Dodonæus in Japan

Dodonæus in Japan

Translation and the Scientific Mind in the Tokugawa Period

Edited by W.F. Vande Walle
Co-editor Kazuhiko Kasaya

1425

Leuven University Press
International Research Center for Japanese Studies, Kyoto

Contents

Preface

This collection of essays is the outcome of an international symposium, jointly organised by the International Research Center for Japanese Studies, Kyoto, and the Section of Japanese Studies of the Katholieke Universiteit Leuven in October 1998. It was the second in a series of three international symposia that the International Research Center for Japanese Studies organised in Europe in conjunction with a European partner. The first one was held in Sheffield (1997), the second in Leuven (1998) and the third in Leiden (1999).

The Leuven symposium, which went under the general title of *Translation of Culture, Culture of Translation*, actually consisted of two parallel sessions. The first one was a workshop on Gender and Modernity in Japan: *Japanese Women's Magazines*, the second one was a symposium devoted to a reflection on *Translation and Adaptation in the Formulation of Modern Episteme: A Reappraisal of Dodoens*. The essays in the present volume are the reworked and elaborated versions of the presentations made at the latter symposium.

It was clear that many of the issues we had to tackle had to do with translation, and that translation was not a phenomenon limited to Japan, but equally prominent in European cultural history, nor limited to texts as such, but involving broader cultural contexts as well. As a result we undertook an investigation of Dodoens's (Dodonæus) importance in Europe as well as in Japan through the prism of translation, transposition, adaptation etc., defined as a moving force in cultural and social development and an indispensable lubricant in the process of functional differentiation. As Japanologists, our main concern was evidently Japan, yet we have deliberately opted for a perspective that kept a certain distance from boundaries. Therefore we have endeavoured to confront experts in the field of Western herbals and botany with historians of early modern Japan.

Hayao Kawai, director-general of the International Research Center for Japanese Studies, delivered the keynote address. In the margin of the symposium a little exhibition including old editions of Dodoens's *Herbarius,* also known as *Cruijdeboek* or *Cruydt-Boeck*, as well as a few books on anatomy by Flemish authors, which found their way to Japan through the agency of the Dutch East India Company, was held at the Central Library of the Katholieke Universiteit Leuven. It was conceived and executed by Dr. Chris Coppens, archivist at the said library.

THE PARTICIPANTS TO THE SYMPOSIUM WERE (IN ALPHABETICAL ORDER):

H.E. Mr. Nagao Hyôdô, ambassador of Japan in Belgium
Prof. Dr. Herman De Dijn, Vice-rector, Katholieke Universiteit Leuven

OF THE INTERNATIONAL RESEARCH CENTER FOR JAPANESE STUDIES, KYOTO

Haga, Tôru, professor emeritus / Inaga, Shigemi, associate professor / Ishii, Shirô, professor / Kasaya, Kazuhiko, professor / Kawai, Hayao, director-general / Matsuda, Kiyoshi, visiting professor / Shirahata, Yôzaburô, professor

OF THE KATHOLIEKE UNIVERSITEIT LEUVEN

Coppens, Hans, researcher / Joos, Joël, researcher / Lambert, José, professor / Matsue, Mariko, lecturer / Sakurai, Naoko, lecturer / Schiltz, Michael, researcher / Takeyama, Ena, lecturer / Van Ourti, Gretl, lecturer / Vanoverbeke, Dimitri, associate professor / Vanpaemel, Geert, professor / Van Put, Ineke, researcher / Vande Walle, Willy, professor / Verplancke, Ingeborg, lecturer / Wostyn, Paul, lecturer

OF OTHER JAPANESE UNIVERSITIES

Crijns, Frederik, researcher, Kyoto University
Kashioka, Tomihide, professor, Kyoto Women's University

OF OTHER EUROPEAN UNIVERSITIES

Ambrosoli, Mauro, professor, University of Udine, Italy / Beukers, Harmen, professor, Leiden University, The Netherlands / Blomberg, Catharina, lecturer, University of Stockholm, Sweden / Lukacs, Gabor, directeur de recherche, Centre National de la Recherche Scientifique, France / Métailié, Georges, directeur de recherche, Centre National de la Recherche Scientifique, France / Screech, Timon, senior lecturer, SOAS, University of London, U.K. / Visser, Robert, professor, Utrecht University, The Netherlands / Wille, Helena, researcher, Ghent University, Belgium.

The international symposium was made possible by the generous support of Honda Europe N.V., Ghent, in commemoration of its twentieth anniversary and by the sponsorship of the International Research Exchange Foundation for Japanese Studies, Kyoto.

We owe a great deal of gratitude to the personnel of the following institutions who offered us their help or their hospitality during the symposium or in the course of editing the present volume: the Central Library of the Katholieke Universiteit Leuven; the Bibliothèque générale et de sciences humaines and the *Collection Couvreur* of the Université catholique de Louvain; Dr. Francine de Nave, *bestuurscoördinator-conservator*, and the staff of the Plantin-Moretus Museum in Antwerp. We thank all the institutions and individuals, those who remain anonymous as well as those who are mentioned, who lent illustrations for reproduction in this book or gave the permission to copy and reproduce illustrations in their possession.

We thank the following individuals for their logistic assistance: Hans Coppens, Mariko Matsue, and the other staff members of the section of Japanese Studies, Véronique Preuveneers of the department of Oriental Studies, Dr. Chris Coppens and Guido Cloet of the *tabularium* of the Central Library, all of the Katholieke Universiteit Leuven, as well as Claude Detienne, librarian of the *Institut Orientaliste* of the Université catholique de Louvain.

To no one do the editors owe a greater debt than to David de Cooman, Japanologist, who devoted countless hours to checking dates and names, titles, spelling and grammar, and to bringing uniformity in the references and style.

INTRODUCTION

W.F. Vande Walle

Department of Oriental and Slavonic Studies, K.U.Leuven

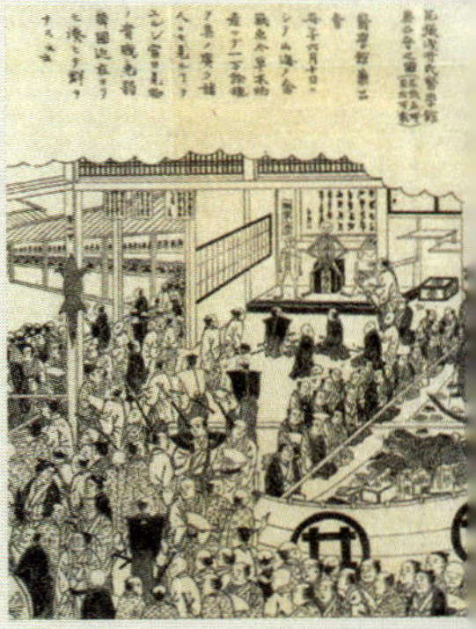

Translation is a vital element of development in any culture. We only have to recall the itinerary Greek classics travelled through Arab translations into Latin, to realise how vital translation has been in the preservation and transmission of an existing culture or legacy as well as in stimulating the birth of new cultural offspring, a 'renaissance'. It is no exaggeration to say that "[a] translator is not someone whose task is to conserve something but to propagate something, to spread and develop it: translators are agents of change. Translators, in fact, make a difference."[1] If until early modern times, however, translation may have seemed of less importance, this was due to the paucity of communication and exchange. Communication across boundaries and borders was carried on in Latin, but actually involved a tiny minority of the people. During the Renaissance, what with the discovery of new continents and cultures and the translation of the Bible into several European literary vernaculars, opportunities for communication were multiplied and European high culture gradually diversified from unilingualism to multilingualism. In the process, translation became vastly more important, a fact of daily life so to speak.

The translation of the Bible constituted a challenge to traditional authority, and at the same time embodied the recognition of the value of the vernacular. It did not only break the hierarchy of authority but also challenged the hierarchy among the languages, for up to that time only Greek and Latin and to some extent the Italian of Dante had been considered languages that were "rational" and therefore having a grammar, in contrast to the "vulgar" languages which were supposed to lack rationality and grammar[2]. Translation into the vernacular languages therefore was a concomitant to the rise of the critical and scientific spirit, of modern episteme. As time went by, the more widespread and economically strong vernaculars were gradually upgraded and in the end became a cornerstone in the process of nation building. Although national cultures are deeply indebted to translation, they have tended somehow to obfuscate its importance, no doubt because it runs contrary to the myth of independence, which is at the heart of the nation construct. Even nowadays, the role of translation and the translator tend to be slighted or ignored. They are viewed as a necessary evil, or something like a mechanical interface, something that ideally could to be replaced by a machine, and that does not colour the originality and the quality of the contents that are being rendered, the one exception by general consensus being the translation of *belles-lettres* and poetry.

In the scheme of things of the nation-state the translation of a particular work, whether scientific or literary, can hardly be ranked among the masterworks of the national cultural heritage. The nineteenth century transformed the nation-state into the norm, creating a fictitious microcosm of self-subsistence and independence. It was supposed to be able to exist on its own, to be independent from others, other states, other nations, other communities, or whatever these others are called in present-day parlance. There was a national literature, a national science, national philosophy etc. In such scheme of things, the translation of a foreign work can never be central to the own tradition. Translation was relegated to the periphery, was something that could supplement, but was not indispensable. Japan may be an exception in this respect, which does not necessarily mean that it has relied more extensively on translation than other cultures, but it has always been strongly aware of the function and importance of translation.

In reality, translations have been essential to new developments in the most varied areas of human endeavour, even in those cultures that claim the highest degree of originality for themselves. *Die Leiden des jungen Werthers* was of tremendous importance for the development of the Romantic Movement in French literature, yet its French translation never featured as an important work. In science the situation was probably worse, for here authorship was less revered than in literature, tended to slip more easily into anonymity or oblivion, while scientific writings were sometimes translated almost surreptitiously. The situation seems to be different in Japan. The *Kaitai shinsho* is the translation of a rather pedestrian Dutch book on anatomy, yet to this day it is considered a masterpiece and a classic. Admittedly, whether it is really a translation in the strict sense is debatable, but what matters here is

解体新書

that it was at least conceptualised as a translation. Does this confirm once more the *cliché* that Japanese culture is derivative? This qualification is more of a value statement than it is a useful standard or criterion, for derivation is central to all cultures. Only, the perspective on derivation is different in different cultures. Ours may be a culture just as derivative as the Japanese, but we are less apt to recognise or perceive this, because the derivation may be hidden behind a screen of distortion or the mist of time. The willingness to acknowledge indebtedness may be a more important distinction characterising cultures than the degree of derivation itself. The so-called derivative cultures, by the very nature of their awareness, tend to stress their own derivation and by the same token the alleged originality of other cultures. Nowhere more than in Japan does the *cliché* of the four great civilisations (Ancient Egypt, Mesopotamia, India and China) hold sway, thus reinforcing the dichotomy between derivative and original cultures. This is a simplistic and reductionist vision of culture, for, borrowing is not passive receptivity, but implies active adaptation and creativity.

Our nation-state has provided us with a powerful framework to write cultural histories in. Each national culture is conceived of as something that is ideally self-contained and self-supporting, while outside influence is acknowledged but perceived as something of a minus, "indebtedness" as something that has to be limited, something that supplements a lack or defect in the own culture. Corollary to this prejudice is the conceit that dominating cultures usually are "inspired" by other cultures, whereas non-dominant cultures are always "influenced" and "indebted". It is evident that we are dealing here with a perspective that is determined by contingent elements: geography, political and economic history, remoteness in time, which allows to "forget" that some or other cultural element was once borrowed etc.

With reference to the situation in the Far East, people have sometimes distinguished between continental and insular cultures. In that frame Chinese culture is continental. Vast and encompassing, it has always projected an image of autonomy and being self-contained. This has not prevented it from taking in many foreign loans, but since the Chinese territory kept expanding, the place of origin, which was initially outside the boundaries of Chinese rule, eventually ended up within the territory of a subsequent dynasty or period, thus feeding the Chinese construct that they were all part of the Chinese cultural legacy. Between the seventeenth and the nineteenth century, when confronted with Western science and technology, many Chinese intellectuals held that they were actually rediscovering ancient wisdom lost from China in antiquity, which the Westerners had merely "augmented and nurtured." Here we are confronted with the ideology of an all-encompassing integrative cultural matrix. Conversely, we could say that insular cultures cultivate the ideology of duality and distinction. In view of the clear-cut geographical rupture it is easier to identify something indigenous from something imported, but as time goes by, something that was once foreign may end up being perceived as something indigenous. Yet, the two never fully integrate, there is always some measure of coexistence of the foreign with the indigenous, in which the foreign continues to be meaningful as foreign. The preceding argument may have gone some way in showing the limited usefulness of a parameter such as the "origin" of a cultural element. Our approach is that all cultures are indebted to others and that exchange, communication, is essential for the vitality of any culture or rather *is* culture. In the final analysis, the existence of derivative cultures is not at issue here, it being obvious that all cultures, great or small, are derivative, and for their own sake better be.

The primacy of language

The aforementioned dualistic typology of cultures is easily correlated with another duality: that of classical culture versus the vernacular and the popular. Leaving the opposition continental-insular- merely a geographical metaphor- aside, it is not difficult to see an analogy between the relationship between Chinese and Japanese culture on the one hand, and the Mediterranean and Germanic spheres of culture in Mediaeval Europe on the other. China and the Mediterranean world respectively embodied classical civilisation. The weight of the "classic" was heavy, so overwhelming that at first what was indigenous was neglected or overlooked. The learned elite looked at the books and not at reality. Within a scholastic tradition, classical texts were the source of authority. In the field of abstract thinking, speculation and theory, it was hard to challenge scholastic authority and there was nothing readily available to falsify or disprove long held speculative theories. In the field of the study of nature however, the conditions were different. There, reality was bound one day to thrust itself much more compellingly upon the scrutiny of students of classical texts. One such area of nature was the world of plants. Since the first century A.D. Dioscorides' *De materia medica* had been the cynosure of botanists and phytographers. When Renaissance botanists tried to correlate Dioscorides' plant descriptions to the reality of their native floras, they found many discrepancies. They thus realised that their compass was not reliable, and classical authority had to be supplanted by empirical scrutiny. German botanists were the first to initiate this methodological shift, which was to have far-reaching implications. Rembert Dodoens, alias Rembertus Dodonæus, the central figure of this collection of essays, followed in their footsteps, but he took their work one step further. While the methodological innovation of his predecessors had mainly been embodied in their illustrations, in Dodoens's case it equally informed his descriptions. Dodoens took a particular interest in the nomenclature, because he realised that vernacular names of plants are closely linked to local flora and are not easily translatable into equivalents in other vernaculars, because each has other local connotations. His approach involved the attitude of a physician, a botanist and a lexicographer. Accurate naming and comparison of the various names was a *conditio sine qua non* to an adequate grasp and understanding of the regional particularities of the plant world. By including the multitude of local flora, i.e. varieties not found in the Mediterranean, into their repertory, the botanists in Germany and the Low Countries veered away from the classical botanical writings. The new development in North-western Europe amounted to a departure from the classical matrix. It was a rejection of authority and went hand in hand with the assertion of the local and the particular. Instead of the illusion of a unified worldview fostered by classical learning, came a vision of diversity and particularism. It will be noted that these new herbalist treatises came out of Lutheran Germany, that had cancelled its subscription to Roman dogma.

The Southern Low Countries, a prosperous area, were not insensitive to the new developments, both in the field of religion and herbal studies. The area boasted a high density of botanical and herb gardens. The demand for botanical knowledge created a market for books on the subject and publishers saw the new opportunities. The first botanical treatise to be published in the Low Countries was *Liber ruralium Commodorum* by the Bolognese agriculturist Petrus de Crescentiis, printed by Jan Veldener in Leuven in 1474. In 1484 Veldener, who hailed from Bavaria, published an illustrated *Herbarius in dyetsche*, which was reprinted in the beginning of the sixteenth century in Antwerp by Willem Vorsterman and Govaert Back. In 1514 the Antwerp-based printer Claes de Graeve published *Den groten herbarius met al sijn figueren die ortus sanitatis ghenaemt is*, which went through reprints in 1526, 1532, 1533, 1538 and 1542. Leonhart Fuchs had barely published his *De historia stirpium* (1542) when a Dutch translation was being prepared and published in Basel (either 1543 or 1445) under the title *Den nieuwen herbarius dat is dboeck vanden cruyden*. The translator has remained anonymous, but it is assumed that it was the young Rembert Dodoens. At any rate, we have to note how through translation, notably into the vernacular, new herbalist and medical knowledge was being absorbed, to be further developed in the Low Countries. It is not hard to see the similarities with the development of herbal studies in Tokugawa Japan.

The new trend was the harbinger of stronger regional trends in culture and religion on the European continent. In the political field it was the dawn of an era marked by the advent of absolutism, and subsequently the formation of the nation-state. We do not mean to say that these changes were caused by the innovations in the study of plants, only that the innovations were part of the broader frame of transformation. If it was true for the world of plants it also applied to the world of culture. Latin, although not supplanted, was no longer the only language worth that name, while gradually literary vernaculars claimed their rightful place on the stage of higher culture. Significantly, Ambroise Paré, the French surgeon who revolutionized European surgery, was not formed in the classical mould. He did not know Latin and therefore wrote all his treatises in French.

An analogous development appears to have taken place in Tokugawa Japan. Herbal studies in Japan, like many other fields of science, were based on knowledge transmitted from China. The Chinese herbal *par excellence* that offered the Japanese herbalists their paradigm was Li Shizhen's 李時珍 (1518-93) *Bencao gangmu* 本草綱目 (Jap. reading *Honzô kômoku*). It actually provided the mould for all traditional East Asian herbal scholarship. In 1607 the famous Confucian scholar Hayashi Razan 林羅山 acquired a copy of this book in Nagasaki and presented it to the Bakufu. From 1638 on the Bakufu laid out two herbal gardens (yakuen 薬園) in the environs of Edo, one in Shinagawa and one in Ushigome. The Confucian scholar and herbalist Inô Jakusui 稲生若水 (1655-1715) was an avid and intelligent student of the Chinese herbal. His edition of *Honzô kômoku* (1714) is considered authoritative and superior to its original, yet all improvements and corrections he added, were based on philological and textual study. His perspective was and remained *identifying* Japanese plants with the ones described in the *Honzô kômoku*. He was still looking at the book instead of at reality. The same attitude prevailed 庶物類纂 in compiling his magnum opus *Shobutsu ruisan*, which in its original 巻 planning was to comprise thousand chapters (maki). Although death prevented him from achieving this Herculean feat, yet he managed to complete 362 chapters. It includes descriptions on animals and plants culled from more than 170 Chinese books. Unfortunately, since the compilation was stored away as a secret book it never rendered service to any students. It is well organised and erudite, yet it contains few novelties or original views.

貝原益軒 In contrast, the polymath Kaibara Ekiken (1630-1714), while following *Honzô kômoku*, at the same time went beyond its limits and enlarged and refined its descriptions on the basis of his own observations. Like the Renaissance botanists who found Dioscorides' *De materia medica* to be at variance with their native floras, Ekiken discovered many discrepancies between the Japanese flora and the descriptions in *Honzô kômoku*. This not only lead him to a practice of juxtaposing Chinese and Japanese varieties, but also to including many varie- 大和本草 ties that were not mentioned in the Chinese herbal. His *Yamato honzô* (1709) diverges in various ways from his Chinese model, adds information drawn from other Chinese herbals and includes numerous indigenous varieties, for which there are no Chinese names and which he specifically marks as Japanese. It contains descriptions of no less than 1366 varieties.

平賀源内 Hiraga Gennai (1728-1779), a man of Renaissance proportions in vision and aspirations, if not in achievement, would take this "departure" one step further by incorporating herbal knowledge from Dutch herbals, notably Dodonæus, and be one of the first to formulate a programme of natural history in Japan. Since first-hand observation was of paramount importance, opportunities to see natural products had to be increased. Herborizing and studying plants in their natural habitat were the answer, but required an enormous amount of time and resources. Exchange among like-minded was a far better and cheaper 田村藍水(元雄) way. Therefore Hiraga Gennai and Tamura Ransui (Gen'yu) organised exhibitions of medicinal and natural products in Edo. This became a popular practice among herbalists, physicians and students of natural history, not only in Edo, but also in Nagoya, where during the nineteenth century the Asai academy of medicine organised yearly exhibitions of medicinal and other useful products. These were highly acclaimed events that drew a large public, as we gather from the descriptions in *Owari meisho zue*. 尾張名所図会

These practices marked an important step forward towards empiricism. This new direction, outside the Chinese matrix, was concurrent with a policy of indigenisation, promoted by the Bakufu, and with an increasing interest in novel knowledge from the West. Indigenisation did not mean at once a stronger emphasis on indigenous culture, but on indigenous nature. Arguably, indigenous popular culture would not have been powerful enough to supplant the prestigious classical Chinese culture in an effort to overcome its constraints, only nature itself could achieve this. Interestingly, the so-called Nativists (*Kokugakusha*) frequently refer to nature in Japan, when they argue the superiority of Japanese "culture" over Chinese. At such a juncture of time precise scientific knowledge about nature based on observation was introduced in Japan from the West. Among that Western knowledge figured Dodonæus' herbal. The first copy known to have been imported in Japan was the 1618 Dutch edition. It was presented to the Shogun Tokugawa Ietsuna in 1659 by Zacharias Wagenaer, the head of the Dutch factory on Deshima. However, it was stacked away in the Shogunal library, where it gathered dust for decades, until it was allegedly "rediscovered" in 1717 by the Shogun Yoshimune, who subsequently ordered it to be studied, thus setting in motion an intellectual current that was later to be labelled *Rangaku* (Dutch Studies). Thus both in Europe and in Japan we meet the Flemish botanist at a crucial juncture in the formation of the scientific mind, the articulation of modern episteme. Just as was the case in Europe, the new developments in Japan were marked by a broadening of the scope from a strictly medicinal interest to a more detached and encompassing view of nature. The shift is also visible in the terminology, which *per se* is derived from Chinese but takes on new meaning in eighteenth-century Japan. The initial term is *honzô*, which corresponds to *materia medica* and hence herbal studies, but as the interest of the Japanese intellectuals shifts towards natural history they use new compounds containing *butsu* and *hin* or com- 物、品
binations thereof such as *buppin* to finally adopt *hakubutsugaku* as the 物品、博物学
standard equivalent to the Western concept of natural history.

Here it would seem that the East and the West for the first time caught up with each other. In Europe Linnaeus published his *Systema naturae* in 1735 and Buffon (1707-1788) his *Histoire naturelle* between 1749 and 1784. In Japan Kaibara Ekiken published his *Yamato honzô*

物類品隲
小野蘭山
本草綱目啓蒙

in 1709, Hiraga Gennai his *Butsurui hinshitsu* in 1763, while Ono Ranzan (1729-1810) was working on the compilation of his *Honzô kômoku keimô* during the eighties of the eighteenth century, although this work was actually published in 1803. The interest of the West in natural history is obviously related to the discovery of the world and the development of international trade, but in Japan it is less easy to explain. To all intents and purposes the country remained closed and isolated from the outside, yet a tiny minority of intellectuals were discovering the world vicariously, through foreign books.

水谷豊文
物品識名
物品識名拾遺、物品
伊藤圭介
泰西本草名疏

The interest for and the discovery of the natural world may have run parallel in Japan and the West, yet the shift from *materia medica* to natural history in Japan took longer and was more gradual and vacillating. This is clear from the self-styling of the herbalists. In the early seventeenth century it was self-evident for Kaibara Ekiken to style his book *honzô*, but one century later, Ono Ranzan, at a time that natural history was riding the crest of the wave, still referred to *honzô* in the title of his book. The first to deviate from this style was Hiraga Gennai who opted for *butsu* (thing, creature) and *hin* (class, category) in a clear reference to classification of the natural world. Mizutani Hôbun echoed this style in his *Buppin shikimei* (1809) and *Buppin shikimei jûi*, (1825). *Buppin* seems to be the first term that was used as an equivalent of natural history. When Von Siebold came to Japan he had in his luggage a copy of Carl Peter Thunberg's *Flora Japonica*, a work that is based on Linnaeus' taxonomy. He presented the book to Itô Keisuke (1803-1901), who in 1829 published his Japanese translation of Thunberg's book under the antiquated title *Taisei honzô meiso*. Not only does the title include the term *honzô* but it also styles itself a gloss in the fashion of Confucian exegetical literature, while the term *Taisei* for West is reminiscent of the seventeenth-century Chinese adaptations of Western works composed by Jesuits and their learned Chinese converts.

It may strike the reader as an anachronism to find Dodonæus in the company of Buffon, Linnaeus, Thunberg etc., because it was precisely the herbal tradition represented by Dodonæus that was superseded by natural history. Yet, in Japan his herbal endured along the new science of natural history. During the twenties of the nineteenth century, while Itô Keisuke was preparing the compilation of what was the first presentation of the Linnaean system in Japanese, the integral translation of Dodonæus' herbal was under way. It was completed around the year 1823 under the title 'Illustrated Herbal of the Westerner Dodonæus' and in 1829, the very year that Itô Keisuke published his compilation, plans were laid out to publish it, but due to the death of the sponsor of the project, Matsudaira Sadanobu, and the loss of the woodblocks in a fire, the plan was never carried through. Conversely, Itô's comprehensive compilation incorporating the Linnaean system, *Nihon sanbutsushi* was only published in 1873.

日本産物志

From natural history sprang botany in the narrow sense of the term and horticulture. This stimulated the cultivation of plant varieties, particularly flowering plants, thus creating the need to inventory cultivated varieties. Representative of this interest is *Honzô zufu*, a botanical album compiled by Iwasaki Kan'en (1786-1842). It consists of 96 *maki*, which were published between 1830 and 1844. It was the first Japanese botanical album produced in colour. It describes about 2000 plants, which are still arranged according to the order of *Honzô kômoku*. Kan'en not only included wild varieties but also cultivated ones, as well as foreign plants and products. Some of the illustrations are reproduced from Johann Wilhelm Weinmann's *Taalrijk register der plaat- ofte figuur-beschrijvingen der bloemdragende gewassen* (1736-1748) (known in Japanese as *Kenka shokubutsu zufu*), a book owned by the famous Rangaku scholar Udagawa Yôan (1798-1846)[3].

本草図譜
岩崎灌園
顕花植物図譜
宇田川榕庵

TRANSLATION

Although the chronological boundaries set by the title of this collection of essays coincide with the Tokugawa period, let me take an example from the Meiji and Taishô periods. I am warranted to do this, because of the teleological slant given to this collection of essays. Since we want to demonstrate the contribution of Dodonæus to the articulation of modern episteme in Japan, the implied teleology is that it reached certain maturity in the subsequent modern era. From that perspective, the Meiji and Taishô periods are the realisation of something that was embryonic in the mid-Tokugawa period, was gestated in the late Tokugawa period, and burst into full bloom after the Meiji Restoration. My example is taken from philosophy since that is an area where the fundamentals of a culture are at work.

Nishida Kitarô is generally recognised to be the first original Japanese philosopher. In this one statement lies hidden a whole set of cultural *aprioris*. It is obviously very hard to prove or to disprove. What we can say is, that he is the first to have mastered the Western philo-

漢文調 翻訳調 善の研究

sophical idiom. His predecessors were still writing in a so-called antiquated style, exemplified by e.g. Nakae Chômin. This is a style that is still very redolent of *kanbun-chô* Japanese, not only in the grammar but also in the vocabulary. It is true that Nishida's Japanese is much less tributary to that kind of style and approximates much better *honyaku-chô*, the style evolved during the Meiji period to translate English, German and French original works of fiction and non-fiction. By the time Nishida was publishing his *Zen no kenkyû*, the Japanese language had already undergone a deep change, equipping it with a syntax that came much closer to that of Western languages. This was in the first instance the result of four decades of frantic translation activities during the Meiji period, but actually we have to go back even further, at least to 1774. That is the date when the *Kaitai shinsho* was published, the first Japanese translation of a Dutch book. It marked the beginning of an era of Dutch studies (Rangaku), in which, as we shall amply see in this collection of essays, the study of language and translation played a central role. Consequently, the scientific Japanese of Meiji and Taishô times equally owes a great deal to the Rangaku translators. In the process, the source languages (Dutch, English, French, German etc.) have rubbed off considerably on the target language (Japanese). In a sense, in order to make translations of Western works into Japanese faithful, the Japanese language had to mimic the source languages. If the early translations of Western works seemed less reliable or faithful, it was perhaps not so much because the translators were poorly qualified, as because the tool was unfit for it. Japanese was simply too distant from many of the Western languages. No individual could single-handedly force the language through such a complex mimetic process; it took several decades to accomplish. When this work of "gutting" had finally been done, what was left was an altogether different language. It was "easy" then for Nishida to write something that sounded convincingly like western Philosophy.

The preceding argument attributes primacy to language over any other cultural element, and although it may strike the reader as fairly radical, there is a good measure of truth in it. If we were to take this line of reasoning to the extreme, we could even deny Nishida any philosophical originality, brand any perception of originality as illusory and attribute it entirely to the impression created by his usage of a Westernised language. In the early phase Japanese was unable to accommodate (Western-style) philosophy because it was so alien, but in the end it caught up with the "alienness" of philosophy. The bottom line is however that we are still left with the doubt whether the confirmation by Westerners that Nishida's philosophy is indeed mature philosophy and universal, is not just another variation of the oft encountered tendency of equating universal with western and qualifying anything that is not Western as deviant or particularistic.

The big gap between Western languages and Japanese is perhaps most eloquently exemplified by an example again taken from the Meiji period. Oda (Niwa) Jun'ichirô (1851-1919) translated Bulwer-Lytton's *Ernest Maltravers* into Japanese as *Karyû shunwa* (A Springtime Tale of Blossom and Willows) and published it in 1879. Obviously, by the time Oda made his translation Rangakusha had already been translating Western texts for more than a century, but except for a few curiosities, they had been texts of a non-literary nature. Now Oda was confronting a far more difficult task: producing a text that combined stylistic grace with fidelity of rendering. The original was intended as literature, and its translation had to be perceived as such by the Japanese. We may assume that Oda, who had studied in England and Scotland, had a sufficient command of English. However, the syntactical structure and the semantics of this language were so totally different from his native as to defy rendition. Therefore the novel had to be "rewritten", nothing more, nothing less. Not only the title was changed beyond recognition but the contents too were drastically altered, if not in the plot, at least in terms of mood and style. He adopted the stiff literary style known as *kanbun-chô*, a variety of Japanese that both lexically and syntactically echoes Classical Chinese and lacks the flavour of contemporaneity that is associated with more vernacular varieties of Japanese.

織田純一郎
花柳春話

This does not mean to suggest that at the time he was writing a more vernacular variety of Japanese would have solved the problem. In fact, Oda was using a tried and tested method: that of substituting the unfamiliar for the unknown. For centuries "overseas" and "foreign" had meant Chinese to the Japanese. When they were confronted with the West, things Chinese all of sudden looked familiar by comparison, for many of the latter the educated at least knew as bookish knowledge, although the social and institutional referent of a great deal of the Chinese vocabulary they knew, was unknown or at least abstract. But, precisely because of that, Chinese terminology and by extension style redolent of Chinese was considered the most suited interface to negotiate the cultural gap yawning between Japan and the West. The way Itô

Keisuke titled his adaptation of Thunberg's book *Taisei honzô meiso* is an example of this method, which we mentioned a while ago. Chinese was traditionally the Japanese paradigm for foreignness and as such familiar enough. That is a procedure that is not uncommon in English or any of the other European languages either. We often borrow from Latin to denote something novel, something that is (at first) not part of our social and political experience. We know the Latin word, but since it is devoid of any of its sociological or political implications, we can easily transfer it. It is at the same time distant and familiar enough. It was only when Japanese social and institutional reality, and in its wake, customs and way of life, were gradually aligned to Western models, that commensurable notions could be articulated. In the process the Japanese used Chinese terms for the Western concepts and notions, terms that had hitherto no sociological and political reality for them, to denote phenomena that were analogous to the Western ones. At that point, these terms, coined to denote novel phenomena that had their counterparts in Western societies, could be "translated" into their Western equivalents without great loss of meaning. Subsequently, by dint of use, the terms became workable translations of Western phenomena, because they were meant to be just that. Here we have an indigenous signification. But before Japanese reached that comfortable stage, they had to traverse a phase of baffling and unnatural literal translation. However great the resistance may have been against this procedure, it proved to be the only way out of the conundrum.

Reference to classical models and paradigms was doomed to remain 朝比奈知泉 unsatisfactory. Asahina Chisen (1862-1939) was perhaps the first to attempt a faithful translation of a literary text, making no concessions to traditional literary conceits and style or to fears that the reader might be baffled by the unfamiliar setting of the plot. His translation of Bulwer-Lytton's *Kenelm Chillingly*, under the Japanese title *Keishi-* 森田思軒 *dan* (The Story of K.C., 1885) was hailed by the critic Morita Shiken (1861-1897) as the first truly exact rendering of a European novel. But a heavy price had to be paid. The Japanese was unnatural, betraying heavy influence of English idiomatic expressions, and sometimes it was hard to grasp the intended meaning. "To pay attention" was translated as "chui o harau" and personal pronouns such as he (kare) and she (kanojo) were explicitly mentioned, a practice then uncommon in Japanese. Similes and metaphors linking phenomena unknown to the Japanese must have baffled many readers. However, these literal transpositions were repeated in subsequent translations until in the end they had imperceptibly acquired citizenship, deeply affecting both the syntax and the vocabulary of the Japanese language[4].

Subsequently, step by step, in the course of a polishing process through thousands and thousands of pages of translation, a new language completely different from what it had been at the outset, came into being. Admittedly, all languages evolve, but Japanese went through a much more incisive transformation in a short span of time. Even the final form of verbs and adjectives at the end of sentences had to be reinvented. Not until the publication of *Ukigumo* was this realised satisfactorily. 浮雲

The complete translation of Dodonæus' herbal was undertaken during the twenties of the nineteenth century, a period when Dutch Studies reached full maturity. Scholarly families such as the Katsuragawa's and Udagawa's took the level of Dutch Studies to an unprecedented level. We must however not exaggerate the influence of Dutch Studies in the Edo period. After all, many of the writings of these scholars were never printed and remain to this day in manuscript form. If influence there was, it was more indirect in the sense that it paved the way for the right attitude, which matured in the Meiji period. However, the importance lies in the fact that we have here a tendency for systematic translation. It is a recognition that systematic translation will help forward society or science. It is without parallel in terms of the sheer effort that was invested into the enterprise. Moreover, scientific terminology was not systematised until the end of the Bakufu period, so that the effect of what we study here came with a time lag, having its full impact only in the Meiji period.

THE ESSAYS

The collection of essays presented here lies at the intersection of two lines of perspective: the study of the translation of science and that of the science of translation. More specifically, it endeavours to trace how the Western herbal tradition, notably the herbal of Dodonæus, was received into the intellectual discourse of Tokugawa Japan, and to demonstrate how it contributed to the articulation of modern episteme, the scientific mind. In writing their essays the authors have more or less had that teleological perspective in the back of their minds. In presenting and grouping the essays, we have tried to marry this vantage point with a chronological order.

The first part contains, besides a bio-bibliographical summary, a group of essays that place Dodonæus in the European context and assess his importance and his contributions to the development of modern science. Robert Visser's essay sets the tone and makes a careful evaluation of his position in the herbal tradition from the viewpoint of the history of science. Although he owed a great deal to the so-called "German fathers of botany", his phytographical descriptions, based on a consistently empiricist approach, were superior to those of his predecessors. Moreover, he was much more concerned with systematics and botanical classification than they were, although Visser points out that perhaps his innovativeness has been exaggerated in the past. Mauro Ambrosoli approaches Dodonæus from the angle of economic and cultural history and deals with the link that existed between botanical practice and agricultural demands in early modern Europe. Moreover he shows how contemporary religious and political cleavages impinged on the path that botanical science took. Helena Wille gives a detailed analysis of the botanical networks that existed in the Low Countries in the time of Dodonæus and highlights the important contribution made to phytography by amateur collectors. Moreover, she has succeeded in identifying a set of albums with drawings of plants, that is presently preserved in the library of the Jagiellonian University at Krakow.

The second group of essays transfers the scene to Tokugawa Japan in an effort to assess the impact Dodonæus' herbal had on the articulation of the modern episteme in that country. This involved a process of transmission, which in its turn was predicated on translation. The fundamental meaning of translation in society is what Michael Schiltz's essay deals with. W.F. Vande Walle focuses on the issue of translation from the viewpoint of the history of linguistics. He addresses the complicated issues involved in translating European languages and concepts into Japanese and Chinese, problems encountered by the Japanese translators of Dodonæus. Shigemi Inaga extends this approach to translation in another area, notably the visual and the artistic, an issue relevant to the translation of Dodonæus, since it also involved a host of plant illustrations. He demonstrates how transmission and rendition were tied in with the (re-) interpretation of the Western linear perspective. Kazuhiko Kasaya's essay on the Tokugawa Bakufu's policies for the national production of medicines establishes the important link between the transmission of Dodonæus' *Cruijdeboeck* to Japan and the broader dynamics of national policy.

PART III is a collection of essays more specifically related to rendering Dodonæus into Japanese. Kiyoshi Matsuda's essay is a careful bibliographical study on the reception and spread of Dodonæus' *Cruijdeboeck* in Japan. Timon Screech has contributed an essay about the visual impact Dodonæus had in the culture of representation in Tokugawa society. Tôru Haga traces the way that Dodonæus contributed to the paradigm shift in Tokugawa culture, in particular as embodied by Hiraga Gennai, who reinforced the departure from the Chinese model and had the ambition to formulate a universal science encompassing Chinese, Japanese and Western elements. Yôzaburô Shirahata takes this evolution even one step further and shows how Japan witnessed the transformation from pharmacopoeia to botany to horticulture.

PART IV deals with issues in the broader context of Rangaku, so-called Dutch studies, that are relevant to the transmission and impact of Dodonæus. Harmen Beukers describes the important role Deshima surgeons played as mediators in the early introduction of Western natural history into Japan. Gabor Lukacs's essay shows that Dutch studies were not limited to science from Holland, but involved the broader frame of European science. He notably provides a well arranged overview of the contributions made by French science to the articulation of the modern scientific mind in Japan. He pays particular attention to the transmission of the French surgeon Ambroise Paré's works through Dutch translations to Japan. Catharina Blomberg's essay introduces Carl Peter Thunberg, who as the direct disciple of Linnaeus, made many descriptions of the Japanese flora and played a pivotal role in introducing his master's system into Japan. He was the harbinger of the new type of botany that supplanted the herbal tradition as exemplied by Dodonæus. Frederik Cryns's essay deals with Japanese translations by Udagawa Genshin and Tsuboi Shindô that remained in manuscript form. Although their influence was of necessity limited, they are fine examples of the level of sophistication Dutch Studies had reached. They highlight the problems involved in translating fundamental concepts of medicine that have a bearing on anthropological views. Crijns shows how central and fundamental the issue of translation was in the development of modern science in Japan, thus making this collection of essays come full circle.

NOTE ON SPELLING AND ROMANISATION

The spelling of personal names in Europe before the Napoleonic period is not uniform. Moreover, before the eighteenth century, most scholars also had a Latin sobriquet. As a result the same person is often referred to by various names. We have made an effort at uniformity but not at all cost. Thus e.g. the Flemish version Dodoens and the Latin equivalent Dodonæus are used without distinction. In addition the name Dodoens or Dodonæus is often used in the way *Laozi* or *Zhuangzi* are used. As is the case with so many "classical" authors, writer and writing have become each other's metonym, so Dodoens can often be interchanged by the *Cruijdeboeck*. This book has gone through many revisions and editions, and in the process the spelling of the title also changed. When we refer to the book in general we use the spelling of the first edition, i.e. *Cruijdeboeck*. When a reference to a particular edition is intended, explicitly or implicitly, we use the spelling adopted in that edition, usually *Cruydt-boeck*.

For exonyms we have used the spelling that is presently in use in the region, city or place concerned or was used at the time under study. Thus Mechelen is preferred over Malines (French) or Mechlin (English). Exceptions are well known places with generally recognised English equivalents such as Antwerp, Bruges, Ghent etc.

Japanese names are arranged in the order used in Japan: family names precede given names. Japanese words are romanised in what is commonly called the modified Hepburn system, as used e.g. in *Kenkyûsha's Japanese English Dictionary*. Common Japanese words, that have entered the English lexicon, are not italicised. Chinese words are transliterated in *Hanyu pinyin*, representing present-day standard pronunciation, except in quotations, where the original transcription is respected, except for those words that have an accepted spelling in the English language. Chinese characters that are included in the list of the *Jôyô kanji* are consistently given in their simplified form.

NOTES

1 Andrew Chesterman, *Memes of Translation* (Amsterdam/Philadelphia: John Benjamins, 1997), 2.

2 Tanaka Katsuhiko 田中勝彦 H. Hâman (Harald Haarmann), *Gendai yôroppa no gengo* 現代ヨーロッパの言語 (Tokyo: Iwanami shoten, 1985), 35-53 (Iwanami shinsho 292).

3 For an in-depth treatment of the history of pharmacopoeia *(Honzôgaku)* and its development into natural history *(Hakubutsugaku)*, see Yamada Keiji ed. 山田慶児編, *Higashi Ajia no honzô to hakubutsugaku no sekai* 東アジアの本草と博物学の世界, 2 vols. (Kyôto: Kokusai Nihon bunka kenkyû sentâ, 1995).

4 Donald Keene, *Dawn to the West: Japanese Literature of the Modern Era: Fiction* (New York: Holt, Rinehart and Winston, 1984), 55-71.

ILLUSTRATIONS

By courtesy of Université catholique de Louvain, Bibliothèque générale et de sciences humaines, Donation Japonaise.

ILL 1 Preface to Inô Jakusui's edition of *Honzô kômoku*, the Japanese edition of *Bencao gangmu*, published in Edo, Kyoto and Osaka, in 1714.

ILL 2 Pair of facing pages from Inô Jakusui's version of *Honzô kômoku*, reproducing on the right hand side the concluding lines of Wang Shizhen's preface to the Jingling edition of *Bencao gangmu* (1596), on the left hand side the preface of Xia Liangxin, the publisher of the Jiangxi edition (1603).

ILL 3 Exhibition of natural products at the Asai medical academy in the province of Owari. The exhibition was organized by the herbalist Asai Shizan in Nagoya in 1835. Woodcut from *Owari meisho zue*, by Okada Kei & Noguchi Michinao, Nagoya, Tenpô 15 - Meiji 13, 7 *maki*, 7 *satsu*; reproduced in Shirai Kôtarô, *Zôtei Nihon hakubutsugaku nenpyô*, Tokyo, Meiji 41.

ILL 4 Page from Kaibara Ekiken's *Yamato honzô* (1709-1715), dealing with the cranes. It contrasts Chinese and Japanese varieties and evaluates the taste of their meat.

ILL 5 Pair of facing pages from Kaibara Ekiken's *Yamato honzô* (1709-1715), supplement, featuring varieties of maple leaves.

ILL 6 Pair of facing pages from Kaibara Ekiken's *Yamato honzô* (1709-1715), supplement, featuring on the left page an ostrich fern in the violet.

ILL 7 Title page of *Butsurui hinshitsu* by Hiraga Gennai, Osaka, Hôreki 13, 6 *maki*, 6 *satsu*, revised by Tamura Seiko.

ILL 8 Effigy of Ono Ranzan, after a drawing by the painter Tani Bunchô. *Chôtei honzô kômoku keimô* by Ono Ranzan, Edo, Kôka 4, 48 *maki*, 6 *satsu*, revised and supplemented by Obara Yoshinao.

ILL 9 Page from Itô Keisuke, *Nihon sanbutsushi*, 11 *satsu*, Meiji 6 (1873).

ILL 10 Page from Itô Keisuke, *Nihon sanbutsushi*, 11 *satsu*, Meiji 6 (1873).

ILL 11 Pair of facing pages from Iwasaki Kan'en. *Honzô zufu*, 96 *maki*, 95 *satsu*, Taishô 10 (1921).

ILL 12 Pair of facing pages from Iwasaki Kan'en. *Honzô zufu*, 96 *maki*, 95 *satsu*, Taishô 10 (1921).

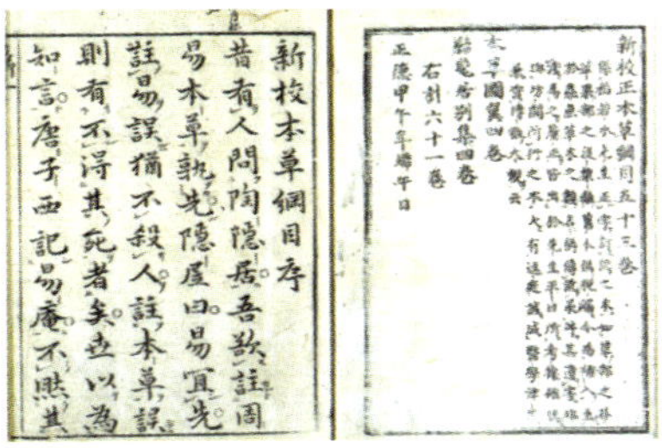

ILL 1

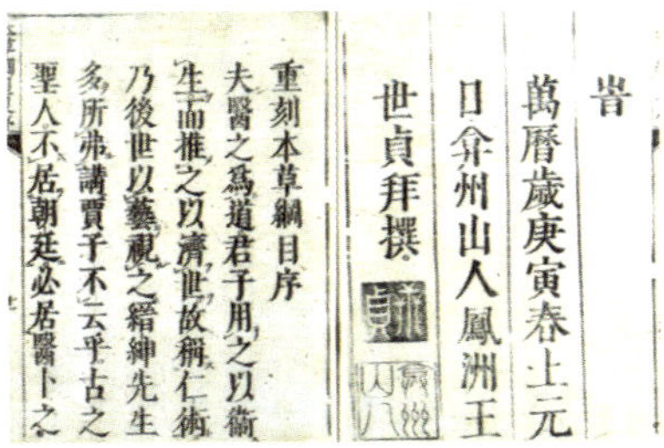

ILL 2

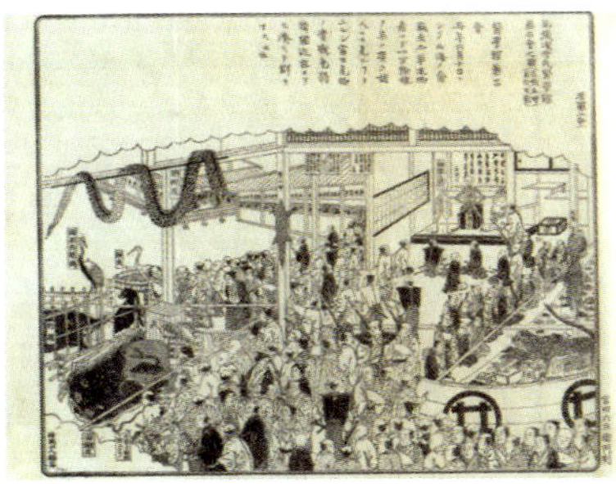

ILL 3

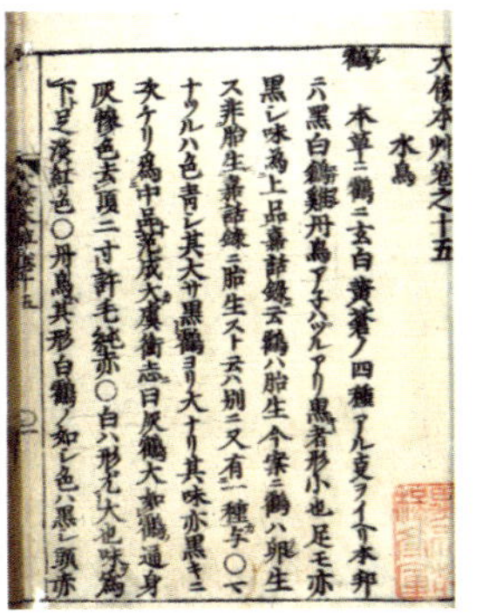

ILL 4

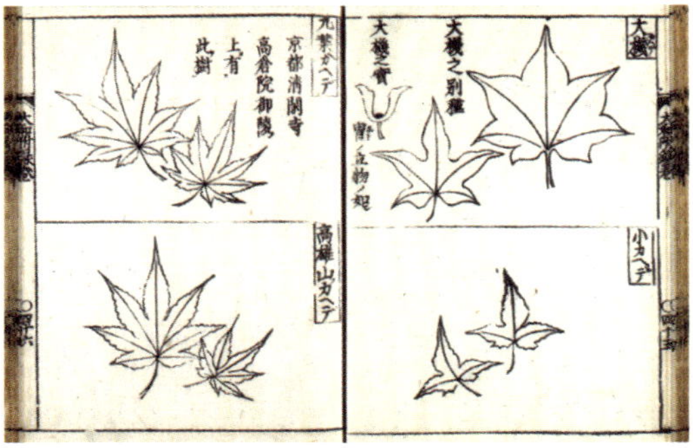

ILL 5

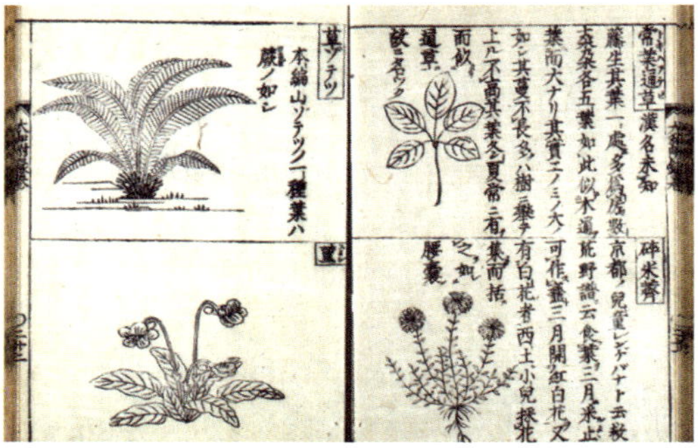

ILL 6

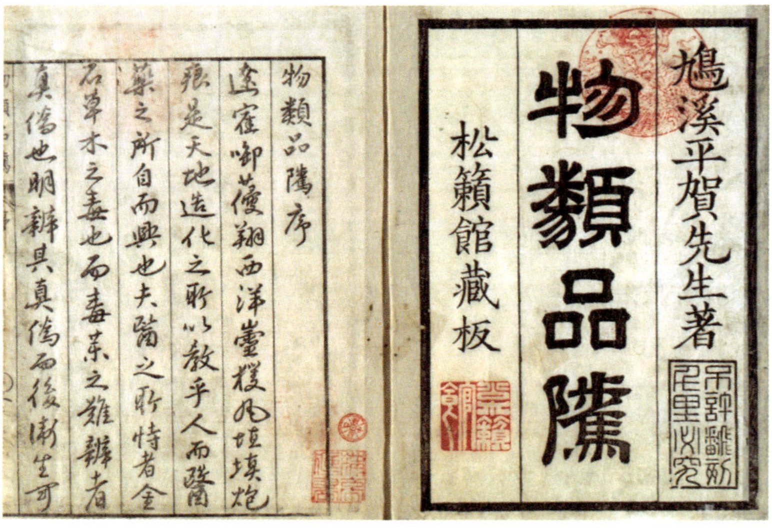

ILL 7

ILL 8

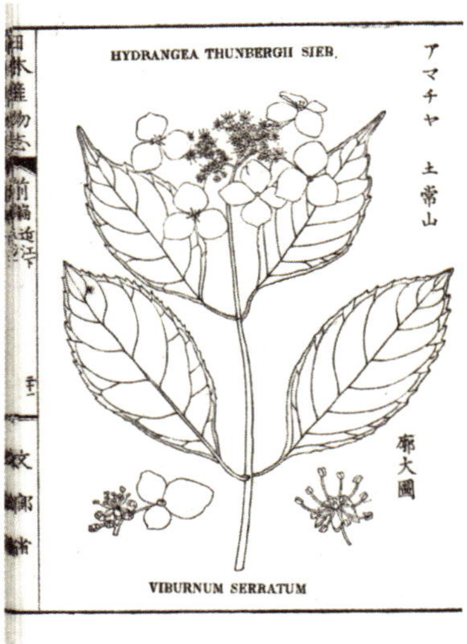

ILL 9

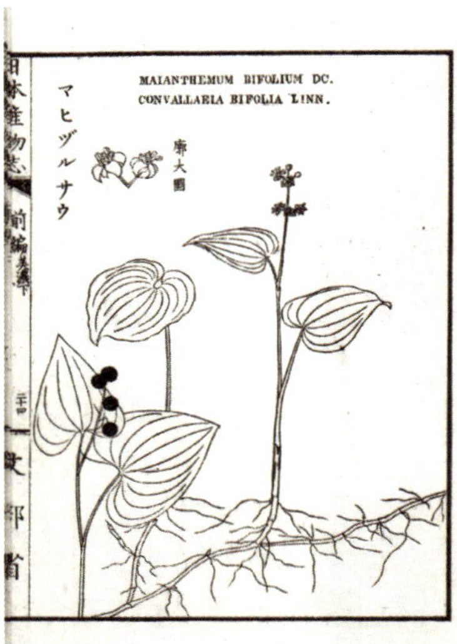

ILL 10

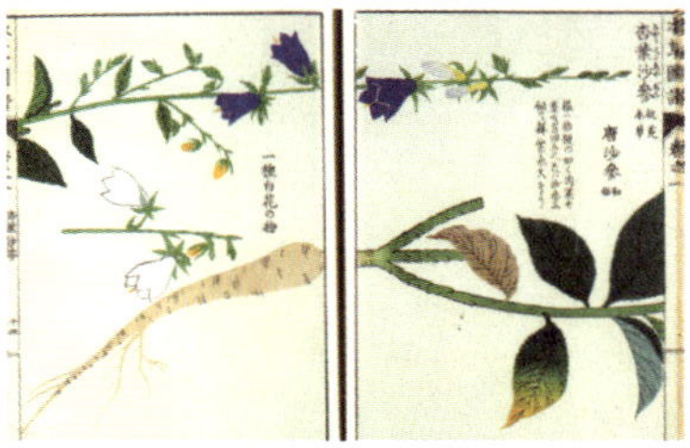

ILL 11

ILL 12

PART I

DODONÆUS IN THE EUROPEAN CONTEXT

I

DODONÆUS:
A BIO-BIBLIOGRAPHICAL SUMMARY

W.F. Vande Walle
Department of Oriental and Slavonic Studies, K.U.Leuven

Cruÿdeboeck.

Rembert Dodoens was born on 29 June of either 1516 or 1517 in Mechelen, a Flemish town in what was then the Spanish Netherlands. His original name was Rembert van Joenckema, a name that testifies to his Frisian ancestry. His father was physician in ordinary to the town government. At the age of thirteen Rembert went to the University of Leuven, where he studied cosmography with the famous Gemma Frisius, as well as physiology and medicine. He graduated as a medical doctor in 1535. From 1541 to 1575 he succeeded his father as physician in his native town. In the sack of Mechelen by the Spanish troops in 1572 Dodoens lost most of his possessions. Between 1574 and 1577 he served as physician in ordinary to the Holy Roman emperor Maximilian II and his successor Rudolph II. He was offered a position as professor of medicine at the University of Leuven, but he declined the honour. In 1578 he also declined an invitation to become physician in ordinary to Philip II, king of Spain. Between 1578 and 1581 he lived as a religious exile in Cologne. In 1582 he stayed in Antwerp at the printing house of Christopher Plantin, preparing the publication of his new edition of the *Cruijdeboeck*. In the meantime he had been appointed professor of medicine at the University of Leiden, where he died in 1585.

In 1533, at the age of sixteen, he published his first book entitled *Nederlands Herbarium*, published by Claes de Graeve of Antwerp. Soon after that, he published three botanical atlases at the printing house of Jan van der Loe in the same city. In the field of cosmography, he composed a work entitled *Cosmographica in astronomiam et geographiam isagoge*, equally published by Van der Loe in Antwerp (1548). Before long, he turned his attention to pharmacology and medicine, two sciences to which he was to devote the rest of his life. In 1554 he published the first edition of the *Cruijdeboeck*, which includes the first scientific description ever of the cauliflower, the Jerusalem artichoke and the potato. For his description of the potato he relied on data collected by his friend Karel van der Sluis, alias Charles de l'é(s)cluse, alias Carolus Clusius (1526-1609). The book also contains the first description of the ailment *peritonitis*. Prior to the publication of the *Cruijdeboeck* he had published two botanical atlases, entitled *Trium priorum de stirpium historia commentariorum imagines* (1553) and *Posteriorium trium historia commentariorum imagines* (1554) resp., which contained the illustrations that were subsequently used in the *Cruijdeboeck*. The book was originally intended as a Dutch adaptation of Leonhart Fuchs's *De historia stirpium* (1542), but it surpassed its model both in the rationale of its arrangement and the number of plants described. The printer had purchased the wood-blocks of Fuchs's book, but for the 200 new descriptions that Dodoens added, he commissioned the Mechelen artist Peeter vander Borcht with the drawings and Arnold Nicolai with the cutting of the wood-blocks. While Fuchs used an alphabetical arrangement, Dodoens adopted a systematic classification, largely based on medicinal criteria. Subsequently he would further expand and refine his knowledge and elaborate a new classificatory system more explicitly based on botanical criteria, which he published in *Stirpium historiae pemptades sex. Sive libri XXX* (1583), his undisputed masterpiece.

The *Cruijdeboeck* consists of six parts. The first part contains a number of plants that have no clear relationship with one another. The second part contains descriptions of flowers and odoriferous herbs; the third is devoted to medicinal and harmful plants, the fourth to cereals, vegetables and fodder plants. The fifth part deals with herbs, roots and fruits that are used in cooking, while the sixth and last part present descriptions of trees and other ligneous plants. In the preface the author deplores that medical practitioners tend to slight herbarism and leave it to "dispensers and other unschooled persons".

Overview of Dodoens's major herbalist and medical writings, arranged in chronological order.

De frugum historia liber unus.: Ejusdem Epistolae duae, una de farre, chondro, trago, ptisana, crimino et alica. Altera de zytho et cerevisia. Antverpiae: ex officina Ioannis Loei, 1552.

Cruijdeboeck: in den welcken die gheheele historie, dat es Tgheslacht, tfatsoen, naem, natuere, cracht ende werckinghe van den Cruyden, niet alleen hier te lande wassende, maer oock van den anderen vremden in der Medecijnen oorboorlijck, met grooter neersticheyt begrepen ende verclaert es, met der selver Cruyden natuerlick naar dat leven conterfeytsel daer by ghestelt. Antwerpen: J. van der Loe, 1554.

The title page has a dedication to Mary of Hungary, Governess of the Low Countries.

Histoire des plantes, en laquelle est contenue la description entière des herbes, c'est à dire, leurs espèces, forme, noms, tempérament, vertus & operations: non seulement de celles qui croissent en ce païs, mais aussi des autres estrangères qui viennent en usage de medecine/ Nouvellement traduite de bas Aleman en François par Charles de l'Escluse. En Anvers: De l'imprimeris de Iean Loë, 1557.
The French translation of the *Cruijdeboeck.*

Cruijdeboeck. In den welcken die geheele historie, dat es tgheslacht, tfatsoen, nae[m], natuere, cracht ende werckinghe, van den cruyden, niet allen hier te lande wassende, maer oock van de[n] andere[n] vremde[n] in der Medecynen oorboorlijck, met grooter neersticheyt begrepen ende verclaert es, met der selver Cruyden natuerlijck naer dat leven conterfeytsel daer by ghestelt ...Van nieuws oversien, ende met seer veel schoone nieuwe figueren vermeerdert. Geprint Thantwerpen: by Jan vander Loe, 1563.
A revised and enlarged edition of the 1554 edition.

Frumentorum, leguminum, palustrium, et aquatilium herbarum, ac eorum, quae eo pertinent, historia ... Additae sunt imagines vivae ... quarum pleraeque novae, et hactenus non editae. Antverpiae: Ex officina Christophori Plantini, 1566.
This work superseded *De frugum historia* (1552). It contains descriptions of grains, leguminous plants and marsh and water plants. Peeter vander Borcht drew the illustrations, while Cornelis Muller, Arnold Nicolai and Gerard Jansen van Kampen cut the wood-blocks. 800 copies were printed of this edition. In 1569 a second edition was printed from the same wood-blocks.

Florum et coronariarum odoratarumque nonnullarum herbarum historia/ Remberto Dodonaeo Mechliniensi Medico auctore. Antverpiae: ex officina Christophori Plantini, 1568.
This book contains descriptions of remarkable floriferous and odoriferous plants. In it the author describes the sunflower, which had been

imported into the Low Countries from South America via Spain. The first specimen grown in the Low Countries could be admired in the garden of a Mechelen plant-breeder. The drawings for the illustrations were made by Peeter vander Borcht and the wood-blocks were cut by Arnold Nicolai and Gerard Jansen van Kampen.

Florum et coronariarum odoratarumque nonnullarum herbarum historia/ Remberto Dodonaeo Mechliniensi Medico auctore. Altera Editio. Antverpiae: ex officina Christophori Plantini, 1569.
This is a revised edition of the preceding work.

Purgantium aliarumque eo facientium, tum et Radicum, Convolvulorum ac deleteriarum herbarum historiae Libri IIII, Remberto Dodonaeo Mechliniensi Medico auctore. Accessit Appendix variarum et quidem rarissimarum nonnullarum stirpium, ac florum quorundam peregrinorum, elegantissimorumque icones omnino novas nec antea editas, singulorumque breves descriptiones continens. Cuius altera parte Umbelliferae exhibentur non paucae, eodem auctore. Antverpiae: Ex Officina Christophori Plantini, Architypographi Regii, 1574.

A nievve herball, or Historie of plantes: Wherein is Contayned the Whole Discourse and Perfect Description of All Sortes of Herbes and Plantes, Their Divers & Sundry Kindes ... and That Not Onely of Those Whiche are Here Growyng in This Our Countrie of Englande, but of All Others Also of Forrayne Realmes Commonly Used in Physicke/ First set foorth in the Doutche or Almaigne Tongue ... and nowe first Translated out of French into English by Henry Lyte. Antwerp: Hendrik van der Loe for Gerard Dewes, 1578.
When Jan van der Loe died in 1563, his son Hendrik inherited the wood-blocks that his father had used for the illustration of Dodoens's *Cruijdeboeck*. When Hendrik was commissioned by the English publisher Gerard Dewes to print the English translation of Dodoens's herbal, he used the same wood-blocks. Henry Lyte's English translation was based on Carolus Clusius' French version (1557) and the second Dutch edition (1563). The English edition, which went through several re-editions up to 1619, contained no less than 872 woodcuts, which made it the most richly illustrated edition of all versions published up to then. Dodoens's herbal exerted a considerable influence on English botany. John Gerarde's *The Herball* (London, 1597), the best-known English herbal of its time, owes a great deal to the *Cruijdeboeck*. The illustrations of the 1633 and 1636 editions of Gerarde's herbal were

printed from the several hundreds of wood-blocks of Dodoens's herbal, which had been shipped from Antwerp to London for the purpose.

Historia vitis vinique et stirpium nonnullarum aliarum; item Medicinalium observationum exempla. Coloniae: Apud Maternum Cholinum, 1580.

Remberti Dodonaei Medici Caesarei Medicinalium Observationum exempla rara, recognita et aucta: Accessere et alia quaedam ... Coloniae: Apud Maternum Cholinum, 1581.

Tractatus de magno animali, quod Alcen nonnulli vocant, Germani vero Elend, et de ipsius partium in re medica facultatibus. Item historia cervi rangiferi et gulonis Filfros vocati/ ... Accessit Remb. Dodonaei ... de Alce epistola. Coloniae: Apud Maternum Cholinum, 1581.

Stirpium historiae pemptades sex. Sive libri XXX. Antverpiae: Ex officina Christophori Plantini, 1583.

Medicinalium observationum exempla rara, Accessere et alia quaedam. Lugduni Batavorum: Ex officina Christophori Plantini, 1585.

A niewe herball or historie of plantes: wherein is contayned the whole discourse and perfect description of all sortes of herbes and plantes: their divers&sundry kindes: their straunge figures, fashions, and shapes: their names, natures, operations, and vertues: and that not onely of those which are here growing in this our countrie of Englande, but of all others also of forrayne realmes, commonly used in Physicke / First set foorth in the Doutche or Almaigne tongue, by that learned D. Rembert Dodoens ... and nowe first translated out of French into English by Henry Lyte. London: Edm. Bollifant, 1595.

Cruydt-Boeck van Rembertus Dodonæus, volgens sijne laetste verbeteringe met biivoegsels achter elck capittel, uut verscheyden cruydtbeschrijvers: item in 't laetste een beschrijvinge vande Indiaensche gewassen, meest getrockenuut de schriften van Carolus Clusius. Leyden: Fr. van Ravelingen, 1608.

Remberti Dodonaei Mechliniensis Medici Caesarei Stirpium Historiae Pemptades Sex Sive Libri XXX. Varie ab auctore paullo ante mortem, aucti et emendati. Antverpiae: ex Officina Plantiniana, Apud Balthasarem et Ioannem Moretos, 1616.

Praxis medica Remberti Dodonaei Mechliniensis. Amsterdami: Impensis Henrici Laurentij, 1616.

Cruydt-Boeck van Rembertus Dodonæus, volgens sijne laetste verbeteringe Met Biivoegsels achter elck Capittel, uut verscheyden Cruydtbeschrijvers: Item in 't laetste een Beschrijvinge vande Indiaensche Gewassen, meest getrocken uut de Schriften van Carolus Clusius. Tot Leyden: Inde plantijnsche druckerije van Françoys van Ravelingen, 1618.

Medicinalium observationum exempla rara, Accessere et alia quaedam. Hardervici: Apud viduam Thomae Henrici, impensis Henrici Laurentii, 1521 [i.e. 1621].

In D. Remberti Dodonaei praxim artis medicae / ... Sebastiani Egberti Cos. scholia, cum auctario annotationum Nicolai Fontani ... Amstelodami: Sumptibus Hendrici Laurentii, 1640.

Cruydt-Boeck Remberti Dodonaei, volghens sijne laetste verbeteringhe met bijvoeghsels achter elck capitel, uyt verscheyden cruydtbeschrijvers Item, in'tlaetste een beschrijvinghe vande Indiaensche ghewassen, meest ghetrocken uyt de schriften van Carolus Clusius. Nu wederom van nieuws oversien ende verbetert. Antwerpen: Inde Plantijnsche Druckerije van Balthasar Moretus, 1644.

Dodoens published his *Stirpium historiae pemptades sex* in 1583, two years before his death. It presents a comprehensive arrangement of the plant world known in his time from a utilitarian perspective. It was the synthesis of all his preceding herbalist publications. The book is divided in six *pemptades*, which consist of five 'books' ('pemptas' refers to 'pente', the Greek word for 'five') each. The first *pemptas* contains an introduction and five 'books' about miscellaneous plants that do not fit any of his classes. The second *pemptas* deals with flowers, odoriferous and umbelliferous plants. The third *pemptas* is devoted to descriptions of medicinal herbs, the fourth deals with cereals, leguminous plants, marsh and water plants, the fifth with vegetables and the sixth with fruit-bearing plants and trees. In all there are 26 groups of plants. The plants are illustrated in no less than 1306 woodcuts. Plantin re-used the wood-blocks that had been cut for illustrating earlier works of Dodoens's, as well as those that he had used for the books of Carolus Clusius and Mathias Lobelius. In 1581 Plantin had purchased the wood-blocks that Jan van der Loe had owned from the latter's widow. This Latin version was translated into Dutch and published in 1608 under the title *Cruydt-Boeck van Rembertus Dodonæus, volgens sijne laetste verbeteringe ... etc.*, which also includes the revisions made by Joost van Rafelingen, brother of the printer Françoys, grandson of Christopher Plantin, and a competent physician in his own right. The Latin version was reprinted in 1616. The Dutch editions of 1618 and 1644 are reprints of the 1608 edition. The 1644 edition is 1492 pages long and contains 1367 woodcuts.

As is testified to by the various editions it went through for over a century, Dodoens's *Cruijdeboeck* remained a standard reference for a long time, and in Flanders even until the beginning of the nineteenth century. He wrote his works both in what was called then *Nederduytsch* (Dutch or Flemish) and Latin. The list of his publications known today includes twenty-five titles on botany, eight on medicine, two on geography and astronomy and numerous almanacs. In his plant descriptions he included the name in Dutch (Flemish), German, "Bohemian", Hungarian, Slovenian, Croatian, French, English, Italian, Spanish, Arabic, Latin and Greek, "Indian" and "a few other foreign languages". The wood-blocks used for printing the woodcuts in the *Cruijdeboeck* are still preserved in the Plantin Moretus Museum in Antwerp. Dodoens's memory lives on in two statues: one in the *Kruidtuin* (botanical garden) of Mechelen and one in the *Kleine Zavel*, a little public garden in the centre of Brussels. Busts of him are found in several universities and museums throughout the Benelux.

REFERENCES

1 Louis, Armand. *Geschiedenis van de plantkunde: de plantkennis in dienst van farmacie en landbouw.* Gent: Story-Scientia, 1977.

2 Nissen, Claus. *Die botanische Buchillustration: ihre Geschichte und Bibliographie.* Stuttgart: Hiersemann, 1951.

ILLUSTRATIONS

ILL 1 Portrait of Rembertus Dodonæus. Engraving by Esme de Boulonnois, 1682, 186 x 138 mm. From: Isaac Bullart (1599-1672), *Académie des sciences et arts*, ed. Jacques-Ignace Bullart, Brussel, 1682, subsequently reproduced in Jan Frans Foppens, *Bibliotheca Belgica, sive Virorum in Belgio vita, scriptisque illustrium catalogus, librorumque nomenclatura*, Brussel, 1739. Collection Katholieke Universiteit Leuven, Central Library, Prentenkabinet PA 2909.

ILL 2 Autograph dedication and signature of Dodoens in the *Album amicorum* of Wilhelm Helbig of Würzburg, fol. 204. The dedication reads: "Nobili ac generoso domino Dño Gulielmo Helbich scribebat Rembertus Dodoneus medicus ad Caesaream maiestatem proficiscens. A° DM LXXIIII.XVIII Septembris" (i.e. Dodonæus, physician, on his way his Imperial Majesty, wrote this for the noble and magnanimous Lord Wilhem Helbig, 18 September of the year 1574). Collection Katholieke Universiteit Leuven, Central Library, Tabularium, BATAB.

ILL 3 Title page of *Cruijdeboeck*, 1554. Collection Université catholique de Louvain, Bibliothèque générale et de sciences humaines.

ILL 4 Title page of *Purgantium aliarumque eo facientium, tum et Radicum, Convolvulorum ac deleteriarum herbarum historiae Libri IIII*, 1574. Collection Université catholique de Louvain, Bibliothèque générale et de sciences humaines.

ILL 5 Title page of *Medicinalium Observationum exempla rara, recognita & aucta*, 1581. Collection Université catholique de Louvain, Bibliothèque générale et de sciences humaines.

ILL 6 Title page of *Praxis medica Remberti Dodonaei Mechliniensis*, 1616. Collection Université catholique de Louvain, Bibliothèque générale et de sciences humaines.

ILL 7 Title page of *Stirpium Historiae Pemptades Sex Sive Libri XXX*, 1616. Collection Université catholique de Louvain, Bibliothèque générale et de sciences humaines.

ILL 8 Title page of *Cruydt-Boeck*, 1618. Collection Katholieke Universiteit Leuven, Central Library, Tabularium, B 3998.

ILL 9 Title page of *Cruydt-Boeck*, 1644. Collection Katholieke Universiteit Leuven, Central Library, Tabularium C1827

ILL 10 Title page of *Cruydt-Boeck*, 1644. Collection Albert Couvreur, Ecole de Pharmacie, Université catholique de Louvain, Bruxelles (Woluwe Saint-Lambert).

ILL 1

ILL 3

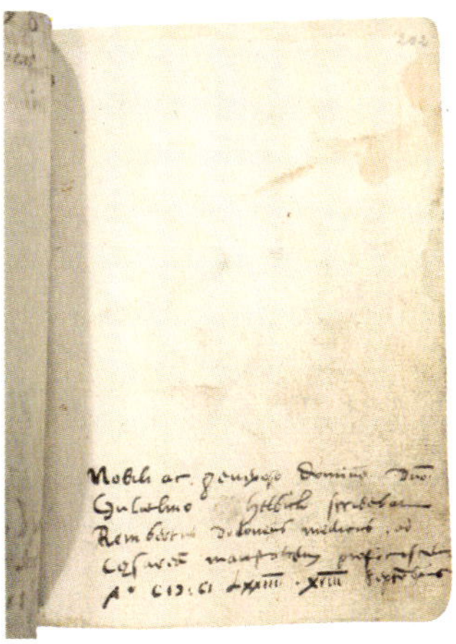

ILL 2

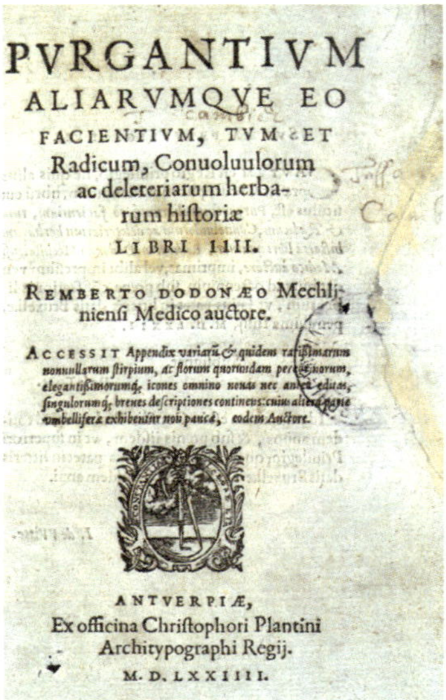
PVRGANTIVM ALIARVMQVE EO FACIENTIVM, TVM ET Radicum, Conuoluulorum ac deleteriarum herbarum historiæ LIBRI IIII.

REMBERTO DODONÆO Mechliniensi Medico auctore.

ACCESSIT Appendix variarū & quidem rarissimarum nonnullarum stirpium, ac florum quorumdam peregrinorum, elegantissimorumq; icones omnino nouas nec antea editas, singulorumq; breues descriptiones continens: cuius altera parte vmbelliferæ exhibentur non paucæ, eodem Auctore.

ANTVERPIÆ, Ex officina Christophori Plantini Architypographi Regij. M. D. LXXIIII.

ILL 4

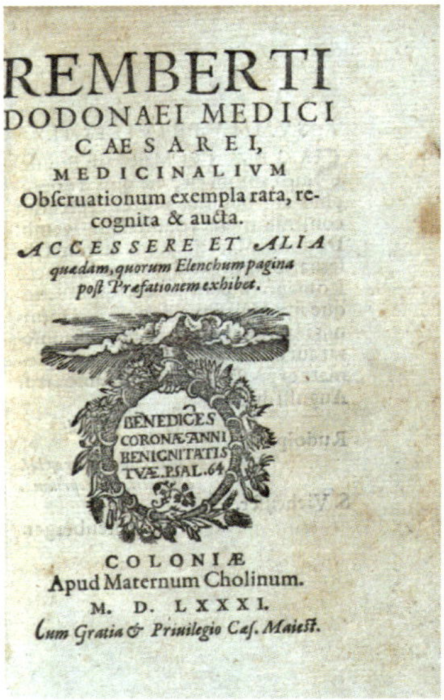
REMBERTI DODONAEI MEDICI CÆSAREI, MEDICINALIVM Obseruationum exempla rara, recognita & aucta.

ACCESSERE ET ALIA quædam, quorum Elenchum pagina post Præfationem exhibet.

COLONIÆ Apud Maternum Cholinum. M. D. LXXXI. Cum Gratia & Priuilegio Cæs. Maiest.

ILL 5

PRAXIS MEDICA REMBERTI DODONÆI MECHLINIENSIS.

AMSTERDAM, Impensis Henrici Laurentij. ANNO 1616.

ILL 6

REMBERTI DODONÆI MECHLINIENSIS MEDICI CÆSAREI STIRPIVM HISTORIÆ PEMPTADES SEX SIVE LIBRI XXX. Variè ab AVCTORE, paullò ante mortem, aucti & emendati.

ANTVERPIÆ, EX OFFICINA PLANTINIANA Apud Balthasarem et Ioannem Moretos. M. DC. XVI.

ILL 7

ILL 8

ILL 9

ILL 10

II

DODONÆUS AND THE HERBAL TRADITION

Robert Visser
University of Leiden and University of Utrecht

The Greek scientist Theophrastus of Eresus, a pupil of Aristotle and his successor as head of the Peripatetic school at the Lyceum, is generally regarded as the father of European botany. The authoritative *Dictionary of Scientific Biography* was very specific when it stated that he laid "the groundwork for modern botany."[1] One cannot deny that the Greek natural philosopher has rather good credentials for this position. Theophrastus was, first of all, the author of the oldest distinctly botanical writings that are still extant.[2] More substantial claims for historical recognition must of course be founded on the actual contents of these writings. His methods and aims look indeed surprisingly modern. Theophrastus attached great importance to a critical and empirical approach. Undoubtedly inspired by his teacher, he recognised the diversity of plant forms as the central problem of botany. He used morphological and physiological data as the principal means to solve it. Besides, we find in his work the first signs of some kind of natural classification. Such traits link Theophrastus closely to the new botany that started to emerge at the beginning of the early modern period. Theophrastus' parenthood is remarkable in the sense that its first recognisable offspring was only born in the first half of the sixteenth century, more than eighteen centuries after his death. During that long interval, the majority of botanists, instead of following his example of a truly scientific study of plants, concentrated their attention on the practical aspects of their science. During the Middle Ages Western botany derived its *raison d'être* first and foremost from the services it rendered or was supposed to render to agriculture and especially to medicine. The medieval botanists drew their inspiration and guidance not from Theophrastus but from the Greek physician Dioscorides.

About the middle of the first century A.D. Dioscorides had written a comprehensive book that became widely known as *De materia medica*, the Latin translation of its title. The book was for the greatest part devoted to plants. Dioscorides manifested himself as an experienced and knowledgeable phytographer. It is quite clear however that he was not primarily interested in plants *per se*. Dioscorides' subject was medical botany and his chief concern was the remedial use of plants and other natural products. His book was written in Greek and therefore inaccessible to most of the European botanists. This did not prevent its author from being the most popular, influential and the most highly esteemed botanist for a very long time.

It is an often-told story how in the Renaissance Dioscorides' work became one of the factors that induced a drastic change in botanical practice. When European botanists tried to correlate his plant descriptions to the species of their native floras, they inevitably discovered considerable discrepancies. The ensuing doubts about the universal applicability of Dioscorides' descriptions were an important stimulus for botanists of the early sixteenth century to reconsider their ways of obtaining scientific knowledge. They soon became convinced that the classical texts and the one produced by Dioscorides in particular should be rejected as the main source of factual information. It was argued that instead of relying on what botanists from preceding ages had written, which until then had been a fairly widespread procedure for collecting scientific data, the botanists should consider it their primary duty to consult nature directly.

It is amazing to observe how easily a practice was abandoned that had reigned supreme for many centuries. In just a few decades there grew a consensus in the European botanical community that only the empirical approach could result in meaningful scientific knowledge. From now on this method would never be regarded as anything less than the guiding principle for all investigations in the field of botany. German botanists are considered, with good reason, to have been the pioneers of this fruitful methodological innovation. The notable names are those of Otto Brunfels (ca 1489-1534), Jerome Bock (Tragus) (1498-1554), Leonhard Fuchs(ius) (1501-1566) and Valerius Cordus (1515-1544). Not without some nationalistic bias the German historian of botany Kurt Sprengel introduced Brunfels, Bock and Fuchs as the "German fathers of botany"[3] and considered their herbals, published between 1530 and 1542, as inaugurating a completely new era in the development of the plant sciences.[4]

The authors glorified by Sprengel wrote books that are clearly not of equal merit. Brunfels's herbal broke new ground with the life-like illustrations made by the artist Hans Weiditz. His text on the other hand was entirely conventional. Brunfels himself indicated that he had extracted it from what he called "ancient and trustworthy authors".[5] Bock and especially Cordus[6] were the full-blooded empiricists among the German botanists. They based their descriptions on a logical and systematic application of the empirical doctrine. By studying the plants with their own eyes they enriched botanical science with quite a number of original observations. In view of the often rather exaggerated views on the revolutionary merits of the German fathers of botany – views expressed by Sprengel and many later historians of botany[7] – it is good to keep in mind that the changes brought about by these botanists, did certainly not constitute a complete break with the past. Their important innovations were primarily of a methodological nature. Most of the German fathers were rather traditional however in regard of the ultimate aims of their research.

Brunfels, Fuchs and to a lesser degree Cordus were also basically medical botanists. They regarded plants first of all as raw material for medicines. For them the new methodological practice was a means to assist the physician in better identifying the plants and thus improving the range and quality of his herbal drugs. Bock on the other hand was more instilled with the philosophical spirit of Aristotle and Theophrastus than with the practical spirit of Dioscorides. He was the only German father of botany who deserves to be called a scientist. Bock's investigations were not guided by the needs of the medical profession. He focused on the plants for their own sake. His botanical studies were driven by a sincere wish to analyse as many biological aspects of his objects as possible.

The developments that were started in Germany were assured a fruitful continuation in the Low Countries, where a flourishing botanical culture originated in the second half of the sixteenth century. One of the most formative influences on this process was the central figure in this collection of essays. Rembert Dodoens (1517-1585) was the first in this part of the world to follow in the footsteps of the German fathers of botany – for Sprengel apparently a reason to honour him with the title of "one of the oldest and most important fathers of botany".[8] Dodoens did not remain the only one for long. Within a few years he received the company of Carolus Clusius (Charles de l'Escluse,

1526-1609) and Matthias de Lobel (1538-1616), who investigated the vegetable kingdom in a similar way as their countryman.

Dodoens had started his career as a physician.[9] He extended his sphere of activities when the Antwerp publisher Jan van der Loe asked him to write a herbal that was apparently intended for a broad readership, since it had to be in the vernacular. Dodoens complied with this request. Preceded by some smaller botanical writings, his voluminous *Cruijdeboeck* was published in 1554. With this substantial contribution to botanical literature, the study of plants became a major object of Dodoens's scientific interests for the rest of his life.[10] His final literary achievement as a botanist was a considerably enlarged (the number of plants was almost doubled) and thoroughly rewritten and rearranged version of the *Cruijdeboeck*. It appeared in 1583 as the *Stirpium historiae pemptades sex* and was published in Antwerp by the famous printing office of Christophe Plantin. Dodoens's herbal met with considerable success. The original version was translated into French (1557), English (1578 and subsequent editions) and Japanese (1790s) and was reissued several times. In 1608 the *Stirpium historiae* version appeared in a Dutch translation that had been prepared by Dodoens himself. A new edition of the *Cruijdeboeck* was issued even as late as 1644. Dodoens's book has been rightly qualified as one of the most popular herbals of the sixteenth and early seventeenth century. It seems that in his native country Dodoens's popularity even lasted well into the nineteenth century. In 1850 one of the members of the Belgian medical community stated that "there is not a pharmacist in the Flemish part of Belgium who does not possess a Dodoens and does not use its pictures everyday for identifying the wild species of medicinal value".[11]

The *Cruijdeboeck* was conceived by the publisher as a kind of encyclopaedia on plants, with special attention to their medical virtues. Van der Loe had urged Dodoens to use Fuchs's herbal as a model and Dodoens had agreed to this suggestion. Hence the fact that Dodoens's book relies heavily on the illustrations, the phytography and the pharmaceutical instructions of his German colleague. However, this certainly does not mean that Dodoens was a mere translator or a slavish copier and that his *Cruijdeboeck* lacks in originality. It is true that most of the illustrations of the *Cruijdeboeck* were borrowed directly from Fuchs, but that was primarily a decision of the publisher, based on commercial considerations. In the medical part Dodoens adheres closely to the original. In the botanical texts we meet with a somewhat different Dodoens. There he clearly shows that he did not lack critical sense. He demonstrated moreover an independent mind whenever he added novel and original observations to Fuchs's descriptions and that is something he did regularly. Dodoens dealt with substantially more species than Fuchs had done, included a lot more details in his descriptions and provided his readers with better information on the places where the plants grow.

All taken together we have to ascertain that, although the first edition of the *Cruijdeboeck* showed numerous and often very promising signs of Dodoens's empirical leanings, the new method played as yet a rather modest role in his fact-finding activities. Therefore, I think that we give Dodoens too much credit when we subscribe to the opinion, advanced first by historians of science of an older generation,[12] but also shared by contemporary ones, that this first version of his herbal is a major example of the fundamental methodological shift that was taking place in the botany of that time.

It was only after the publication of the first edition of the *Cruijdeboeck* that empirical research developed into an essential element of Dodoens's investigative practice in botany. It really became a prominent activity from the early sixties onwards, when he was preparing what would become his *magnum opus*, i.e. the *Stirpium historiae pemptades sex*. Dodoens's activities in this period make it quite clear that he now rated the investigation of the plants themselves as the principal source of phytographical data. His methodological position had changed considerably and when his life's work was published, Dodoens had become one of the leading empiricists in botany. It is interesting to note that while until then the methodological innovation of the sixteenth-century herbals had been mainly embodied in their illustrations drawn from nature, with Dodoens it shaped the verbal descriptions as well.

Several resources were used by Dodoens to give his botany the necessary empirical foundation. There are strong indications that he studied material in herbaria. It seems however that these collections of dried plants were only of minor importance. He was much more interested in investigating living plants. The numerous ornamental gardens in the Low Countries were an ideal place for pursuing this line of research. Dodoens was a frequent visitor of such gardens and amply availed himself of the opportunities they offered the botanist. The majority of the more than hundred new species he described were garden plants. Field trips near the places where he lived were another important means of gathering information. Dodoens was a pioneer in exploring the local floras of the Low Countries. Even in his old age he was still taking time off to herborize and study plants in their natural habitat.[13] These activities provided him with a rather extensive knowledge not only of floristics but also of the ecology and especially the sociology of plants. His varied experiences found their way into the *Stirpium historiae*, where he referred time and again to his own observations of these topics.

Dodoens was indeed a keen observer. The descriptions in the *Stirpium historiae* are a testimony to his abilities in this respect. They capture the characteristic morphological properties of the plants, which are often emphasised by comparing them to those of related species, and give a clear and minute, although not too detailed, picture of their appearance. Dodoens made it easy for his readers to form an idea of the habitus of the plants. The verbal descriptions in the *Stirpium historiae* are almost unparalleled as an aid in identifying plants. If we compare Dodoens with his colleagues we have to conclude that he was among the best and perhaps even the best phytographer of his time. The popularity of his writings made Dodoens an influential propagandist of the empirical practice in botanical science. There is no doubt that its rapid spread in the Low Countries owed much to him.

The methodological changes sketched above were one of the major developments that left their mark on sixteenth-century botany. The growing attention to taxonomical matters was another determinant of great importance. Here too, Dodoens made an important contribution. Before dealing with it, let us have a closer look at what his predecessors and contemporaries had achieved in botanical classification. Theophrastus had already studied the possibilities of arranging plants in groups. He distinguished four primary *divisions*: trees, shrubs, half-shrubs and herbaceous plants. At a lower level he ranged the species, especially those of the fourth division, in subgroups, which coincide more or less with some of our natural families, like grasses and *umbelliferae*. This rather sophisticated arrangement remained unsurpassed for many centuries. Between Theophrastus and the Renaissance, attempts at botanical systematisation usually resulted in classifications of a rather pragmatic nature. They displayed a striking lack of consistency and of naturalness in the criteria for establishing the relationships between plants. Plants could be ordered by means of utilitarian principles, with the consequence that we find in one and the same system disparate groups like "ornamental plants", "aromatic plants", "medicinal herbs" etc. A favourite method was to range the plants alphabetically, according to whatever name in whatever language the author chose to give them. It will be clear that in neither case much light was thrown on the morphological affinities between plants.

On the other hand we find several botanists in the sixteenth century who showed an awareness of the existence of a natural order in the plant world and who seemed to have been convinced that it was their duty to try to reveal it. Of course a full-blown classificatory science did not develop overnight. Explicit and elaborate views on the discipline's aims, principles and diagnostic procedures were still lacking. Moreover, the concrete attempts at classification were generally restricted to the *species* and *genus* level. Nevertheless, classification was again on the botanical agenda and this time it was there to stay. The revived interest in a more natural classification was undoubtedly tributary to Theophrastus, whose botanical writings had become better known in the West towards the end of the fifteenth century, when the first Latin translation was published (1483). Among the German fathers of botany Jerome Bock was an outspoken supporter of the new systematics. He emphatically rejected the utilitarian and alphabetical classifications as unscientific and emphasised the need to try and find a natural system of plants. Bock practised what he preached. His classification was primarily concerned with the *genera*. Seen against the backdrop of his time, his classificatory attempts were meritorious in every way. His colleague Valerius Cordus worked in the same spirit and went one step further. He was the first who made substantial efforts to unite related *genera* in the same family.

In view of the fact that Leonhard Fuchs provided the model for the first edition of Dodoens's herbal, it is useful to see what he achieved in the field of systematics. In the historical literature on Dodoens, Fuchs is usually pictured as an old-fashioned scientist without any notable interest in the new trends of botanical systematics. This judgement is not entirely fair and perhaps inspired by too great an admiration for the pioneering qualities of Dodoens. It is indeed true that Fuchs arranged his plants alphabetically. But it is essential to keep in mind – and that was missed by the majority of the historians referred to above – that it was an arrangement of *genera* and not of *species*, and that in quite a lot of these genera Fuchs brought together plants that according to the then accepted criteria resembled each other. It is undeniable that

Fuchs was much less concerned with classification than for instance Bock, Cordus and later Dodoens, but this does not mean that he had no part in the quest for a natural order among plants, albeit in a modest way. This leads us to the conclusion that as for botanical classification he may have been of greater importance to Dodoens than is commonly suggested.

Dodoens demonstrated right from the start an involvement with classificatory matters. The more than 1000 plants in the first edition of the *Cruijdeboeck* were split up in five main divisions. The creation of these divisions was an aid to structuring his book in a way no printed European herbal had been before. Dodoens was however rather traditional in his choice of characteristics to differentiate these divisions. He based his classification on the useful properties of the plants. Contrary perhaps to what one might expect after such an orderly beginning, the arrangement within the divisions was for the greater part chaotic and without apparent order. It was only in a few occasions that he grouped together those plants that showed morphological similarities. Although we can ascertain in Dodoens an actual interest in plant systematics, it is obvious that his first results, as published in the *Cruijdeboeck* in 1554, were in no way ahead of his time.

However, during the following decades Dodoens developed his systematics considerably, as appears from the *Stirpium historiae*. The book presented a classification that was much more elaborated than what any of his predecessors had produced. It contributed to its recognition as a pioneering work. There is a kind of consensus that it is one of the landmarks in the history of the plant sciences. According to the nineteenth-century historian of botany Ernst Meyer it was "a first crude attempt at a scientific arrangement of plants".[14] In the *Stirpium historiae* Dodoens started by dividing the plants in the same five major groups he had used 30 years earlier in his *Cruijdeboeck*. Not insignificantly, he now called them *pemptades* and equated them with *classes*. Entirely new was their further subdivision into a total of 25 subgroups. In doing so Dodoens gave an original extension to the hierarchical structure of the botanist's classificatory schemes. The definition of these groups was less innovative. Dodoens had again recourse to utilitarian characteristics. Only in a few cases did he employ more natural criteria. As a consequence the composition of these groups was on the whole rather heterogeneous, at least from our point of view.

Dodoens's real merits as a systematist are to be found in the way in which he handled groups at the level of *genera* and *families*. Here he showed an acute insight into the purpose of botanical classification. By these groups Dodoens put into practice what he had announced in the short introduction to the *Stirpium historiae*, namely that he would bring together those plants that resemble each other and separate the ones that did not, and that he would do this on the basis of their "forma et figura" (= morphology and habitus). This procedure enabled him to recognise many of the larger genera and families and even to establish relationships between plants that do not show much resemblance at first sight. While most of his predecessors had regarded genera as elementary taxa, Dodoens was one of the first to treat them as a composition of species and that gave his systematics a distinctly modern touch. The same holds for his nomenclature. Another result of his focusing on genera was Dodoens's almost strict adherence to a binary generic nomenclature, a type of nomenclature that the eighteenth-century Swedish botanist Carolus Linnaeus would make into a cornerstone of biological taxonomy, which it still is today.

The views expressed in the preceding paragraphs are based on a reading of the *Stirpium historia* as a botanical text. It is good to realise that in this way we only get a part of the story that can be told about this important and influential book. To do full justice to Dodoens's intentions, our story should also pay attention to its medical and pharmaceutical aspects. There is, or at least there was in the past, some debate about the relative importance of the botanical and the medical parts of the *Stirpium historiae*. During the Dodoens celebrations of 1917, Hunger, who later became known as the author of an impressive monograph on Clusius, defended the thesis that in writing his herbals Dodoens had turned from a physician into a botanist.[15] Louis, another specialist on sixteenth-century botany and writing much later, was not quite sure if such transformation had taken place. He was ultimately inclined to the opinion that Dodoens had always remained a physician and that his herbals were intended for medical purposes.[16]

If we consider the contents of the *Stirpium historiae* quantitatively, we have no reason to doubt that we are dealing with a book that is primarily botanical. The descriptions of a purely botanical nature, including the classificatory discussions, take up by far the greater part of the book. Here Dodoens went to far greater lengths than any other author of a herbal had done before. The assumption that he may have had some kind of botanical agenda and that his plant descriptions were no means to medical ends but an end in itself seems moreover to be vindicated by his statements about classification in the introduction of the *Stirpium historiae* to which I referred above. We cannot but regret that Dodoens never presented unambiguously his own viewpoint on the aims he had in writing this book. Whatever Dodoens's intentions, his *Stirpium historiae* undoubtedly has a place in the history of botany. He is one of the exceptions that Frank Egerton may have had in mind when he wrote in the introduction to Edward Greene's *Landmarks of Botanical History* that "most of the history of botany before 1700 was really the history of pharmacy".[17] Before making a few final comments on Dodoens's place in the botany of his century, I would like to stress that any serious attempt at assessment is hampered by a lack of really thorough and up-to-date historical analyses of his voluminous botanical production. We are even less well informed about its influence, especially in Western Europe. Most of what has been written about Dodoens is fragmentary and dates from many years back. Besides, not a few of these studies are coloured by strong hagiographic tendencies. In view of this situation my concluding remarks on his place in the history of sixteenth-century botany are necessarily of a tentative nature.

In the broad perspective of European botany Dodoens attracts first and foremost the attention because of his systematics. He certainly was one of the pioneers in this field. Dodoens's historical significance is not merely determined by the actual results and the scale of his classificatory activities. Classification was at that time more than just a new botanical specialty. It was also being propagated as a means to sever botany's ties with medicine and give it the status of an independent scientific discipline. In the same year that the *Stirpium historiae* appeared, the Italian physician and botanist Andrea Cesalpino published an important and influential theoretical study entitled *De plantis libri XVI*. In this book Cesalpino voiced his regrets that botany had fallen in the clutches of medicine. He proved himself to be a zealous advocate of an autonomous botany. He was convinced that botany could win a position of its own if and when its practitioners concentrated on taxonomy. This branch of botany was in his view pure science and was supposed to be of no relevance to medicine or any other utilitarian pursuit. Already in the sixteenth century the movement started by Cesalpino gained considerable momentum. It led to lasting results in the following centuries, when scholars like Tournefort and Linnaeus completed the process and gave botany definitely a place of its own among the life sciences.

If Dodoens had any ideas similar to those of Cesalpino, he expressed them neither explicitly nor implicitly. Therefore we cannot simply connect his attempts at systematics with the pursuance of botany for botany's sake. On the other hand it is hardly conceivable that there was not any interaction between the theoretical and the practical dealings with taxonomy. This certainly applies where Dodoens is concerned. He was one of the most seminal practitioners of plant systematics of the late sixteenth century. In particular, he demonstrated how to do classificatory research and also that it actually worked. The *Stirpium historiae* showed that comparative research can reveal meaningful patterns in the plant world. Dodoens's text could easily be interpreted as proof that classification yielded results that could be accepted as science, thus giving concrete foundation to the claims of Cesalpino and his allies with regard to the emancipatory function of taxonomy. We can at least credit Dodoens with an indirect and supportive role in the movement towards an independent science. Besides empiricism and classification, to which he also contributed, this was another characteristic of the modernisation process that European botany underwent during the sixteenth century.

NOTES

1 See J.B. McDiarmid, *Dictonary of Scientific Biography*, vol. 13 (1976). s.v. "Theophrastus."
2 I.e. *Historia plantarum* and *De causis plantarum.*
3 Kurt Sprengel, *Geschichte der Botanik* (Altenburg/Leipzig, 1817), vol. 1, 258.
4 The books concerned were: Brunfels, *Herbarum vivae icones* ... (1530); Bock, *New Kreütter Buch* ... (1539); Fuchs, *De Historia stirpium* ... (1542).
5 Brunfels, *Herbarum*, dedicatory epistle.
6 His botanical writings were published posthumously by Conrad Gesner in 1561.
7 E.g. K.F.W. Jessen, *Botanik der Gegenwart und Vorzeit in culturhistorischer Entwicklung. Ein Beitrag zur Geschichte der abendländischer Völker* (Leipzig, 1864), 176 seq.
8 Sprengel, op. cit. (see note 3), 307.
9 The only book-length biography of Dodoens is the rather obsolete study by P.J. Van Meerbeeck, *Recher ches historiques et critiques sur la vie et les ouvrages de Rembert Dodoens (Dodonæus)* (Malines, 1841).
10 The best analysis of Dodoens's botany is to be found in Edward Lee Greene and Frank N. Egerton, ed., *Landmarks of Botanical History* (Stanford: Stanford University Press, 1983), vol. 2, 847-876.
11 P.-J. D'Avoine, *Eloge de Rembert Dodoëns, médecin et botaniste Malinois du XVIe siècle* ... (Malines/ Bruxelles, 1850), 53-54.
12 As e.g. A. Louis, "Over het leven en het botanisch werk van Rembert Dodoens (1517-1585) ...", *Biologisch Jaarboek* 21 (1954): 262-264 and passim.
13 See A. Louis, " Critische beschouwingen bij de Dodoens-brief van 26 november 1583," *Mededelingen van de Koninklijke Vlaamsche Academie voor Wetenschappen, Letteren en Schone Kunsten van België* 16 (1954), no. 13.
14 E.H.F. Meyer, *Geschichte der Botanik* (Königsberg, 1857), vol. 4, 348.
15 F.W.T. Hunger, "Dodonée comme botaniste," *Janus. Archives internationales pour l'Histoire de la médecine et la géographie médicale* 22 (1917): 156.
16 Louis, op. cit. (see note 12), 244.
17 Greene, op. cit. (see note 10), 5.

III

BOTANICAL PRACTICE AND AGRICULTURAL DEMANDS IN EARLY MODERN EUROPE: THE CONTRIBUTION OF REMBERTUS DODONÆUS

Mauro Ambrosoli
Università di Udine

This paper will discuss the works and the activities of Dodoens in respect of (1) his study and travels in continental Europe associated with de l'Escluse, (2) Dodoens's and L'Obel (de Lobel)'s book production and botanical practice in England, (3) the English translation of Dodoens's *Cruijdeboeck*, practical information and the English agricultural demands, and (4) the impact of Dodoens's books on seed production and the interaction with a developing agriculture in the early seventeenth century.

1 Preliminary remarks

To evaluate the contribution of Rembert Dodoens to the construction of botanical discourse we must keep in mind that he belonged to the second generation of European botanists. Since the formal teaching of botany took place in the faculties of medicine, these scholars were mostly interested in medical botany. They were doctors of medicine themselves and eager to obtain top positions as private doctors of the Emperor, the Pope and other monarchs. Others, more simply, were active in the major hospitals of their days. Dodoens was a contemporary of scholars such as de l'Escluse, L'Obel, Anguillara, Mattioli, Cesalpino, d'Aleschamps, Gesner and the brothers Bauhin, to name but a few. As a result of their work as botanists, they all collected regional floras out of local specimens, which became the base of European botany and provided the general information for what was due to become botanical taxonomy. Dodoens himself made a valuable attempt towards a natural classification and de l'Escluse's contribution was particularly important in respect of regional European floras thanks to his extended travels from Portugal to Hungary.[1]

Practical research was pushing botanists and scholars to exchange written information and samples of specimens, plants and seeds, to find a solution for the many new problems of taxonomy, faced by Renaissance scientists, who were operating at an international level. At the same time new fractures were opening in mid-sixteenth century Europe even in the small world of men of science.[2] To appreciate the role of Rembert Dodoens and his contribution to the organisation of modern botany, we must keep in mind the social dimension of botanical discourse. I wish to illustrate how political and cultural divisions, which occurred in sixteenth, and early seventeenth century Europe, made the creation of botanical discourse more difficult and delayed the adoption of an international classification.

The growing demand for a complete botanical treatise was a common feature of the European book market. The botanists mentioned above performed a very useful task: they offered the European public general works and comments upon the plants that had been described in Greek and Latin (and later Arab) medical tradition, to which they added a mass of specialised information collected from direct experience. Furthermore, the public for the big (and expensive) botanical treatises was growing after 1550: whilst first editions were generally written in Latin and published everywhere in Europe, numerous translations into Italian, French, German and Dutch, Czech, Polish, English and Spanish frequently appeared in print. Although these numerous editions contained many differences vis-à-vis the original works, the translations did enhance the chance of European botany becoming a fully-fledged science rapidly. These translations helped some plant names to become more widely used than others in the description of old and new plants. In this process, famous institutions, such as the botanic gardens of Padua or Leiden, the Hôtel Dieu in Lyon or the Vatican Garden in Rome became better known and were visited more often by scholars and gentlemen who had acquired an interest in botany.

A solution to the twin problems of information (the great number of new plants never described before) and nomenclature (which names had to be given to plants unknown to the ancients) might possibly be found thanks to practical experience and some kind of agreement among scientists.[3] Although various social groups throughout Europe (male and female members of the religious orders, gentlemen, scholars, but also gentlewomen and housewives) were developing new interests in the natural environment, which were founded on practical experience, scientific unity was a long way to come. Local and regional economic and cultural forces quickly responded to this scientific demand even in the midst of a very awkward political situation: the Emperor Charles V and his opponent Francis I of France were throwing every available resource into their epoch-making contest. As a result of this political polarisation, economic and scientific competition grew stronger around regional centres of power.

Furthermore, Dodoens's contribution endured into the following generations thanks to the numerous editions printed by Plantin. Many of the scientific problems connected with Rembert Dodoens's activities, and somehow brought to light by his work, found a solution in the following decades thanks to the interests and the exacting demands made by a larger public of consumers. Rembert Dodoens's achievements were a good representation of the possible solutions, that were available to a developing discipline in an age of deep crisis. My first acquaintance with Rembert Dodoens dates back to some years ago when I was working on the agricultural revolution in Western Europe and pointed out that the growth of botany preceded and aided agricultural modernisation during the sixteenth century. Rembert Dodoens's contribution to the creation of a botanical discourse was founded on his botanising journeys, on his collections of fresh and dried specimens sent to him by others, on his travels in Central Europe, but also on the position of Antwerp in the world economic system, and particularly Antwerp's place in the European grain trade.[4]

In the introduction to *De frugum historia* (1552), Dodoens dedicated the work to Viglius Zuichem, president of the State Council of the Netherlands, to whom he had been introduced by his cousin Joachim Hopper, and declared that he was complying with the requests of many Belgian and foreign friends to publish the part about cereals and legumes, which he was preparing for the *Cruijdeboeck*. In this first work, Dodoens remained faithful to the medieval tradition and, among other things, made clear the value of cereals to keep one's health, but warned his readers against eating rotten wheat, because of the hazard of getting scurvy. His descriptions of lesser-known grains, such as panic, sorghum, rice and maize were full of details and useful contemporary nomenclature. In the same category of healthy foodstuffs for human and animal consumption were also listed various kinds of beans and peas, both cultivated and wild varieties. Dodoens was in a position to make valuable remarks about the garden cultivation of many of these imported grains and legumes by the Flemish peasantry and keen herbalists, thanks to the multiple trading connections of Antwerp with Spain and hence Spanish America on the one side, and Northern Europe and the Baltic on the other. Unfortunately, Rembert Dodoens lived in a time when Western Europe was split by religious and political divisions and the Low Countries were at the heart of these divisions. I believe that these religious and political cleavages even manifested themselves in the scientific solutions proposed by botanists.

II A brief genealogy of European botany in the sixteenth century

I shall approach Rembert Dodoens's contribution from the viewpoint of cultural history. Botany benefited greatly from improvements in printing and engraving in the early decades of the sixteenth century. The circulation of Rembert Dodoens's scientific achievements was made possible by the development of printing in the late sixteenth century. Plantin could easily sell the botanical books of Dodoens, although some works came out during very turbulent years.[5] Plantin printed 800 copies of Rembert Dodoens's *Stirpium historiae pemptades sex* and illustrated books usually reached issues of about 400 to 800 copies in the same century (probably this number remained fairly stable until the eighteenth century).[6] Although the first generation of botanists-humanists such as Ermolao Barbaro (1454-1493) and Jean Ruel (1474-1537), held different opinions on particular points, they fulfilled the function of discussing classical (i.e. Greek and Roman) botany and made it available to the following generations of scholars. It was the German books printed in the 1530s, which enlarged the market for the herbals thanks to the superior quality of their illustrations and to their vernacular nomenclature. But the works of Brunfels, Bock, Fuchs, Cordus etc., came out of Lutheran Germany (for example, Brunfels supported his work as a botanist with the income from the sale of his own protestant pamphlets[7]) and were therefore looked upon with suspicion, especially in Roman Catholic Italy.

During the same decades, Italian universities at Pisa, Padua and Pavia, as well as Cosimo de Medici in Florence created the first botanical gardens to support medical teaching;[8] yet Italian botanists remained firmly attached to Dioscorides' texts for the identification of the European flora. In the short run this strict conformity to the classical model reinforced the division with botanists of northern Europe. Most illustrious botanists throughout Europe looked to the classical heritage with respect, and a visit to Italy or stay at one of its major universities, Bologna or Padua, was an obligation for anyone who aspired to a scholarly career. Dodoens himself is said to have travelled to Italy (as well as France and Germany), at least briefly after his graduation from university: these were the same years when Luca Ghini lectured on the *Simples* at the university of Bologna. Many European botanists considered Luca Ghini a master (for example: Valerius Cordus from Saxony spent two years studying with Ghini in Bologna, then travelled to Padua and Ferrara, 1542, went as far as Rome, visited everything and died in the city in 1544;[9] often the Englishman Turner referred to Ghini as "my master").[10] Although many botanists from northern European regions thought that Theophrastus' and Dioscorides' writings (which were the foundation of botanical teaching at the universities) contained very important observations on the natural history of plants and pharmacopoeia, yet they found Dioscorides' selection of plants biased and related to a precise region, Asia Minor, where he had served as a surgeon to the Roman imperial army (first century A.D.).

In the second half of the sixteenth century German and northern European botanists were discovering families of plants that were not known to the ancients. Brunfels realised that certain plants were not known to Dioscorides, so he maintained the name German peasants gave them, such as *Gauchblumen* i.e. *Cardamine pratensis*, L.[11] Northern European botanists were not ready to adopt any nomenclature that found its origin in the Mediterranean regions simply on the grounds that it was founded on the classical experience. This 'scientific option' would have maintained botany under the domination of the Italian universities and reinforced the commercial value of Italian botanical handbooks (especially Mattioli's, which contained fewer engravings compared to German and Flemish block prints) to the detriment of a larger book market in continental Europe.

III Beyond the regional floras: P. A. Michiel and R. Dodoens

There was one way to overcome these difficulties and that was to produce regional floras for quick reference books, which had a larger market than expensive folio volumes about medical botany. These were the first books produced by Rembert Dodoens with the publisher Van der Loe in 1552, 1553, 1554 (and collected as *Cruijdeboeck* in 1554) and later with Plantin in 1565, 1566, 1568, 1574, 1580 (and collected as *Pemptades* in 1583, 1616). Smaller octavo books had a larger market and certainly answered better the demands of landowners and householders in search of practical information about the management of plants as medicines, about the conservation of grains as foodstuffs, recipes for winemaking, and flower cultivation for personal pleasure. Furthermore, since the 1550s the geographical boundaries of urban conspicuous consumers were extending dramatically and products from the New World became available, all sorts of new information on new plants were in demand.

Rembert Dodoens's collection was also founded on the labours of others[12] and was not dissimilar to that large unpublished collection of plants put together by P. A. Michiel, a patrician from Venice, who refused active service for the Republic of Venice and preferred to work quietly in the shadow of the Padua botanical gardens.[13] The latter received information and dried specimens from Europe and the Mediterranean area thanks to the Venetian consular network. Parallel to the exchange of plant collections, letters were being exchanged between the most open-minded botanists, such as Gesner, Aldrovandi, d'Aleschamps, de l'Escluse, all discussing common problems of nomenclature. It was Jean de Brancion, director of the botanical garden at Mechelen, and a personal friend of Dodoens, who supplied the illustrious scientist Ulisse Aldrovandi of Bologna with small parcels of seeds and handwritten information about their management in the years 1566-1568 (in those years Dodoens was about to leave Mechelen). Among other items, in these parcels Brancion made clear the differences between the *Trifolium* and the *Medicago* genus and supplied the right seeds of lucerne (alfalfa), a cultivation that had been lost in most European regions (see below).[14] This is only one example of the way botanists behaved in mid-sixteenth century: amidst quarrels there was also scientific co-operation. However, these attempts to find solutions, which would have soon reduced the differences within the botanists' community on both sides of the Alps by comparing practical information, were hampered by one major disaster, the revolt in the Low Countries.

IV Politics, war and Rembert Dodoens's life

Rembert Dodoens's father, Denys, was born into a family of civil servants from Friesland: he moved to Mechelen when he was appointed to the post of 'médecin juré de la ville', and married a widow, Urselle Rollands. Junius Rembert had been born there in 1517 out of her first marriage.[15] At the age of eighteen he took a degree in medicine:

his studies also covered geography and cosmography, classical languages, and, obviously, botany. He was also a student at the *Collegium Trilingue* of Leuven University. After his graduation he purportedly made extensive travels in France, Germany and Italy, where he visited *'multas universitates ibidem'*, according to his personal friend Petrus Suffridus.[16] It is not clear whether Rembert Dodoens's travels and his friendship with de l'Escluse took him away from Mechelen and the Low Countries at an early age or later.[17] Yet, after he had made Mechelen the centre of his early career, Dodoens was in the position to benefit from the journeys and the experience of others. He was on friendly terms with Jean de Brancion, a gentleman who directed the botanical gardens there and who supplied him with plants and seeds. De l'Escluse studied at the medical school of Montpellier, and travelled the length of Europe, from Portugal to Hungary; L'Obel frequently visited England and had good relations there. In the course of these travels Dodoens, de l'Escluse and L'Obel exchanged a great deal of information. In the *Pemptades* (1583 and 1616) Rembert Dodoens acknowledged de l'Escluse's and L'Obel's contributions to his own plant collection. However, the centre of Dodoens's well rewarded professional activities as a physician remained Mechelen, as long as he was not forced to move. Soon afterwards (1568) he was offered the vacant post as physician to Philip II of Spain, thanks to his cousin Joachim Hopper (by then secretary of the same king).

His refusal has to be understood in the light of the difficult political situation of those years of rebellion and civil strife. Dodoens remained in Mechelen for a few more years, until the Spanish troops of the Duke of Parma besieged and sacked the city, along with other cities, in 1572: Mechelen's population dropped from 30,000 inhabitants in 1550 to 11,000.[18] Dodoens, by then a wealthy man, lost all his possessions in this tragic event. Whilst Viglius Zuichem insisted again that he take up the post in Madrid, he accepted Maximilian II's invitation and settled in Vienna, where he stayed from 1574 to 1578, as imperial physician (to Maximilian II and then Rudolf II), in the position previously held by his personal friend de l'Escluse.[19] He remained in Vienna until the 'Religious Peace' had been proclaimed in Brussels and Mechelen (1578); but news that in his native Mechelen the Protestant and Catholic factions were at war again (1579) and that the Catholic patriciate had entered into negotiations with the Duke of Parma, halted him on his way back home. Peace talks were even held at Cologne, the imperial and episcopal city, sponsored by Rudolph II in 1579.[20] Dodoens spent some years in Cologne, which had become a temporary home for many refugees from Flanders who did not want to take the Calvinist side yet were hostile to Spanish rule. He successfully practised medicine there until 1581. Between 1581-83 he lived in Antwerp to supervise Plantin's edition of the *Pemptades* (1583). As Parma's army was getting ready to besiege that city, he accepted de l'Escluse's invitation and settled in Leiden in the years 1583-85, where he was offered the chair of medicine by the newly founded University. His numerous academic successes did not prevent Dodoens from pursuing his major botanical interests relentlessly until the last moment: in 1583 Plantin arranged for a long list of seeds to be sent to him, on Dodoens's behalf, from father Arias Montanus, the curator of the Escorial Library in Madrid.[21]

v Nomenclature and botanical discourse

In the perspective of political divisions that divided the major European regions during the second half of the sixteenth century, even European botany was organised around centres of cultural and political influence. The quarrel over nomenclature (classical names against popular and vernacular ones) was also reinforced by the tough competition in the publishing world in a bid for commercial supremacy in the European book market. In the mid-sixteenth century some printers, notably in Venice or Lyon, became more able than others to produce smaller and cheaper editions: this improvement also affected the production of illustrated botanical books. For example, Valgrisi put together and produced 562 illustrations of plants for the edition of P.A. Mattioli's *Commentari* (Venice, 1554). Of that edition and the following reprints, Valgrisi published a total of 32,000 copies. Still, he could not compete with Plantin's entrepreneurial skills. Van der Loe and later Plantin had engraved, collected and later on bought most of the print blocks already available on the market (in Antwerp and London) and produced 1300 illustrations for Dodoens's *Cruijdeboeck* (1554 and following editions) and around 2000 tables for the *Pemptades*.[22] Among these illustrations 600 descriptions and prints were taken out of Fuchs (1539) and Brunfels (1532). To these Dodoens added 300 new plants, never illustrated nor described before. In the 1563 edition Dodoens increased them to 1200 plants: it took him and Plantin another twenty years to prepare the illustrations and text for the *Pemptades* of 1583.[23]

The increased number of plants described since the 1550s forced all botanists to abandon the alphabetical order, a remnant from the medieval herbal tradition, which still was the usual manner to organise plant lore. Henceforth other external characteristics were taken as criteria to group plants into families. In 1583 Andrea Cesalpino (1519-1603) arranged plants in a very sensible manner (according to fruit and seed) and prepared a binary nomenclature that, unfortunately, did not find supporters. Rembert Dodoens (and later Caspar Bauhin in 1623) believed that the leaves presented good enough characteristics to support a general classification of plants. Yet these characteristics were too limited and the example of the genus *clover*, in Latin *Trifolium*, proved that the class of plants with trifoliate leaves had become too large and unmanageable for practical use.[24]

Once again the question of nomenclature revealed the sensible position on the subject held by Rembert Dodoens (and the Flemish school), who followed the local nomenclature when available: the botanist could not expect that unknown plants revealed their 'true' names by some 'inner virtue'. Pandolfo Collenuccio (who belonged to the humanist tradition of Padua, 1444-1504) was the first to suggest that botanists should ask peasants about the names and use of unknown plants: his suggestion was taken up only by German botanists (Brunfels published Collenuccio's botanical work, 1532-1539). Jean Ruel (1536)[25] had clarified the difference between wild and cultivated plants and demonstrated that wild plants existed before cultivated ones, that some existed at both states and that wild plants usurped the place of cultivated ones (and this was the case of *Medicago*).[26] It became the task of the botanist to record folk names and local uses of plants known to peasants and peasant women, great experts in plant medicine. For example, Euricius Cordus (1486-1535), son of a farmer, gardener and botanist, took great delight in outings into the country and learned a great deal from illiterate women and peasants, but also from the comparison of his findings with other scholars such as Dioscorides or Brunfels.[27] Rembert Dodoens followed the same way: when he brought some order into the genus of leguminous plants, he was first to use the name *Onobrychis* (instead of the older name *Polygala*, which went into disuse since) to describe what is now known as O. *sativa* (Lam.) or O. *viciaefolia*, L. To this he added one local spelling of the Flemish name of the plant (today known as *hanekam*), which he translated as O. *crista gallinacea*. Clusius reinforced the statement and added that this plant was known in Brabant as *Hanecammekens*.[28] It was a sensible thing to keep Latin nomenclature also in vernacular textbooks, to help identification of less familiar and new plants.

vi Rembert Dodoens, L'Obel, Lyte: a growing market for botanical information

L'Obel (1538-1616)[29] and Pierre Pena had travelled in England for the first time in 1569. They stayed in London to publish the *Adversaria nova* with a dedication to Elizabeth I in 1571. Pena from Provence and L'Obel from Arras had met at Montpellier University and they remained on good terms since: their journey to England took place when the Duke of Alva's Council of Troubles (Tribunal de los revoltosos) was engaged in the repression of the rebellion in Flanders. On their way through France, they stopped at La Rochelle, another instance of a friendship that shared religious sympathies for the reformed churches. Plantin followed the London edition of the *Adversaria nova* for his edition, which came out in 1576.

The English translation of Rembert Dodoens's *Cruijdeboeck* was undertaken in this circle. It was Henry Lyte who translated it from the French version of the *Cruijdeboeck* published in Antwerp by de l'Escluse in 1557. He had once been a student at Oxford and became an amateur botanist, a cultured landowner in Somerset, who had travelled in the Continent. The copy he used has come down to us, with his corrections and marginal notes.[30] It is difficult to establish whether Henry Lyte wrote all the notes we find in this copy himself. This translation faced a problem that is central to this collection of essays: English, French, Flemish and Latin plant names had to agree with the illustrations in an age when formal nomenclature was still in its infancy. If the translator wanted that his book was useful, he needed precision and on some crucial points it was not easy even between languages with many common points such as French, Flemish and English. Unfortunately, Rembert Dodoens-de l'Escluse somehow confounded the translator and the reader about one particular fodder plant, which appeared in places in Brabant but, more generally, was grown in the gardens of herbalists. This plant is sainfoin, *O. viciaefolia* L. Whilst Rembert Dodoens had acquired experience on chosen plants growing in gardens, de l'Escluse had studied plants in their natural environment: the latter was better acquainted with grasslands and pastures in central Europe, where *O.viciaefolia* grew wild. Translating a text that had been originally written in Flemish by Dodoens, but published in French by de l'Escluse (Antwerp, 1557), Lyte was somehow confused and added that the plant in question was the famous 'Medick' of the ancients (luzerne today, *M. sativa* L.), which had been described by Jean Ruel (1536).

Sainfoin and luzerne are similar fodder plants, which have, very unfortunately, different soil requirements. Since the 1550s landlords in Italy, France, Germany and England had begun to look again at plants not purely for their medical value. I have pointed out above that the first printed book by Dodoens was, not incidentally, a small treatise on cereals and legumes (1556) for human and animal consumption. In their quest to improve production, landlords and botanists rediscovered forgotten plants, which in the Roman and Greek farming experience had been highly valued. Although leguminous fodder plants are naturalised plants in Western Europe, the best strains of cultivated clover, sainfoin and luzerne had all but been lost and what remained of them was hidden in only a few areas (Southern Spain and Provence for instance), whilst their place had been taken by weeds and undomesticated plants of the same species.[31] In the following century enterprising English landlords took these grasses out of their semi-wild context and sowed them – an innovation in their capitalist farms. Because of the opportunities created by the dissolution of monasteries (1536), English landlords became extremely interested in developing their agricultural methods and read anything that dealt with the economic use of plants more eagerly than other social groups in Europe. Yet, it took about seventy years for English landlords, who read botanical and agricultural textbooks by the letter, to overcome the inaccuracy, when they wanted to put into practice Dodoens-Lyte's description. Some years later, L'Obel, who was very active in England at that time, clarified these and other inaccurate names that were featured in the Flemish books on fodder plants. But unfortunately his efforts never reached the printing press and remained unpublished among his papers.[32]

By the beginning of the seventeenth century, botanists had brought together so much information that their discipline required a new sort of treatise. By then, the book market was capable of sustaining the production of a great number of plant engravings with the desired precision, in order to illustrate a rather elaborate and complex text, which contained information for the landlord, the botanist, the physician, the householder, the corn merchant, and eventually the housewife. By the standards of the time all this information could not be compressed in one volume. When the Bauhin brothers printed the largest collection of plants ever produced by Renaissance botany (1623, 1650-51, 1671), they had to suppress much of the practical information and to introduce abbreviations.

vii Seed production in the early seventeenth century

Rembert Dodoens's writings and activities produced their major effects in the following generation, in a less learned *milieu* and abroad. It was in England that innovative landlords created an international market for agricultural seeds and benefited from the recovery that Flanders and Brabant were enjoying in the early 1600.[33] Recent research does show an upward movement of land rents in the Spanish Low Countries as soon as the war moved towards Maastricht, at the end of the Twelve Years' Truce (1609-1621).[34] Once again, intensive agriculture became the option chosen by the Flemish peasantry to pay high rents: their noted (by L'Obel for instance) ability to produce garden seeds provided a new source of income. A regular trade of garden seeds (especially onion!) did exist from late medieval times (and earlier) between London and the Low Countries. Irregular parcels of clover seed appear in merchant records to anticipate a regular trade after 1690: these seeds, known to Rembert Dodoens and L'Obel, became the object of an intriguing obsession for inquisitive English landlords, who were looking for new methods to overcome their own agricultural and economic crisis. During the years of the English Civil War, an English exile, Sir Richard Weston, travelled extensively through Flanders and Brabant and did not fail to notice the new wave of economic and farming activities, which had made a quick recovery possible in these regions. From 1650 onwards a growing interest in intensive agriculture gained momentum in England. And although clover, flax, linen, rape and turnip were all native English plants, their successful cultivation required fresh and good quality seeds, not easy to obtain by the English peasantry, who were not specialised in this industry.[35]

More mundane landowners had tried their luck with fruit plants and flowers: e.g., on 6 April 1639 one particular landowner, in his London garden was growing flowers from 153 tulip bulbs, all different.[36] His was not an isolated case: agricultural and garden seeds were soon quoted and advertised in farming books and later newspapers. Clover seed especially became the object of a trade, which went alongside with the major grain trade. Nurserymen, gardeners and seeds men in England had always been strongly connected with the Continent, especially Flanders and France (and some English gentlemen with Italy).[37] In 1567 and later, a number of refugees from Flanders settled in southeast England developing horticulture and running nurseries.[38] Regions such as Flanders, Holland, Zeeland and later Friesland, provided seeds and trees to English gardens and improved a botanical patrimony that was naturally poorer than the rest of Europe. Between 1550 and 1650 (and later during the eighteenth century) the position of England in respect of the oceanic trade gave new impetus to the diffusion of non-European plants in the continent: yet English agriculture depended heavily on seeds which European peasants provided in regular quantities. In the mid-seventeenth century the complex social and economic transformation, which we usually refer to as the agricultural revolution, was, for the botanical part at least, founded on the diffusion of books, information, and seeds from Flanders and Holland. Lesser men, like peasants and gardeners, and important botanists, like Rembert Dodoens and L'Obel, all made their valuable contribution to make this transformation possible.

viii Conclusions: Rembert Dodoens's irenic position

The newly founded University of Leiden was not a place for strict Calvinist orthodoxy: Lipsius and other professors openly professed their Catholic faith, and Catholic students quickly moved there to listen to the most prominent teachers in many disciplines. During the next few years following Dodoens's arrival, the new Republic invested lots of money to make Leiden into a leading university in Northern Europe, with institutions that included a student college, a large library and a botanical garden (1587). It was de l'Escluse, a professed protestant, who turned the garden into an efficient institution. It has been suggested that Rembert Dodoens shared the religious views of other Erasmian scholars.[39] His personal experience seems very much tied up with the political events that framed the lives of his fellow countrymen. Dodoens's dedications to Hopper and Viglius probably suggest his conciliatory position towards the court of Madrid and those at home who wished to win the elite in Flanders to the Spanish side after the Troubles. Dodoens twice declined the position offered to him in Madrid, which probably put him on difficult terms with the Spanish government. This might explain why he preferred to stay in Vienna at the court of that famous patron of scientists and artists, Rudolf II, instead of looking after his endangered possessions in Mechelen. It is very difficult to approach and to discuss the problems of the scientific community in the sixteenth-eighteenth centuries, without keeping in mind these sharp religious and political divisions, which tore Europe apart. Writing to his good friend, father Arias Montanus, in 1585, among other news, Plantin spoke of the death of Dodoens and clarified his public position in respect of his post at Leiden: the professors of the Catholic faith were free from taking an oath of allegiance to the States of Holland in political matters. On other occasions (1572, 1575, 1583) Plantin had informed father Arias about Dodoens's necessities and troubles.[40] Arias had been chaplain of Philip II (1566) and in 1568 had been sent to Antwerp to supervise the 'Bible royale et polyglotte', to be printed by Plantin. It seems that Arias, Plantin, Hopper and Viglius interceded on behalf of Dodoens with the Spanish government either in Flanders or in Madrid: yet he did not commit himself to the Spanish side during the rebellion and acted freely on important occasions.

During the second half of the sixteenth century a very interesting group of botanists arose in Normandy and Flanders: Belon from Mans, d'Aleschamps from Caen, Jean and Caspard Bauhin from Amiens, Rembert Dodoens from Mechelen, de l'Escluse from Arras, L'Obel from Lille. One thing they had in common: they all travelled throughout Europe to a greater extent than other European botanists did. Furthermore, they had studied at the university of Montpellier, in clear opposition to the medical school of Paris. The contribution of the Flemish botanists to the development of botanical discourse was very important. They were men of the North (at least in respect of Greek and Latin botanical studies). They all shared a competence in current spoken languages and an interest in peasant and local folk plant names; yet they were all versed in classical botany without following it strictly as Mattioli did in Italy. It seems to me that Rembert Dodoens practised a kind of irenic botany, which took in consideration the ideas of other authors without overwhelming them. The seventeenth-century developments of this botanical school, which were carried on at Leiden University by Commelin, Rumphius and Boerhaave, enabled Carolus Linnaeus, active in Leiden between 1735-1738, to achieve a botanical classification thanks to a binary nomenclature.

In respect of the formation of a botanical discourse, the following major steps took place roughly between 1450 and 1550. When Greek plant names were translated into Latin they became binaries (Leuconion – viola alba, Melanion – V. nigra etc)[41]; at the same time, individual readers added hand-written plant names in their vernacular language to the few printed herbals available.[42] The corrected versions of Dodoens's Latin and German herbals, which he bequeathed to Plantin in 1585, reveal a great similarity of practice between botanists and their public.[43] This uniformity made scientific communication easier between the high and the low sector of society and it promoted the creation of a botanical discourse. Latin and vernacular nomenclature were both present in Rembert Dodoens's work: if we take into consideration that the *Cruijdeboeck* was also printed in different languages, we easily understand that it appealed to a larger public, which had primarily practical interests. At the same time the men of learning had the opportunity to acquaint themselves with a very large body of popular nomenclature, which became the foundation of any subsequent classification.

NOTES

1 E.L. Greene and F.N. Egerton, ed., *Landmarks of Botanical History* (1909), 2 vols. (Stanford, Cal.: Stanford University Press, 1983), passim.
2 Mauro Ambrosoli, *The Wild and the Sown. Botany and Agriculture in Western Europe, 1350-1850* (Cambridge: Cambridge University Press, 1997), chap.3; M. Findlen, *Possessing Nature. Museums, Collecting and Scientific Culture in Early Modern Italy* (Berkeley, Cal.: University of California Press, 1994), passim.
3 Thomas S. Kuhn, *The Structure of Scientific Revolutions* (Princeton: Princeton University Press, 1970).
4 Herman Van der Wee, *The Growth of the Antwerp Market and the European Economy (Fourteenth-Sixteenth Centuries)*, 3 vols. (The Hague: Nijhoff, 1963).
5 Leon Voet, *The Golden Compasses. A History and Evaluation of the Printing and Publishing Activities of the Officina Plantiniana*, 2 vols. (Amsterdam: Vangendt, 1969-1972), 1:46, 90.
6 M. Rooses and J. Denuncé, *Correspondence de Christophore Plantin*, 8 vols. (Antwerp, 1883-1918), III:6; L. Febvre and H. Martin, *Histoire du livre* (Paris: A. Colin, 1958); E. Eisenstein, *The Printing Press as an Agent of Change* (Cambridge: Cambridge University Press, 1979).
7 Greene, *Landmarks*, 1:240 ff.
8 A. Chiarugi, "Le date di fondazione dei primi orti botanici nel mondo", *Nuovo Giornale Botanico Italiano* 60 (1953): 785-839.
9 Greene, *Landmarks*, 1:369 ff.
10 Ambrosoli, *Wild and Sown*, 263.
11 Greene, *Landmarks*, 1:264.
12 Armand Louis, "La vie et oeuvre de Rembert Dodoens (1517-1585)", *Bulletin de la Société Royale de Botanique de Belgique* 82 (1950): 281-82.
13 Ambrosoli, *Wild and Sown*, 101, 106, 115-17.
14 Ibid.,116.
15 Louis, "Vie et oeuvre", 272.
16 P. Suffridus, *De Scriptoribus Frisae decades xvi. Lovanii* (1593).
17 Rembertus Dodonæus, *Cosmographica in Astronomiam et geographiam isagoge*, 1548. Facsimile of the original Latin text, with an introduction by A. Louis (Nieuwkoop, NL: B. De Graaf, 1963).
18 Jonathan Israel, *The Dutch Republic: Its Rise, Greatness, and Fall, 1477-1806* (Oxford: Clarendon Press, 1998), 308.
19 Biographisch Woordenboek der Nederlanden (1852) ii, R. Dodonæus.
20 Israel, *The Dutch Republic*, 195-96.
21 M. Rooses and J. Denuncé, *Correspondence de Plantin*, VII: 1011.
22 See paper by Helena Wille in present volume.
23 Dodonæus, *Cosmographica in Astronomiam*, 1548. Intr by Louis (1963), 41.
24 Greene, *Landmarks*, 1:108-9.
25 Ibid., 2:636.
26 Ambrosoli, *Wild and Sown*, passim.
27 Greene, *Landmarks*, 1:366.
28 Woordenboek der Nederlandsche Taal, sub nomine.

29 Ibid., 2:877 ff.
30 London, British Library, Printed Books Dep.,442h9.
31 Ambrosoli, *Wild and Sown*, CHAP. 6 passim.
32 Ibid., 279-80.
33 Ibid., 335-356.
34 Vandenbroeke, *Annales* ESC,39 (1984) and Israel, *The Dutch Republic*, 412-13.
35 Mauro Ambrosoli, *"The political context of the agricultural activities of the Hartlib circle"*, in *Politics Religion and Economy in the Hartlib* Papers, ed. S. Mandelbrote (Oxford University Press), forthcoming.
36 Sloane Ms 95.5. British Library,London.
37 F. Harvey, *Early Nurserymen* (London Chichester: Phillimore, 1974) and M. Thick, *"Market Gardening in England and Wales"*, in J. Thirsk, *The Agricultural History of England and Wales* (Cambridge: Cambridge University Press, 1990), AHEW,V,II.
38 AHEW, vol V, 1988.
39 Blok, *Janus. A Review of the History of Medicine* 22 (1915).
40 M. Rooses and J. Denucé, *Correspondence de Plantin*, III:421, V:664, VII: 1011, 1045.
41 Greene, *Landmarks*, 108 ff.
42 Ambrosoli, *Wild and Sown*, CHAP.2-3, passim.
43 M. Rooses and J. Denucé, *Correspondence de Plantin*, VII: 1045.

IV

BOTANICAL COLLECTORS AND COLLECTIONS IN THE LOW COUNTRIES

Helena Wille
University of Ghent, Archives Departement

Collectors

In Renaissance Europe a growing number of people were fascinated by plants. All over Europe the collections gathered by these plant lovers became centres of a new scientific culture that was conducive to the development of botany. Their gardens, mostly full of plants, were used as a sort of depot for storage and distribution of new varieties. In these gardens all kinds of plants were introduced, collected and cultivated, so that their numbers were constantly growing. This led some plant lovers to try and catalogue the plants and find the 'natural order' that governed them. Expanded travel and comparison of different samples laid the basis for a complex network of plant lovers who exchanged information about new specimens. The social background and the accomplishments of many of the plant lovers and collectors during the Renaissance received ample treatment by several authors (Reeds, 1991; Olmi, 1992; DaCosta Kaufmann, 1993; Findlen, 1994). Unfortunately, not many of the plant collectors in the Low Countries have been studied, so that we know very little about them.

Who were these plant enthusiasts called "cruydeliefhebbers" in Flemish, what was their profession and which plants did they cultivate in their gardens? In order to reconstruct to a certain extent the history of plant collecting, letters, catalogues, lists of plants and of course printed herbals constitute indispensable sources. During the Renaissance herbals appeared all over Europe. In the Low Countries they owed much to Christophe Plantin of Antwerp. He published the works of the triumvirate of botany: Rembert Dodoens, Latinised as Dodonæus (Mechelen 1517 - Leiden 1585), Charles de l'Escluse, Latinised as Clusius (Atrecht i.e. Arras 1526 - Leiden 1609) and Mathias de Lobel, Latinised as Lobelius (Lille 1538 - Highgate 1616). Each of them, in his own way,

Acc. 9590
S P Q A
Kruydtboeck
OFT
Beschryvinghe
Van allerleye Ghewassen,
Kruyderen, Hesteren, ende
Gheboomten:
DEVR
MATTHIAS DE L'OBEL
Medecyn der Princ. Exc.tie
LABORE ET CONSTANTIA
ARA SALVTIS.
T'ANTWERPEN
By Christoffel Plantyn
M. D. LXXXI.

ILL 1

succeeded in offering new perspectives for the development of botany. Dodoens set about cataloguing the indigenous flora in particular. De l'Escluse paid more attention to plant geography and was the first to study the flora peculiar to certain regions. Lobelius built on the systematic work of Dodoens. In terms of his classification, which he illustrated clearly in synoptic tables, he was ahead of most of his contemporaries.

1 Lobelius' Kruydtboek (1581)

Lobelius' works, especially his *Kruydtboeck*, have an extraordinary significance in the history of plant collecting in the Low Countries. He studied medicine in Leuven and travelled abroad (Italy, England, France). He was appointed personal physician to William of Orange and later to Lord Zouche, an English plant lover. In 1581 Christophe Plantin published his *Kruydtboeck* in Antwerp. This herbal, an adaptation of his *Plantarum seu stirpium historia* (1576), came as an impressive folio volume, lavishly illustrated with 2189 woodcuts. At the end of the book, arranged in two sections, a supplement and an extensive index of plant names in Flemish, Latin, German, French, Portuguese and English was appended, followed by an index of medicines. The plant descriptions do not only make reference to classic authors, in particular Dioscorides, but also to contemporaneous plant collectors, mostly recorded as "most learned, extraordinary friend ..." A perusal of Lobelius' *Kruydtboeck* has allowed us to conclude that many of these plant collectors in the Low Countries belonged to the nobility and the civil or ecclesiastical elite.

For noblemen and patricians the ownership of a garden often amounted to a valuable asset in the game of social and political power, in which even a few women, such as Marie de Brimeu (a personal friend of Clusius), Radegonde Duquesnoy (Bruges) and Catharina van Eeckeren (Antwerp) played a modest role (table 1).

TABLE I

NOBLEMEN AND PATRICIANS	CITY	LOBELIUS' *Kruydtboeck* (1581)
Boisot, Jean	Brussels	29(II), 132, 141, 178, 591, 123, 162, 168, 146, 640
Bousbecq, Augier van	Lille	74
De Brancion, Jean	Mechelen	8, 37, 41(II), 52, 54, 90, 122(II), 141, 144, 145, 146, 147, 151(II), 152, 162, 167, 184, 202, 207, 208, 222, 249, 270, 342, 343, 344, 353, 358, 386, 459, 465, 467, 591, 597, 639, 644,

		673, 689, 697, 725, 743, 800, 804, 819, 854, 944, 953, 982
De Brimeu, Maria	Antwerp, Lier	29(II), 89, 92, 129, 154, 158, 162, 179, 202, 204, 208, 342, 651, 770
Duquesnoy, Radegonde	Antwerp	162, 177
Van Sint Omaars, Karel	Bruges	851
Marnix van St Aldegonde, Philips	Brussels	162, 207-208, 531, 753
Rym, Charles	Ghent	71, 152
Schetz, Maria	Antwerp	89, 209
Utenhove, Jacob	Ghent	74, 116, 118
Van der Delft	Antwerp, Lier, The Hague	89, 92, 105(II), 129, 151(II), 162, 168, 194, 207, 208, 209, 217(II), 222, 226, 241(II), 248(II), 338, 342, 386, 619, 639, 672, 770, 814, 815, 817, 819, 870, 919
Van Eeckeren, Catharina	Antwerp	141, 162
Van Praet	Bruges	51
Van Rye, Joris	Mechelen, Lier	40, 139, 141, 146, 151(II), 155(II), 162, 207, 208, 343, 808, 982

Collectors mentioned in Lobelius' *Kruydtboeck*
(Figures in the right column of this and the following tables refer to page numbers in the 1581 edition)

Also wealthy and learned *bourgeois* stimulated the development of botany by their delight in plants. Some of them were rich merchants; others were councillor, treasurer, registrar, bailiff, canon, professor or even painter (table 2).

TABLE II

Learned and wealthy Bourgeois	profession	city	Lobelius' *Kruydtboeck* (1581)
Arsseliers, Philip	merchant	Antwerp	29(II)
Boone, Ambrosius	councillor	Brussels	106, 959
Breugel, Peeter	professor	Leuven	380, 452, 497
Coxie, Raphael	painter	Mechelen	132, 162, 177, 162, 201, 207, 208, 258
Douwe alias Voghelsanck, Jan	merchant	Antwerp	72, 97, 296
Fourmenstraux, Andries (Niclaes) de	merchant	Lille	275,294*
Gemma, Cornelius	professor	Louvain	142, 220, 222
Lobel, Mattheus	canon	Lille	532
Martini, Willem	registrar	Antwerp	162, 202, 241
Plantin, Christophe	printer	Antwerp	158, 914
Pluym, Cornelius	treasurer	Antwerp	207, 208, 926
Quickelberghe, Pieter	merchant	Antwerp	78, 274
Roelofs, Gaspar	registrar	Leuven	54(II)
Scholier, Jeronimus	merchant	Antwerp	77(II), 785
Van Hoboken, Jan	registrar	Antwerp	152(II), 158, 241(II), 207, 208, 135, 139, 141, 152, 158, 169, 211, 344, 658, 767

Merchants and members of the free professions mentioned in Lobelius' *Kruydtboeck*

Nevertheless the practice of collecting plants in the Low Countries, as in other regions in Europe, was first and foremost linked to the practice of medicine. Medical practitioners were the most significant group of collectors, and it is safe to say that the appearance of important plant collections in the Low Countries was closely connected to the transformation of the medical profession. Apothecaries collected specimens for professional interest. The knowledge of plants was necessary for them as an evident component of their professional activities, because plants and other products were ingredients used and sold in pharmacies. According to Findlen (1994: 246) apothecaries were the first collectors to limit their museums consciously to the natural world because of their professional interest in the subject. Collecting enhanced the status of these men, who publicised their possession of the most exotic ingredients that nature could supply (table 3).

In the Low Countries some apothecaries gained an international reputation. Thanks to Gesner's *Horti Germaniae*, published in Cordus' *Annotationes* (1561), we have some descriptions of the six hundred thriving and blossoming exotic plants in the botanic garden of Peeter van Coudenberghe (Vandewiele, 1993: 24).

TABLE III

APOTHECARIES	City	LOBELIUS' *Kruydtboeck* (1581)
Beyerlinck, Adriaen	Antwerp	252, 710
Coene, Jan	Antwerp	162, 651, 915
Donrez, Walerant	Lille	3, 18(II), 91, 99(II), 101, 174(II), 175(II), 188(II), 201, 219(II), 267(II), 275(II), 462, 580, 622, 940, 956, 959
Driesch, Olivier	Antwerp	99, 131, 138, 162, 170(II), 178(II), 206, 215(II), 253, 640, 770, 831, 870, 926, 946
Pennin, François	Antwerp	90, 138, 176, 311, 508, 940
Van Coudenberghe, Peeter	Antwerp	247, 273, 497, 556, 588, 710
Van Zinnich, François	Brussels	106(II), 174(II), 959
Mouton, Jean	Tournai	14, 26(II), 37, 56, 64, 65, 91, 105, 132, 137, 153, 154, 155, 157, 162, 169, 191, 253, 296, 298(II), 310, 365, 398, 474, 497, 588, 602, 604, 619, 622, 630, 640, 665, 721, 736, 750, 809, 851, 920, 926, 959, 978

Apothecaries mentioned in Lobelius' *Kruydtboeck*

Also the physicians emphasised the importance of a better knowledge of the *materia medica* but from a different perspective. They collected natural specimens and plants to strengthen their status as observers of nature and practitioners of medicine. Moreover, they used their training in the *materia medica* to expand their regulation of apothecaries, just as the study of anatomy increased their supervision of surgeons and midwives (Findlen, 1994: 247) (table 4).

TABLE IV

PHYSICIANS	CITY	LOBELIUS' *Kruydtboeck* (1581)
Cobelgiers, Seger	Antwerp	272, 610
De Clerck, Hermes	Tournai	450, 546, 644
Eerssel, Lucas	Louvain	716
Espillet, Niclaes	Lille	134, 262
Favolio, Hugo	Antwerp	310, 767
Mitens	Brussels	236(II)
Monncel, Jan	Tournai	182
Ruer, Alart	Lille	97, 750
Thomas, Denys	Antwerp	105(II)
Van Maude, Davi	Antwerp	47(II), 619

Physicians mentioned in Lobelius' *Kruydtboeck*

II WOODBLOCKS

The activity of growing new plants in the gardens was reflected in print and a large number of botanical treatises were published during the sixteenth century. These books were mostly written for an audience of plant lovers and practitioners of medicine, to whom their prefaces were addressed. The information they provided was essentially of a botanical and medical nature: names of the plants, descriptions, habitats, medicinal virtues and properties of drugs, the opinions of the Ancients, including Pliny, Galen and Theophrastus, and polyglot indices. Often medicinal glossaries were added because most of these herbals were written by physicians who were interested in the medicinal properties of the plants. The chief motive behind the books of Dodoens and Lobelius, both physicians, was medicinal. Their books were written in order to aid students and colleagues in the correct identification of plants with curative properties. Illustrations of plants were considered to be very helpful in recognising the plants.

The woodblock collection of the Officina Plantiniana

It was through the copiously illustrated herbals of the threesome Dodoens, Clusius and Lobelius that the *Officina Plantiniana* in Antwerp became a centre of botanical publications. Together these three authors compiled about forty botanical treatises (De Nave, 1993). The considerable number of woodblocks used for the illustration of these herbals remains the pride of the Plantin-Moretus Museum to this day.

Woodblocks purchased by Plantin

As was customary in the sixteenth century, Plantin did not only commission and supervise the production of a number of new botanical woodcuts, but he also purchased a large number of woodblocks that were already in circulation. In 1581 he acquired a job lot of 500 woodblocks from Van der Loe's widow, which comprised the woodcuts used by Joannes Van der Loe for the publication of Dodoens's first herbal, the *Cruijdeboeck* of 1554. These woodcuts were engraved copies of the ones used in the octavo edition of Fuchs's *Primi de stirpium historia* (1545) and they were of excellent quality. In most cases the copies can only be detected by means of a magnifying glass. Most of the woodblocks show indeed only very minor differences: some line engraved slightly thicker or thinner, printed a little longer or shorter. Some plants were less precisely copied, while other small details were changed. Only in exceptional cases do the alterations strike the eye, while in two cases a flower was added to the illustration. [1]

These 500 illustrations were not copied mirror wise. Nissen (1966:60) speaks of "seitenrichtige" illustrations, so that it may be assumed that the tracing method was used here. This means that the drawing was fitted or stuck onto the block as a reverse image or that it was traced. Maybe the new blocks were cut from illustrations that were printed from "worn" blocks, and were made transparent by means of water or oil. In this group of woodcuts almost all hatching is missing. The lines are very fine, the contours of the plants are very clear and the illustrations were presumably meant to be coloured. The lines would then have functioned as the boundaries of the plants. According to Nissen (1966:61) and Delen (1934: 78) the illustrations of the *editio princeps* of the *Cruijdeboeck* (1554) were drawn by Peeter van der Borcht and engraved by Arnold (Arnaud) Nicolai. One should however be careful with these attributions, as not one single author has thus far produced sound evidence for these claims. The presence of the monograms A and PB on the title page would seem insufficient to conclude that Arnold Nicolai and Peeter van der Borcht designed and engraved all plant illustrations of the herbal. Not one monogram can be found in the illustrations themselves and it therefore would seem unwarranted to speculate about the identity of the artists. In 1581 Plantin also purchased

250 woodblocks from the English publisher Thomas Purfoot, who had used the blocks for the publication of the *Stirpium adversaria nova* by Lobelius and Pierre de Pena in 1571, a book later published under the title *Plantarum seu stirpium historia* (1576).

Woodblocks added by Plantin and his successors

To the 750 woodblocks of Van der Loe and Purfoot, Plantin added another 2,847 botanical woodblocks and thus contributed to the creation of favourable conditions for the botanical sciences to flourish in the Low Countries. Arnold Nicolai, Janssen van Kampen and Anton van Leest cut most of these blocks. In these illustrations an obvious preference for more relief and plasticity is striking. The artists worked very carefully and added small fluid parallel lines, especially in stalk and leaf. The hatchings are more frequent and systematically fill the planes of leaf and stalk. Even the root was sometimes filled up with strictly parallel hatchings. The outline is strikingly heavier and the use of cross-hatching imparts to some of the illustrations a heavy, sometimes almost overloaded impression because every available square inch of the block's surface is filled up.

The iconographic care which Plantin and his successors lavished on the botanical works, combined with the high scientific level of the contributions made by Dodoens, Clusius and Lobelius, ensured that the Flemish botanical works of the day became a standard throughout Europe. Few other European printers besides Plantin had access to a comparable collection of blocks needed for the production of these impressive botanical volumes. When Plantin died in 1589 the *Officina* had acquired, through commissioning and purchase, one of the largest collections of botanical woodblocks, numbering nearly four thousand items. When in 1618 the Leiden *Officina Plantiniana* closed down, the van Ravelingen (Raphelengius) family transferred the stock of botanical woodblocks to their relations in Antwerp. It is a remarkable collection which allowed not only Plantin to publish his botanical editions, but also his successors, the Moretus and the van Ravelingen families of Antwerp and Leiden respectively: eight botanical books came off the presses in Antwerp, and four off the ones in Leiden.

Until now, Peeter van der Borcht (Antwerp 1535/40 – Antwerp 1608) has been considered the principal artist who provided the drawings for the house of Plantin. He worked for Plantin more or less continually from 1564 on. He designed the title page of Dodoens's *Cruijdeboeck* (1554) and made more than 80 drawings for the illustrations in Dodoens's *Frumentorum* (1566). For Clusius' works as well he made 52 botanical designs from fresh samples provided by the botanist of the garden of his friend Jean de Brancion. Van der Borcht also worked from dried specimens and even from rough sketches made by the botanist himself during his travels (Depauw, 1993:51).

In the field of botany the *Officina Plantiniana* succeeded in publishing *herbarii*, which through their original layout and their approach to the material, belong to that group of scientific, illustrated books that set a new trend in science. Until the mid-seventeenth century the collection continued to furnish other publishers with illustrations, a testimony to the influence it had on the herbals of that time. For the publication by Thomas Johnson in 1633 (reprinted in 1636) of *The Herball or generall historie of plantes*, a translation by John Gerarde of the *Stirpium historiae pemptades sex* (1583), 2,765 woodcuts from the Plantin collection were used. In 1647 the Plantin woodblock collection was used for the last time for illustrating Simon Paulli's *Den Danske urtebog*, published in Antwerp by Balthasar II Moretus (Lemli, 1993: 57).

Inventory

Between 1618 and 1643 Franciscus II van Ravelingen made an inventory of the collection of woodblocks. He compiled two manuscripts inventorying the illustrations of plants in the botanical treatises of Dodoens, Clusius and Lobelius. These two manuscripts constitute the best source for the study of the collection of botanical woodblocks, which is stored permanently in the Plantin-Moretus Museum in Antwerp. In 1659 Balthasar II Moretus had the intention of selling the collection, but fortunately the negotiations ended in stalemate, so that the woodblocks remained in the *Officina Plantiniana*.

III Drawings

Clusius: a key figure in the history of sixteenth-century botany

Especially Clusius, more than Dodoens and Lobelius, had a wide circle of friends with a passion for plants. For the botanist who did not have his own garden or the necessary resources for the upkeep of the plant life he could import from many foreign countries, it was the ideal solution. According to Luis Laca (1998: 135), Clusius made use of materials he received from a network of correspondents distributed throughout Europe. In Spain he kept up correspondence with two Sevillian physicians, Simon de Tovar and Juan de Castaneda, who sent him regular shipments of seeds and bulbs. But Clusius also furnished other plant collectors with plant material. Again, Lobelius' *Kruydtboeck* (1581) reveals the names of the plants Clusius sent to his best friends in the Low Countries, especially during his stay in Vienna (table 5).

TABLE V

PLANT	COLLECTOR	LOBELIUS' *Kruydtboeck* (1581)
iris lutea variegata Clusij	Jean de Brancion	90
ahouay Theveti	Jan Van der Delft	217(II)
orchis minor latifolia	Jan Van der Delft	222
calceolus Mariae	Jean de Brancion and Jan Van der Delft	386
sedum medium teretifolium	Jean de Brancion	459
scabiosa rubra Austriaca	Jean de Brancion and Jan Van der Delft	639
ptarmica Austriaca	Charles de Houchin	646
paralytica Alpina sanicula angustifolia	Jean de Brancion	673
viola	Jean de Brancion	725
convulvulus peregrinus Clusij	Jean de Brancion	743
periclymenon rectum fructu caeroleo	Marie De Brimeu and Jan Van der Delft	770
geranium fuscum flore livido purpurante	Joris van Rye	808
absynthium album	Jan Van der Delft	919
tussilago montana minima	Jean de Brancion	697

Plants sent by Clusius to his friends in the Low Countries

Sometimes his botanical friends would send him drawings that he could use in his publications. Working with an experienced artist was no doubt very expensive. It is therefore quite likely that Clusius sought the backing of wealthy friends to acquire illustrations. A case in point is that of the London pharmacist Jacob Garet, who sent him a drawing of the *piper caudatum* in 1590, when he was living in Frankfurt. Clusius used the drawing for an illustration in his translation of the *Aromaticum* (1593) by Garcia ab Orto. The London apothecary regularly provided Clusius with exotic fruits such as those mentioned in his *Exoticorum* (1605).

Equally remarkable are the coloured drawings, such as the study of the 'taratoufli', which is the earliest known illustration of the potato, sent to Clusius in 1588 by Philippe de Sivry, the governor of the town of Mons, who had himself received the plants from the papal nuncio. Eventually, Clusius would use another study of the plant to illustrate his text. Philippe de Sivry was an important plant collector, who imported colourful flowering plants for his garden. He was one of Clusius' correspondents (Van den Wijngaert, 1947: 41). Some of Clusius' botanical friends stimulated the creation of sumptuous picture books, in which flowers were represented with technical brilliance by professional artists who were able to reproduce paintings with scientific subject matter.

Very interesting is the close friendship between Clusius and the Hungarian magnate Boldiszar de Batthyani (Balthasar Batthyany), who on several occasions invited the scholar to his castle in Nemet-Ujvar, and paid for the coloured illustrations Clusius needed for his study of Hungarian fungi. This collection, now in the Library of Leiden University, contains 87 watercolours of mushrooms and toadstools, made by the French miniaturist Essaye Le Gillon for Balthasar Batthyany, who was bailiff at the Viennese court. Clusius never used the watercolours as illustrations because the album was misplaced for some time in Plantin's workshop.

The "Centuriae plantarum rariorum" of Karel van Sint Omaars

One of the most interesting collections of botanical drawings that are attributed to Clusius and are undoubtedly connected with the woodblocks used by the *Officina Plantiniana* are the *Libri picturati* A16-31 now in the Library of Jagiellonian University, Kraków. These sixteen albums of watercolours were first discovered by Hans Wegener (1936), who attributed the collection to Clusius. We do not agree with this attribution because we find it highly improbable that Clusius, who constantly suffered from financial worries, would have had a large number of hand-coloured drawings made, of which only one third was used for the illustration of his work. Here again, the perusal of Lobelius' *Kruydtboeck* (1581) was of great help. The comparison of the watercolours with this herbal led us to a surprising discovery: the patron turned out to be Karel van Sint Omaars, and one of the principal artists Jacob van den Coornhuuse, both of whom had been lost from sight for about three centuries (Wille, 1998). We learn that the grasses in the first chapter of the *Kruydtboeck* (1581: 26) were observed by *Mijnheer de Reynoutre* and painted by *Jacques van Corenhuyse*, a highly skilful artist in this field. *Mijnheer de Reynoutre* is, as was customary at the time, another way of writing *Charles de Saint Omer, dit de Moerbeke, seigneur de Moerkercke, Dranoutre (aka Ranoutre or Reynoutre)*, the French equivalent of *Karel van Sint Omaars*, Latinised as *Carolus a divo Odomaro*. *Jacques van Corenhuyse* is obviously the same as Jacob van den Coornhuuse.

Equally very interesting is the illustration of the "archangelica met dobbel uutghespreyde croonkens" (Lobelius, 1581:851), which was undoubtedly drawn from the watercolour on folio 19 verso in album A29 of the *Libri picturati* A16-31. In the accompanying text Jacques van Corenhuyse is named as the artist who painted the plant on the commission of Karel van Sint Omaars. It is mentioned cursorily that it was the intention of the patron to publish this illustration, a reference to the publication of the "*Centuriae plantarum rariorum*" that this plant-lover had in mind. The name of the plant collector was thus clearly mentioned.

Karel van Sint Omaars (1533-1569) owned a castle in the district of Moerkercke (Moerkerke) near Bruges. He was a man-at-arms by profession, just like Batthyani, but when the strains of war had weakened his health, he withdrew to his domain in order to devote himself entirely to art and science (Hunger, 1927: 86). According to Guicciardini (1641: 11) the nobleman owned a collection of curiosities and he had a garden laid out in which was bred every possible sort of beautiful and useful plant, both native and of foreign origin. Clusius not only helped him in arranging the garden, but also advised him in the compilation of a large illustrated book which he intended to publish as the "*Centuriae plantarum rariorum.*" We learn all this from a letter Karel van Sint Omaars sent to Clusius in 1567.[3]

In that same year Clusius was still staying at the castle in Moerkerke, where he was writing his Spanish flora (Clusius, 1576). Karel van Sint Omaars's untimely death in 1569 (at the age of 36) also meant that the planned publication of his illustrated book was aborted and explains why a number of the drawings remain unfinished.
The watercolours came into the possession of Antonius Sanderus (1624:32), who according to Hunger (1927:88), bequeathed them to the Collegium Medicum in Leuven. After his death, Karel van Arenberg acquired the collection in 1595 and supplemented it with drawings of plants and flowers not found in the Low Countries. In 1663, the

Elector of Brandenburg and later King of Prussia, Friedrich Wilhelm, acquired the collection from Daniel Weinmann, so that it came into the possession of the Royal Library in Berlin, which was renamed the Prussian State Library in 1919. From Berlin, the collection was evacuated to Grüsau, Silesia (Poland), in 1941, after which time it disappeared, only to resurface in Kraków as late as 1977 (Whitehead, P.J.P., Van Vliet G., Stearn, W.T., 1989), where it is still preserved in the Library of Jagiellonian University as the *Libri picturati* A16-31.

In 1644, one of the last herbals was printed by Balthasar II Moretus under the title *Cruydt-Boeck*. In the commentaries of the plant descriptions (Dodoens, 1644: 512) the numerous plant collectors received no credit, although they had stimulated and practised the study of nature and played an important role in the network of scientific communication. In the text accompanying the *archangelica* (Dodoens, 1644: 512), Karel van Sint Omaars's name was no longer mentioned, so that with the passage of time his collection came to be attributed to another person, notably Clusius. Fortunately, Karel van Sint Omaars has been immortalized in the plant species *Reynoutria Hout* (Houttuyn, 1773-83: 64, 639-640). Most of the other collectors, with the exception of Peeter van Coudenberghe, recently honoured with a statue in the botanical garden of Antwerp (De Munck: 1996), remain unknown. I hope to rediscover them in the course of my continuing research on the *Libri picturati* A16-31.

I would like to express my gratitude to Prof. Dr. Jan Roegiers who allowed me to conceptualise and complete the study of the Libri picturati A 16-31 as a doctoral dissertation at the University of Leuven.

REFERENCES

PRIMARY SOURCES

1 Ab Orto, Garcia. *Aromaticum et simplicium aliquot medicamentorum apud Indos nascentium historia.* Antwerp: Widow C. Plantin and J. I Moretus, 1593.
2 Clusius, Carolus. *Exoticorum libri decem.* Leiden: Fr. Raphelengius, 1605.
3 Cordus, Valerius. *Annotationes in Pedacii Dioscoridis Anazarbei de materia medica libros V.* Strasbourg, 1591.
4 Dodoens, Rembert. *Cruijdeboeck.* Antwerp: J. van der Loe, 1554.
5 Dodoens, Rembert. *Cruydt-Boeck.* Antwerp: Balthasar II Moretus, 1644.
6 Dodoens, Rembert. *Frumentorum, leguminum, palustrium et aquatilium herbarum, ac eorum, quae eo pertinent, historia.* Antwerp: C. Plantin, 1566.
7 Dodoens, Rembert. *Stirpium historiae pemptades sex.* Antwerp: C. Plantin, 1583.
8 Fuchs, Leonhart. *Primi de stirpium historia.* Basel: M. Isengrin, 1545.
9 Gerarde, John. *The herball or generall history of plants.* London: Th. Johnson, 1633.
10 Guicciardini, Lodovico. *Description de touts les Pays Bas autrement appellez la Germanie inférieure ou*

Basse Allemagne. Amsterdam: Henry Laurents, 1641.
11 Lobelius, Mathias and Pierre Pena. Stirpium adversaria nova. Londen: Thomas Purfoot, 1571.
12 Lobelius, Mathias. *Kruydtboeck*. Antwerp: C. Plantin, 1581.
13 Lobelius, Mathias. *Plantarum seu stirpium historia*. Antwerp: C. Plantin, 1576.
14 Paulli, Simon. *Den Danske urtebog*. Antwerp: Balthasar II Moretus, 1647.
15 Sanderus, Antonius. *De scriptoribus Flandriae*. Antwerp: Gulielmus a Tongris, 1624.

SECONDARY SOURCES
16 Dacosta-Kaufmann, Thomas. *The Mastery of Nature: Aspects of Art, Science, and Humanism in the Renaissance*. Princeton: Princeton University Press, 1993.
17 Delen, Adriaan Jan Jozef. *Histoire de la gravure dans les anciens Pays-Bas et dans les provinces Belges des origines jusqu'à la fin du XVIIe siècle*. Paris: Editions d'Art et d'Histoire, 1934.
18 De Munck, Guy. *Peeter van Coudenberghe. Apotheker-botanicus (1517-1599) en tijdgenoten*. Antwerp: Koninklijke Apotekersvereniging van Antwerpen, 1996.
19 De Nave, Francine, ed. *Botany in the Low Countries (end of the 15th century-ca.1650)*. Antwerp: S. Ducaju & Son, 1993.
20 Depauw, Carl. "Peeter vander Borcht (1535/40-1608): de kunstenaar als inventor of creator van botanische illustraties?" In: *De Botanica in de Zuidelijke Nederlanden (einde 15de eeuw- ca 1650)*. Antwerp: Museum Plantin-Moretus, 1993.
21 Findlen, Paula. *Possessing Nature. Museums, Collecting, and Scientific Culture in Early Modern Italy*. Berkeley: University of California Press, 1994.
22 Houttuyn, Martinus Fr. Zn. *Handleiding tot de plant- en kruidkunde (1773-1783) II*, 7: 64, 639-640.
23 Hunger, F. W. T. *Charles de l' Escluse, Carolus Clusius, Nederlandsch Kruidkundige*, 1526-1609. 2 vols. The Hague: Nijhoff, 1927-43.
24 Laca, Luis-Ramon. "The Spanish and American plants in Clusius' correspondence." *Polish Botanical Studies*, Guidebook Series 20 (1998): 135-160.
25 Lemli, Jozef. "De Officinae Plantinianae te Antwerpen en te Leiden en hun botanische edities van 1589 tot 1647" In: *De Botanica in de Zuidelijke Nederlanden (einde 15de eeuw- ca 1650)*. Antwerp: Museum Plantin-Moretus, 1993.
26 Meier Reeds, Karen. *Botany in Medieval and Renaissance Universities*. New York/London: Garland, 1991.
27 Nissen, Claus. *Die botanische Buchillustration. Ihre Geschichte und Bibliographie*. Stuttgart: Hiersemann, 1951.
28 Olmi, Giuseppe. *L' inventario del mondo. Catalogazione della natura e luoghi del sapere nella prima età moderna*. Bologna, 1992.
29 Van den Wijngaert, Frank. "De botanische tekeningen in het museum Plantin-Moretus." *Gulden Passer* 1 & 2 (1947).
30 Vandewiele, Jules. "Wat groeide er in de tuin van Pieter van Coudenberghe?" In: *De Botanica in de Zuidelijke Nederlanden (einde 15de eeuw- ca 1650)*. Antwerp: Museum Plantin-Moretus, 1993.
31 Wegener, Hans. "Das grosse Bilderwerk des Carolus Clusius." *Der Preussische Staatsbibliotheek. Forschungen und Fortschritte* 12 (1936): 374-376.
32 Whitehead, Peter J. P., G. Van Vliet and William T. Stearn. "The Clusius and Other Natural History Pictures in the Jagiellonian Library, Kraków." *Archives of Natural History* 16 (1989): 15-32.
33 Wille, Helena. "The Albums of Karel van Sint Omaars (1533-1569) (Libri picturati A 16-31, in the Jagiellonian Library in Kraków)." *Archives of Natural History* 24, no. 3 (1997): 423-437.

NOTES

1 It concerns the woodblock nr. 7628 of the MPM collection used for the illustration of the Caltha palustris L. in Dodoens's *Cruijdeboeck* of 1554 on folio BBr and the woodblock nr. 7285 of the MPM collection used for the illustration of the Nerium oleander L. used for the illustration in Dodoens's *Cruijdeboeck* of 1554 on folio KK4v.
2 Both manuscripts are still preserved in the Library of Leiden University (BPL 948 and BPL 949).
3 This letter, published in its entirety by F.W.T. Hunger in his bibliography of Carolus Clusius (1927-1943), is preserved in the "Codex Vulcanius 101" of the Library of Leiden University.

ILLUSTRATIONS

ILL 1 Title page of Lobelius' *Kruydtboeck*, 1581: *Kruydtboeck oft Beschrijvinghe Van allerleye Ghewassen, Hesteren, ende Gheboomten: deur Matthias De Lobel Medecijn der Princ. Exc*[en] T'Antwerpen By Christoffel Plantyn, 1581

ILL 2 **Marrubium Hispanicum**
Matthias Lobelius, *Kruydtboeck*, Antwerpen, Christoffel Plantin, 1581, p. 613.
Collector: Jean Mouton (Tournai).
Plantin-Moretus Museum Antwerp, woodblock collection nr. 4466

ILL 3 **Scorsonera (viperaria)**
Lobelius, *Kruydtboeck*, 1581, p. 651.
Collector: Jan Coene (Antwerp)
Plantin-Moretus Museum Antwerp, woodblock collection nr. 6357

ILL 4 **Guanabanus Scaligeri**
Lobelius, *Kruydtboeck*, 1581, p. 216.
Collector: Willem Driesch (Antwerp)
Plantin-Moretus Museum Antwerp, woodblock collection nr. 6864

ILL 4B **Guanabanus** van Iulius Scaliger.
Dodoens, *Cryudt-Boeck*, 1618 edition, p. 1419

ILL 5 **Peruviana mechoaca**
Lobelius, *Kruydtboeck*, 1581, p. 747.
Collector: Peeter van Coudenberghe (Antwerp)
Plantin-Moretus Museum Antwerp, woodblock collection nr. 4557

ILL 6 **Solanum somniferum Clusij.**
Botanical name: *Physalis somnifera L.* English name: *Nightshade*
Lobelius. *Kruydtboeck*, 1581. p. 320

ILL 7 **Mala Armeniaca**
Libri picturati album A 20 folio 76
Collector: Karel van Sint Omaars (Bruges)
Library of Jagiellonian University, Kraków

ILL 8 **Herba Doria** Libri picturati Album A 18 folio 38
Botanical name: *Erigeron acer L.* English name: *Fleabane*
Library of Jagiellonian University, Kraków

ILL 2

ILL 3

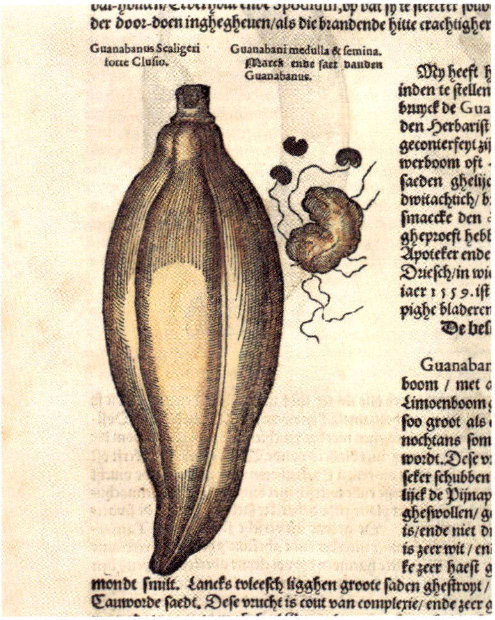
Guanabanus Scaligeri
forte Clusio.
Guanabani medulla & semina.
Marck ende saet vanden
Guanabanus.

ILL 4

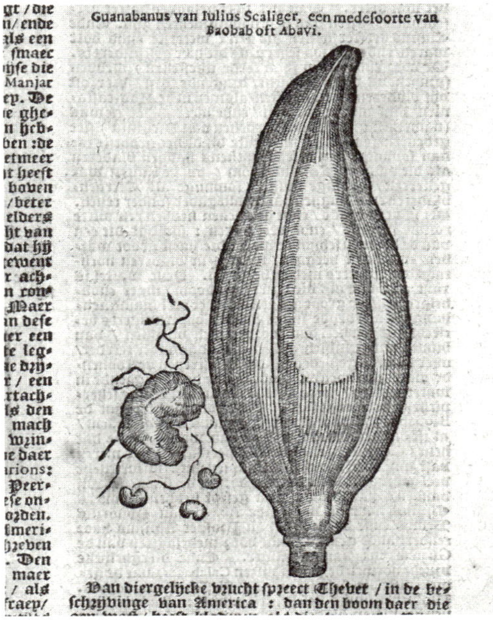
Guanabanus van Iulius Scaliger, een medesoorte van
Baobab oft Abavi.

ILL 4B

Peruuiana Mechoaca Mechoacæ Prouinciæ planta Bryoniæ similis.
de welcke ons ghedaen heeft in Engellandt de zeer vermaerde ende

ILL 5

Slaepmaeckende Naschape van Clusius. In Latijn/ Solanum somniferum Clusij Hyoscyami lutei foliis, ende is Somniferum Verticillatum Matth. In Spaensch / Yerua mora que acarrea locura.
My ghedenckt dat ick ghesien hebbe in Nederlandt inde houē vande

ILL 6

ILL 7

ILL 8

PART II

TRANSLATION AND THE ARTICULATION OF THE MODERN EPISTEME IN JAPAN

V

BREAKING BOUNDARIES: ON TRANSLATION AND THE CONCEPT OF SOCIETY

Michael Schiltz
research assistant of the Fund for Scientific Research (FWO) (Flanders), Department of Oriental and Slavonic Studies, K.U. Leuven.

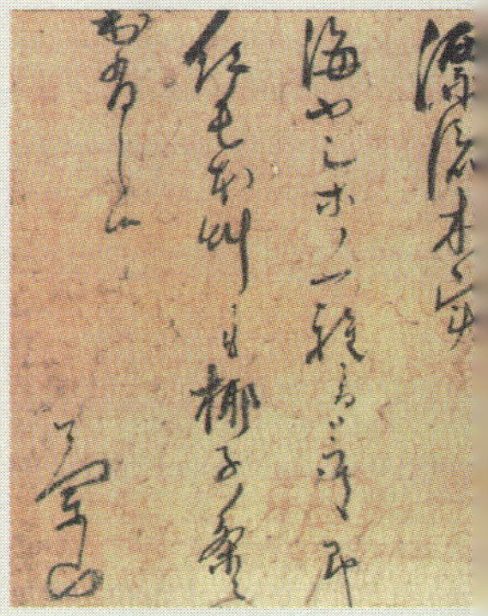

I Introduction

Let me start this paper with a rather ironic observation: the study of translation is an uneasy topic within contemporary Japanese Studies. The reason for this, as can easily be demonstrated, is to be found in the concept of society they embrace. Everybody will agree that translation is communication; and that it is communication crossing regional boundaries, 'trans-regional' communication so to speak. It is the 'trans-' in the above description that indeed poses a problem for current notions and theories of society, whether they are explicitly thematised or implicitly underlie sociological research. At the end of the twentieth century, the notion of society is still primarily *regionally* oriented. One views society (or societies) as a territorially defined entity – Japanese society, Belgian society and so on – and one assumes that, as different but equivalent entities, they can be compared. It must be stressed here that this observation does not merely concern earlier, 'pre-reflexive' literature; even the more mature, reflexive literature – of which Japanese Studies has produced quite a bit in the last decade [1] – questioning the very premises of Japan research, does not manage to 'deconstruct' the regionalist undertone governing sociological research on Japan.

It will be clear that, precisely for regionally oriented Japanese Studies, the topic of this collection of essays is rather 'exotic'. When regional differences are viewed as primordial, communication transcending these differences (e.g. translation) is like a logical *tertium non datur*. Consequently, translation studies are a kind of twilight zone. It is not clear how the study of translation fits into the current frame, how it can be thematised; and then it is left to other branches of science. That this is true, can be corroborated empirically from the sum total of research Japanese Studies have produced and still produce: translation studies occupy only a minor part of the large bulk. Indeed, the impact of translations on Japan's social order for instance, has tended to be treated as a marginal phenomenon.

Consequently, the foremost aim of this paper is to propose a new concept of society, one that allows for a full appreciation of communication independent of regional boundaries – among others, translation. The framework we pick is the systems theory of the German sociologist Niklas Luhmann. Highlighting some key concepts of the general framework provided by his theory of autopoietic social systems (paragraphs II), we will show how a surprisingly distinctive concept of society – '(global) society as the encompassing social system' – is developed (paragraph III). Special attention will be devoted to 'functional differentiation' (III 3), as this constitutes, in Luhmann's view, the central characteristic of modern society and remains the dominant social structure at the end of the twentieth century. In the second part of the paper, the topic of translation is taken up and linked to the concept of a functionally differentiated society (paragraph IV). Contrary to the current paradigm, we will stress the profound importance of translation for societal structure.

II Toward a new concept of society?

We may ask ourselves whether it is worthwhile to cling to existing conceptualisations of society, and the area studies's predilection for what could be called national cultures. Does e.g. the philosophy of Jacques Derrida not teach us that precisely the margins deserve our particular interest?[2] Or, formulated in the terminology of second-order cybernetics: is not the blind spot of an observation important as it makes the observation possible at all?[3] Apart from that, the existence *per se* of transregional communication is empirically incontestable. There simply *exists* a large amount of Western works translated into the Japanese language. And especially the twentieth century provides overwhelming evidence of social 'events' that contradict the assumed primacy of territorial boundaries. A phenomenon known in Japanese as *kokusai kekkon* (international marriage) demonstrates that the domain 国際結婚 of love does not seem hampered by geographical or political frontiers; the global economy has been reality for ages; scientists meet at international conferences and thwarts, *by definition*, attempts to reduce it to regional specificities (in the sense of 'American econometrics', 'French sociolinguistics', etc.). To cite a striking example, the Japanese mathematicians Taniyama and Shimura made crucial contributions to the proof of Fermat's last theorem (Singh 1997), obviously unimpeded by regional differences. Indeed there seems to lurk something highly problematic behind the regionalist concept of society. As Harumi Befu has once pointed out, one risks a mere caricature if conscientiously clinging to it; stating that one does not feel not be bound by the consequences of the regionalist view clearly implies surrendering in the face of more rigorous conceptualisations; and even if we express our reservations vis--vis the regionalist paradigm, it does not necessarily safeguard us from theoretical absurdities.

In any case, for Niklas Luhmann the aforementioned facts suffice to designate the regionalist concept of society as an *obstacle épistémologique* (after Gaston Bachelard). In his monumental study of society *Die Gesellschaft der Gesellschaft* ('The Society of Society,' 1997), the regionalist assumption is placed alongside three other obstacles that are contrasted with his own theory of society, which he proceeds to define as an attempt at "radically anti-humanist, radically anti-regionalist, and radically constructivist concept of society" (Luhmann 1997: 34-35). For this chapter one question is of particular importance: is it possible to develop a theory of society that is able to take into account both the trivialisation of regional boundaries by communication on the one hand, and the obvious and equally empirical existence of regional differences on the other? I am inclined to answer positively. How this can be done *concretely* is again explained by means of Luhmann's sociology. *Soziale Systeme: Grundriß einer allgemeinen Theorie* (Luhmann 1984) shifts to a more *abstract* level of investigation. "Our flight must take place above the clouds, and we must reckon with a rather thick cloud cover. We must rely on our instruments. Occasionally, we may catch glimpses below of a land with roads, towns, rivers, and coastlines that remind us of something familiar [...]. But no one should fall victim to the illusion that these points of reference are sufficient to guide our flight" (Luhmann 1995 [1984], L). This calls for some elaboration, as it obviously cannot imply abstraction as pure artistry, and certainly not abstraction with a mere analytical relevance. In spite of the often invoked distinction between theory and practice, systems theoretical concepts do refer to things that undeniably exist in the real world: meaning, time, process, and so on. However, the necessity for abstraction lies in the object of inquiry: society. Societal complexity requires a theory capable of formulating a complex analysis of complex (social) phenomena; and *abstraction may prove extremely useful in analysing some societally produced paradoxes, including the aforementioned apparently simultaneous processes of globalisation and region-*

alisation. The explicit merit of this shift towards a higher level of abstraction clearly lies in the wider range of application abstraction allows for; it may be possible to put things into a different perspective, and experiment with certain problems and the related problem solving methodologies. In 1984, this approach to what constitutes a sociological problem, and how theory building can respond to that resulted in the formulation of a 'general framework': not so much a theory of society in its own right, but a self-supportive construction built of concepts that acquire their meaning in reference to each other (Luhmann 1984: xiix). And precisely because of this, it is necessary to work with concepts that are defined very clearly, very rigorously, in order to achieve a high condensation (systems theory would prefer to speak of 'self-reference') – exactly as in mathematics, in which axioms, definitions, theorems etc. connect to each other.[4] On the one hand, this elaborate attention towards matters of conceptualisation is largely responsible for the esoteric nature that is allegedly characteristic of Luhmannian systems theory.[5] On the other, it illustrates the highly constructivist nature of the theory, as concepts are indeed defined in relation to each other, irrespective of any question about the 'real meaning' of a concept. In the end, questioning the commonsensical use of notions may the appropriate guideline for the design of concepts: one poses a concept, defines it and tests its explanatory potential. "More than any other sociological theory, systems theory does not accept taking anything, be it the world or notions, for granted. Luhmann adds that this is an absurd premise; yet as an absurd premise it avoids the danger of error and moreover forces the theory to reconstruct in its own terms anything it wants to describe" (Baecker 1997: 39).[6] The consequences for the presentation of the general framework as a coherent whole are surprising. As even the more experienced theorists will agree, *Soziale Systeme* is a difficult book, characterized by a pedagogical structure/ arrangement reminiscent only of Wittgenstein's Tractatus or comparable treatises on (formal) mathematical logic. Aware of the fact that its formulation can only take place in reality and in the world, the theory gradually unfold. Whereas it is initially concerned with a presentation of the difference between system and environment as its main thrust, it is prepared to rework this theme throughout the following chapters, and relate it to wellknown topics and methodologies of sociological theory. And just as in case of the *Tractatus*, self-reflection is an ultimate keystone of the venture. The framework is therefore concluded with remarks on the status of its knowledge (epistemology) as produced within the world, and subjected to the very premises the theory committed itself to. The theory thus re-enters itself as one of the objects it has taken to be 'out there' for sociological analysis. It explains the necessity to realize itself as a (social) system in an environment, in concordance with the epistemological presuppositions *unrealised* at the outset but inevitably 'present' in order to support the theory in the course of its composition.

III A radically anti-regionalist concept of society

1 The theory of social systems and society

Let us now consider the way by which Luhmann arrives at formulating a theory of society. In the first place, the theory of society is directly linked to (and must be linked to) the more general conceptual framework, which is the theory of social systems (Luhmann 1984).[7] The distinction between system and environment being its central paradigm, the new systems theory claims to be capable of catching every social phenomenon in terms of either system or environment. This would make it a universalistic – albeit not exclusivist or solipsist – theory. The word 'every' in the sentence 'every social phenomenon' should therefore be taken very literally: it concerns small, relatively simple social systems such as 'interactions', but just as well much more complex systems, e.g. organisations, or very large systems such as science or religion, or even 'society'. Ultimately, as already mentioned, it concerns systems theory *per se* as well, as it comes to view itself as a system in an environment. This 'autological' component need not be reexamined here. Careful to avoid an ontological or definitional muddle, it may be considered wise to rely on the accepted description of (autopoietic) systems as formulated by the Chilean biologist Humberto Maturana: here, systems are defined as "unities, as networks of productions of components, that recursively, through their interactions, generate and realise the network that produces them and constitute, in the space in which they exist, the boundaries of the network as components that participate in the realisation of the network" (1981: 21). When discussing the specific case of *social* systems, this requires the following specifications:

1 Social systems produce *communications* as their elements, i.e. communications are the atomic, not furthermore decomposable elements (*Letztelemente*) that represent the unity of three selections: *information* (one must pick something among alterna-

tives to communicate about), *utterance* (a selected information must be uttered), and *understanding* (distinguishing between information and utterance). One cannot sufficiently stress the importance of the latter selection. Understandings, including misunderstandings, mark a full distinction between communication and the mere speech 'act'. At last, the understanding determines the way in which a communication will be conceived in the communicative process, regardless of how it may have been designed by the 'sender'. Strictly spoken, the above conceptualisation not only renders the metaphor of transmission obsolete, as it destroys the distinction between a sender and a receiver; more importantly, it allows for an autonomous sphere of the social (social systems) by drawing attention to the way in which the communicative process itself assigns meaning to itself and its communications.[8]

ii As a corollary to this, the unity the system achieves for itself via its elements is made *communicatively*, and in no other way. Consequently, if the system differentiates, differentiation too will be executed according to communicative 'attractors'.[9] And the system's evolution too will follow the autopoiesis of communication (Luhmann 1997: 2). This way of operating is indeed not merely limited to the achievement of a system's unity, its differentiation, or its evolution: *everything the system does, is done autonomously, via systemic operations* (= communications) (Luhmann 1984: 35-38; idem: 48). Briefly: "the system does what it does" (Luhmann 1997: 88). That is what we mean by the concept of 'autopoiesis'.

iii Last but not least, these remarks are valid for *all* kinds of social systems. In the case of interactions, for instance, the unity will be achieved by means of themes; for gigantic societal subsystems such as science, on the other hand, communications connect to each other by means of a 'code' available to the respective functionally differentiated systems (Luhmann 1997: 743-776). Important for our discussion, it is argued that scientific communications are regulated and classified by means of the difference between 'true' and 'false'.

The concept of society is brought to the fore for a special reason: "Sociology must have a concept for the totality of what is social – whether one calls this (depending on theoretical preferences) the totality of social relations, processes, actions, or communications" (Luhmann 1984: 408). 'Society' is probably the only serious candidate for this role. Equally total signifiers as e.g. 'the world' are too much of an incoherent mass, i.e. they should be studied as aggregates of meanings rather than possessing a systemic unity. In more characteristically systems theoretical words: society is the social system encompassing all other social systems (*Gesellschaft als umfassendes Sozialsystem*). With every communication, society grows, society is altered. Its function is the (re)production of communication itself, and it thus represents the social system *par excellence*. This has some serious consequences. Society is put into a very special, if not exclusive, perspective: as it encompasses all social systems, it is the only social system possessing an environment in which there exists no other system of the same (i.e. social) kind – but this does not mean that society is a system devoid of environment! This condition reveals the particular status of the social sciences as a scientific discipline. In view of the above definition, society cannot be observed from outside: sociology, and sociological theory must refer to themselves as a product of the autopoietic operation mode of the societal system. Different sociologies, including the lareger bulk of the area studies, are not bothered by the implications for theory building this clearly entails. If one is inclined to believe that society is a territorially defined entity, it makes sense to believe it can be observed from outside as well. But if one is truly willing to adopt the view of society as the all-encompassing social system, sociology must account for its peculiar epistemological status (and its currently insifficient realisation thereof). And it will need to incorporate the aforementioned and often contested 'autological' component, i.e. it must understand it is subject to the very processes and operations it inquires into. It is for this reason, Luhmann argues, that his theory of society is – somewhat paradoxically – entitled 'The Society of Society': it is a description of society through the eyes of the social system 'society'. Above all, this implies the realisation of the contingency of this description, as it is one (and merely one) among different possibilities. Subjected to the historically contingent processes and evolution of the social system society itself, sociological descriptions of this society cannot simply establish themselves as observations of an observer that is 'objectively' there; rather, the self-referential circularity (and thus: contingency) of society's communications will have to be reflected in the description. The acclaimed universalism of the theory is thereby shown to be the opposite of solipsism *on logical grounds*.

II SOCIETY AS GLOBAL SOCIETY

But for our discussion, there is a far more important consequence implied in such theory of society. If we are really willing – and the line of argument indicates that we do – 1) to think of society as a social system realising its unity in autonomously produced communications, and 2) to define society as the all-encompassing social system, one conclusion imposes itself upon us: *society cannot be but global society* (*Weltgesellschaft*). "The definition of society as the encompassing social system has as a consequence that there can be only one societal system for connective communication. Strictly speaking, different societal systems may exist, similarly as in the past they used to talk of a plurality of worlds; but if one does so, then only if communication between these societies is non-existent, or, if seen from the viewpoint of the respective societies, communication with the others is either impossible or without consequences" (Luhmann 1997: 145). One may object here that this is a tautology [10]: the notion of 'global society' is already implied in the aforementioned specifications. And eventually, several systems theoreticians seem to be bothered by this 'problematic' conceptualisation (Blom 1997: 217-220). However, we stress that tautologies need not be theoretical deficiencies. [11] Society as global society is a mere logical implication of a theoretical choice made at the outset. At this point, one must evaluate the fruitfulness of that choice. When we defined society as a social system, we did so *because we believed it useful to do so* – and we must accept the consequences this entails. Denying or being unwilling to accept them would imply embracing the very essentialism Luhmannian systems theory wants to avoid. We remind the reader here that concepts are not to be judged in terms of their 'essential' value, but in terms of their explanatory potential – systems theory would prefer the term connectivity, *Anschlußfähigkeit*, here – that is gained out of the overall construction of concepts referring to each other (cf. supra). In any case, and in contrast to the regionalist paradigm, systems theory seems to be able to account for the empirical fact of globally communicating society; and furthermore it is able to bring a phenomenon such as 'translation' into perspective – *because it was willing to test the definition of society as a social system*!

III The differentiation of society

Does 'global society' then imply a globally homogeneous unity of communications? Certainly not. Defining society as an undifferentiated mass of undifferentiated communications would clearly be a theoretical mistake. The necessity for abstraction indicated in the above paragraphs was obviously only intended in order to enable further specification, and new distinctions. Above all, systems theory wants to be a theory of differences, of distinctions; distinctions that even mark the conceptual apparatus (system/environment; form/medium …) itself.[12] Accordingly, differentiation is one of the most central concepts of the theory. And as is the case with most systems theoretical concepts, its definition is not self-evident. Rather than he breaking up of the system into several pieces, it is defined as "the repetition of system formation within systems. Further system/environment differences can be differentiated within systems" (Luhmann 1984: 18). Hence, when we talk of the differentiation of society, we do not hint at the dissolution of society into subsystems.[13] Rather, society copies itself within itself, thereby producing at the same time more complexity and stability for itself. In Die *Gesellschaft der Gesellschaft*, Luhmann distinguishes four 'types' of differentiation (Luhmann 1997: 609-618):

- *segmentary differentiation*, i.e. differentiation into similar/equal subsystems (e.g. families);
- *differentiation according to centre/periphery* (e.g. the differentiation of society into equivalent units of which some function as 'spheres of influence'; characteristic of cultural empires),
- *stratificatory differentiation*, by which a fundamental inequality of rank is produced (e.g. feudal society);
- functional differentiation, by which both an *inequality* (of function) and an *equality* (of status) of differentiated systems is produced (e.g. the functional subsystems of modern society: science, the judicial system, the economy, politics …)

For our discussion – and this will be elaborated later in this paper – functional differentiation, or rather *the shift to* the functionally differentiated society, is of particular interest, as it concerns the often discussed and often misunderstood transition from feudalism to modernity. Clearly, the form of society characterised by differentiation according to functional points of orientation is *modernity*. Later on, we will discuss translation in relation to this notion. To be more precise, 'functionally differentiated society' means that communication is directed by the communicative beacons of science, the economy, politics, the judicial system, religion, art, and so on…. Or put in other words: it is the communicative trajectories laid out by those systems differentiated from each other that have primacy over other forms of differentiation, define societal structure, and take on primary importance for the continuation of communication. This entails a peculiar combination of both *specificity* and *universality*. Specificity on the one hand, as the systems view reality according to their specific code[14]; but universality on the other, as the entire social reality is grasped in the terms of each system.[15] Science, the functional system that is most of our concern here, tends to view the whole world in terms of the difference between true and false, and thereby delineates its reach from e.g. religion or politics; and the universality of the 'code' is even extended to include even the distinction between true and false itself (epistemological questions). Therefore we say that functionally differentiated systems are *equal in their inequality* (Luhmann 1997: 613). In turn, this has important consequences for the semantics of society. With the primacy of functional differentiation established, society forsakes the unified standpoint of observation, or an 'Archimedean point'. Instead, it is distinguished by an irreconcilable multiplicity of standpoints, by a lack of controlling instance: in short, by 'polycontexturality' (Günther 1976-1979 II: 283-306). This may well be a unique instant in world history: at no other point in time have societies forsaken to appeal to a unifying viewpoint premodern societies obviously needed. Empirical evidence clearly supports this claim. 'Modern' descriptions of society prefer alienation over identity; multiplicity over uniformity,... As stated above, differentiation is, just as all system's achievements, a product of systemic operations, i.e. communications. The shift from the stratificatory differentiation of feudal society to the functional differentiation of modern society, for instance, can therefore not be inferred as taking place out of the blue, a blind change of societal structure. It is a product of systemic operations, of communication: society produced this change by itself. Or more precisely, as has been argued by historians (Eisenstadt 1999 [1983]), philosophers of median and culture (e.g. McLuhan 1964), philologists (Havelock 1963; Ong 1982) anthropologists (Goody 1996) and sociologists (again: Luhmann 1997), it is a product of an evolutionary problem: the enormous *increase in communication*, propelled by the 'media of dissemination' (*Verbreitungsmedien*) that are writing and, especially, printing.[16] We cannot enter into an elaborate discussion of the communication-specific aspects accompanying written communication.[17] One may, for instance, point out the enormous upgrading of capacity for social memory (Assmann 1999) this entails. Writing, being a kind of self-produced memory independent of the co-evolution of human organisms, enables society to remember and forget much more at the same time. But of overarching societal importance, is the fact that writing makes use of symbols.

Symbolising should be understood here as the capability to *pre*sent the *ab*sent as a possible object of communication. Especially Elisabeth Eisenstadt's work has been seminal in describing the importance of printed materials as an 'agent of change'.

Again, this shift in societal may have been historically contingent, but it can never be described as an arbitrary process. Interestingly, the roots of this shift in societal structure are to be found in the preceding, stratificatory society: "For stratificatory differentiation enables the concentration of resources in the upper-stratum of the system, and this not only in the economic sense, but also in the medium of [political and judicial] power and that of [scientific] truth" (Luhmann 1997: 708). The concentration of resources in the upper-stratum of society paradoxically paves the way for its own progressive destruction! We stress here that this has never been a uniquely European matter. Japanese history provides us equally good examples of such dramatic societal shift. Shogun Yoshimune's policy of *kokusanka*,[18] the 'indigenisation' of foreign products, had obviously to be accompanied by the introduction of 'modern episteme', i.e. science:[19] his policy contained the seeds of the far-reaching differentiation that is inescapable in modern global society. The point is that it did not suffice to merely import foreign products or technology. Equally important were the introduction and acquisition of the 'body of knowledge' associated with the relevant technology, and a sufficient amount of societal will and chance to experiment with this novel way of dealing with written material… For this purpose, one needs people able to read. And when one is able to read, one may very well read literature that is damaging to the ruling elite, highly critical of religious dogma's etc. As a reaction to self-produced complexity, society must develop a form of differentiation which is capable of giving itself positive feedback. Functional differentiation, the form of differentiation that enables societal subsystems to manage communication according to specific codes, proved to be the most effective form, even up to this day. From this perspective, as we will show later in this paper, translation studies are of central importance to the understanding of Japan's modernity.

吉宗, 国産化

In any case, it seems undeniable that, from the end of the feudal period on, an increasing amplification of function-oriented communication came to stand out, in the domains of science, as well as the economy (Ooms 1998, 23-47) and politics (Maruyama, 1974).[20] We already mentioned Shogun Yoshimune's policy of *kokusanka* or indigenisation of foreign products, simultaneously introducing modern knowledge. A most interesting and very clear example may however be noted in the context of military history. The introduction of modern 'Western' military technology – together with Japanese willingness to accommodate it – into sixteenth-century Japan revolutionised the conduct of warfare, and the battle of Nagashino (1575) is illuminating in this context. 長篠 Whereas earlier warfare was still characterised by an emphasis on differences of 'rank' – one may think of the key notion of 'honour', or of stylistic peculiarities referring to differences of rank – Nobunaga's for- 信長 eign technology literally blew the traditional code of honor apart, and placed the focal point on the visibilisation of the difference between power and force (societally recognizable 'political power' Luhmann 1975; 2000, 18-68), fundamental for an autonomously functioning modern political system.[21] Admittedly, one must not assume an abrupt turn: the battle of Sekigahara (1600) for instance, is one example of a 関ヶ原 typically feudal battle, with still a very strong emphasis on formalism and stylistic prescriptions (Kasaya 1994). But it also was the last battle of this kind. The gun would play a crucial part in subduing the warring clans at the end of the Age of the Warring States, and it was both its visibility as a new weapon and the bakufu's conscious and skillful employment of it as a threat against future insurrections that had radically changed the face of political power (the so-called Pax Tokugawa). Whereas pre-Tokugawa warfare may well have been functional as preserving equalities among the members of the upper-stratum by means of the prevention of the accumulation of wealth and power, the mere possibility of war (that can be avoided if bakufu power is yielded to) served in Tokugawa Japan as the reflection and representation of a centralized political power claiming *legitimacy* for itself -and for itself only. In this differentiation of power and force sociologists may detect the origination of modern political power, and its alter ego, the nation state (Parsons 1967; Willke 1996 [1992] 31ff.).

iv Regionalisation as conditioning

Does the above conceptualisation unequivocally lead to a view of society as devoid of regional differences? Clearly, this question goes straight to the heart of a regional studies discipline like Japanese Studies. Does Luhmann indeed hint atthe destruction of a theoretical ground for area studies? Are regional differences and the concept of culture shut out of the picture? We are inclined to answer negatively. Like Luhmann, we regard the thesis of convergence as highly premature. Regionalisation need not contradict globalisation; they both rather represent two sides of the same coin, two sides of one distinction.[22] The globalisation of functional differentiation may very well

need regionalisation in order to strengthen its perspective for worldwide dissemination. The reason for this is to be found in the evolutionary improbability of a phenomenon such as functional differentiation: "In the context of the theory of evolution, it must be accepted that the societal differentiation of specific function systems [...] and especially the shift of the encompassing social system "society" to the primacy of functional differentiation is an *extremely improbable happening* [...]" (Luhmann 1997: 707). At last, the shift from rank dominated societies to societies with a primacy of functional orientation is at least without a historical precedent. For the predecessors of the modern function systems, differentiating in a the latter days of feudal society but still strongly under its strains the main task is thus developing some kind of self-recognition, an 'identity' or an 'Eigenwert' to which future operations can be directed – an operation that has historically been accompanied by numerous evolutionary successes and failures, and the obvious coming into being of regional differences. Hence, globalisation need not contradict regionalisation. The latter can fulfill the function of *conditioning* (*Konditionierung*): it backs up the dissemination of functional differentiation, so to speak, by providing the (regional) conditions for dissemination by means of (regional) experiment. Conversely, the dominant pattern of functional differentiation offers an impetus for the development of (regional) distinctions in dealing with it. Such easily misunderstood intricate arrangement of is this which explains e.g. the existence of differences between Japanese and 'Western' styles of management; regional differences in religious practices; differences in the Japanese and the Chinese emperor system; or differences in the execu-
和魂洋才 tion of power... Semantically, the notion of *wakon yôsai* ('Japanese
明治時代 spirit, Western technology'), coined in the Meiji period (*Meiji jidai*,
富国強兵 1868-1912), illustrates our point. Along with slogans as *fukoku kyôhei*
尊王攘夷 ('rich nation, strong army'), *sonnô jôi* ('respect the emperor, expel the
barbarians') etc., it gained a mantra-like power at the time, invoking the modernisation of the Japanese state (Samuels 1994). As no other, it expressed the profound understanding of the Meiji reformers that
洋才 the introduction of foreign technology (*yôsai*) was both desirable and
inevitable. At the same time, because of its overarching dominance, it provided room for experiment with different degrees of functional differentiation, experiment with research and development, manufactur-
和魂 ing, and so on (*wakon*). In twentieth-century Japan, this bivalent process of globalisation and regionalisation seems to be captured in the ideology of ***kokusanka*** (indigenisation) (Samuels 1994; Green 1996). As Richard Samuels explains, the circular relationship between key

technologies and ideas on development, adaptation of such technologies constitute 'national systems of innovation', technoeconomic paradigms, responible for the relative succesfulness with which technologies can be introduced and implemented. In Japan, more specifically, a long lived ideology and beliefs on the importance of high technology for the preservation of national security may have given rise to the most successful production and research methods at the end of the 20th century (Samuels 1994). But regional differences remain of a different order than functional differentiation; the primacy of functional differentiation is undeniable. Regional differences, as they serve the purpose of *conditioning*, facilitate differentiation. Experiments obviously leave room for difference, but do not destroy the unifying thread among them. It follows that we can identify the difference between Japanese and Western styles of management only if we can project 'economic organisation' as the unifying standard to which the differences can be measured. And for the same reason, it is absurd to compare Japanese management style with Western religious practice.

There exists, however, a kind of regionalisation coinciding with the globalisation of functional differentiation and still different from the above distinguished possibility of regional conditioning. This kind of regionalisation does *not* take place on the level of societal structure, but on the *semantic* level.[23] In other words, it is related to the way societal change and structure are registered and 'meaningfully' processed in communication that is possible only in view of the dominant form of societal differentiation – a level, indeed, on which structural 'data' are (un)consciously referred to. Specifically, we mean the simultaneous rise and interrelation of functional differentiation and the semantics of the *nation*. Notions as 'revolution', still strongly associated with the nation state, play a key role here, as they are intended at expressing an innovation for society as a whole. Just as can be said of regionalisation that it has a (conditioning) function for functional differentiation, we attribute a similar *function* to the notion of 'nation'. We view it as a resource of *identification*, not in the least the identification of state and society: "A nation distinguishes itself from other nations (and not from aristocracy or living in the country or the economy or science). It makes it possible to oppose to the *universalisms* of functional orientations the *particularisms* of regional communities as more valuable. [...] In brief: the notion of nation offers a concept of inclusion that is not dependent on the particular conditions of single function systems, and that forces even politics to treat all members of the own nation as equal" (Luhmann 1997: 1051-1052). In order to be successful, i.e.

unhampered by the remnants of premodernity, the various functional systems need to be 'fine-tuned', i.e. all of them must incorporate the semantics of the nation as an 'attractor': "A nation is a means of regulating the political, economical, educational, and semantic reproduction of a society, implementing such a regulation by imposing cross-references from the political to the economical, from the semantic to the educational, and from the educational to the political" (Baecker 1997: 46).[24] The attentive reader may have noticed that our analysis can equally be applied to the regionalist paradigm we initially deemed inappropriate for contemporary Japanese Studies. While it is possible to deny it much explanatory potential (as we repeatedly did), it is clearly more fruitful to try and explain this paradigm as an historically contingent body of thought accompanying modernity. Its analytical potential is, as illustrated by the topic of translation, questionable – but one should not overlook its function for society!

IV Translation as a propeller of functional differentiation

In the preceding pages we have outlined a frame of reference that is prepared to grant societal relevance to all communication. By delineating a distinct sphere of 'the social', distinguished from the psychic, in short, by allowing for 'social systems', systems theory circumscribes the fuzzy terrain of concepts needing prefixes as 'trans-' ('transmission') or 'inter-' (as in 'intercultural'). Instead, its theory of society centres on the focal point of global communication; and translation must thus be treated as highly relevant for societal structure. But what could be the ramifications for the study of translation? Is this concept of society capable of specifying the phenomenon of translation? Clearly, translation must be put into perspective. More assumptions will have to be developed in order to test the value of our frame *vis-à-vis* empirical reality.

One possibility for outlining a systems theoretical treatment of translation could be the analysis of technical difficulties characteristic of translation. But as we are primordially guided by a sociological in casu systems theoretical interest, we consider it more appropriate to conduct such discussion in relation to the above remarks on societal structure and differentiation.[25] More specifically, we assume a strong relationship between the dissemination of translations and functional differentiation. In Japan, this relationship may be considered most prominent in the field of science, or science-related developments – this is, indeed, the thrust of this book. Shogun Yoshimune's policy of indigenising foreign drugs and the concomitant introduction of Western pharmacopoeia have been mentioned repeatedly in our paper. The concept of *Rangaku* 蘭学 as Western learning equally illustrates the development of and interest in the clearly functionally oriented system of science. In this context, one should reserve the necessary attention for the considerable time lag in the introduction in Japan of primordial scientific literature. Dodoens's *Cruijdeboeck* is exemplary. It did not reach Japan before the first half of the seventeenth century; and even then, it did not attract attention until the eighteenth century. Generally speaking *Rangaku*, literally 'Dutch Studies', were in a way a scientific fossil – as Japan's enlightened moderniser Fukuzawa Yukichi 福沢諭吉 was to discover. But once the Japanese leaders firmly decided to modernise the country and 'open up' to Western science and technology, one may even be tempted to regard that as the start of *systematic* functional differentiation.

One can trace such societal changes and the shift towards functional differentiation through the development of technical issues in the translation process. Especially in the first phase of confrontation with Western books and treatises, the clumsiness of the translation is striking; and so is the lack of conceptual refinement, the great number of mistakes, etc. Well into the Meiji period, the enlightened Japanese ruling elite, understanding the importance of 'catching up' (*oitsuki*, *oikose*) 追い付き、追い越せ with the West, could not mobilise enough resources for a smooth incorporation of Western knowledge. At least not immediately: the great number of missions abroad would gradually make good the backlog. Interestingly, this lack of foreknowledge ironically explains the paradigmatic value of Dodoens's *Cruijdeboeck* and *pharmacopoeia* for the introduction of Western episteme and the further development of science in Japan. Herbals, *pharmacopoeia*, or eventually books on fortress building are indeed *illustrated* (in the case of Dodoens they were furthermore said to be excellent!), and that facilitates understanding – as illustrations do not carry the translation-specific difficulty of 'difference'. Language functions by means of differences expressed by tokens[26], and mastering them is obviously the central difficulty in the 'art of translation'.

There is an interesting corollary to be deduced from the above remarks. When translation can indeed be said to have an amplifying value for the dissemination of functional differentiation (*in casu* functionally differentiated science), it contains some embarrassing consequences for the ruling classes of feudal societies. In the final analysis, the shift from feudal society to functionally differentiated society is a shift in societal *structure*, i.e. a change of communicative expectations. Communications are no longer primordially directed by differences in rank, but by functional points of orientation (scientific truth, political power, judicial justice …). Consequently, the upper class loses the lucrative position of collecting resources (cf. supra): "[Societal] changes affect above all the nobility; not in the form of competition from another upper-stratum, but through the slow devaluation of the difference that distinguishes nobility from the masses. […] For all the stress on accustomed differences of rank, [the nobility] will gradually experience that the newly arising functional systems do not depend on nobility, and that their full differentiation cannot be realised by the nobility" (Luhmann 1997: 712-713). Paradoxically, the ruling elite will have to hedge against the self-induced changes! We may borrow a notion from the philosophy of Jacques Derrida here: the idea that writing is subversive and dangerous, as it can 'get out of hand' (Derrida 1972).[27]

Although suppositions about harsh Tokugawa censorship must certainly not be exaggerated (both in terms of legislation and enforcement censorship was unsystematic until after the Meiji Restoration) (Kornicki 1998, 320-362), the Japanese upper class indeed attempted to safeguard its status by shrouding writing and knowledge in secrecy. This
江戸時代 was already the case in the early Edo period (*Edo jidai*, 1600-1868),
幕府 when the Bakufu had the monopoly of e.g. calendars. But especially the establishment and growth of Dutch Studies (*Rangaku*) seems to
出島 have been cloaked in secrecy. Foreigners residing in Deshima were not allowed to travel through the country; studying Japanese was for-
阿蘭陀風説書 bidden to foreigners. But the deliberate holding back of the *Oranda fûsetsu-gaki*, reports written by the Dutch on the geopolitical situation, and medical knowledge, is illuminating. A similar story can be told
通詞 about the knowledge of the *tsûji*, the interpreters. This knowledge was monopolised by a few families, for an obvious reason.

v Conclusion

As will be clear to many commentators, the topic of translation is not a self-evident one for Japanese Studies. With our paper, we have tried to explain that the identification of the concept of society with the geographically delineated nation state may be the crucial stumbling block for putting the importance of translation into perspective. Hence, we analysed the dominant conceptualisation and pointed out that exactly the alleged importance of regional boundaries 1) did and does not allow for the recognition of a globally communicating society, and 2) thus cannot but attribute a marginal status to the study of translation. At the same time, the explanatory potential of a different concept was tested: the concept of society as a social system. We demonstrated that it is very well possible to assign a place to translation studies, if one is willing to cast off 'common-sense' (i.e. semantical) notions, and test the explanatory potential of others instead.

REFERENCES

1 Alberts, Davis S. and Czerwinsky, Thomas J. 1997. *Complexity, Global Politics and National Security.* Washington DC: National Defense University.
http://www.mnsinc.com/cbassfrd/CWZHOME/CWZBASE.htm

2 Assman, Jan. 1999. *Das kulturelle Gedächtnis -Schrift, Erinnerung und Identität in frûhen Hochkulturen* München: Beck.

3 Baecker, Dirk. 1997. "The Meaning of Culture." *Thesis Eleven* 51 (November): 37-51.

4 Barnes, Barry. 1977. *Interests and the Growth of Knowledge.* London: Routledge and Kegan Paul.

5 Bassford, Christopher. 1998. *Clausewitz Homepage.*
http://www.mnsinc.com/cbassfrd/CWZHOME/

6 Blom, Tannelie. 1997. *Complexiteit en Contingentie – een kritische inleiding tot de sociologie van Niklas Luhmann.* Kampen: Kok Agora.

7 Clausewitz, Carl von. 1989. *On War* (ed. and trans. Michael Howard and Peter Paret). Princeton, New Jersey: Princeton University Press.

8 Derrida, Jacques. 1972a. *Marges de la philosophie.* Paris: Minuit.

9 _____. 1972b. *La Dissémination.* Paris: Seuil.

10 _____. 1988. "Letter to a Japanese Friend," in David Wood and Robert Bernasconi (eds.), *Derrida and Différance.* Evanston III.: Northwestern University Press.

11 Eisenstadt, Elizabeth. 1979. *The Printing Press as an Agent of Change – Communications and Cultural Transformations in Early Modern Europe.* Cambridge: Cambridge University Press.

12 Fuchs, Peter. 1998. "Sommerseminar 1".
http://sti1.uni-duisburg.de/Luhmann/mail15

13 Goody, Jack. 1986. *The Logic of Writing and the Organization of Society.* Cambridge: Cambridge University Press.

14 Green, Michael. 1996. *Arming Japan: Defense Production, Alliance Politics, and the Postwar Search for Autonomy.* New York: Columbia University Press.

15 Gluck, Carol. 1985. *Japan's modern myths: ideology in the late Meiji period.* Princeton, N.J.: Princeton University Press.

16 Günther, Gotthard. 1976-1979. *Beiträge zur Grundlegung einer operationsfähigen Dialektik – Zweiter Band.* Hamburg: Felix Meiner Verlag.

17 Harris, Victor. 2000. "Arms: The Balance of Peace," in Robert T. Singer (ed.), *Edo: Art in Japan 1615-1868* (catalogue of exhibition held in the National Gallery of Art, Washington).

18 Havelock, Eric. 1963. *Preface to Plato* Oxford: Basil Blackwell.

19 Hofstadter, Douglas R. 1997. *Le Ton Beau de Marot: In Praise of the Music of Language.* New York: Basic Books.

20 Howard, Michael. 1983. *Clausewitz.* Oxford & New York: Oxford University Press.

21 Hutter, Michael. 1998. "Wie der Unterschied zwischen Ornament und Figur in die Welt kam," in *Zur Diskussion gestellt: Niklas Luhmann, "Die Kunst der Gesellschaft"* (internet-discussie)
http://www.uni-bielefeld.de/sozsys/kunst.htm

22 Jokisch, Rodrigo. 1996. *Logik der Distinktionen: Zur Protologik einer Theorie der Gesellschaft.* Berlin: Westdeutscher Verlag.

23 Kasaya Kazuhiko. 1994. *Sekigahara-gassen: Ieyasu no senryaku to bakuhan taisei* 「関ヶ原合戦・家康の戦略と幕藩体制」. Tôkyô: Kôdansha.

24 ____. "Tokugawa Bakufu's Policies for the National Production of Medicines and Dodoens's Cruijdeboeck" (See article in present volume).
25 Kiel, L.Douglas and Elliott, Euel. 1996. *Chaos Theory in the Social Sciences – Foundations and Applications.* Ann Arbor: The University of Michigan Press.
26 Klagenfurt, Kurt. 1995. *Technologische Zivilisation und Transklassische Logik: Eine Einführung in die Technikphilosophie Gotthard Günthers.* Frankfurt am Main: Suhrkamp Verlag.
27 Kornicki, Peter. 1998. *The Book in Japan: A Cultural History from the Beginnings to the Nineteenth Century.* Leiden-Boston-Köln: Brill Publishers.
28 Lamers, Jeroen. 1997. "Oda Nobunaga (1543-1582): A Japanese Tyrant." *IIAS Newsletter* 13 (Summer): 36-37.
29 Luhmann, Niklas. 1971. "Sinn als Grundbegriff der Soziologie." p. 25-100 in Jürgen Habermas and Niklas Luhmann (eds.), *Theorie der Gesellschaft oder Sozialtechnologie: Was leistet die Systemforschung?* Frankfurt am Main: Suhrkamp Verlag.
30 ____. 1975. *Soziologische Aufklärung 2: Aufsätze zur Theorie der Gesellschaft* Opladen: Westdeutscher Verlag.
31 ____. 1988. *Die Wirtschaft der Gesellschaft.* Frankfurt am Main: Suhrkamp Verlag.
32 ____. 1990. *Die Wissenschaft der Gesellschaft.* Frankfurt am Main: Suhrkamp Verlag.
33 ____. 1992. *Beobachtungen der Moderne.* Frankfurt am Main: Suhrkamp Verlag.
34 ____. 1993a. "Die Paradoxie der Form." p. 197-212 in Dirk Baecker (ed.), *Kalkül der Form.* Frankfurt am Main: Suhrkamp Verlag.
35 ____. 1993b. "Zeichen als Form." In Dirk Baecker (ed.), *Probleme der Form.* Frankfurt am Main: Suhrkamp Verlag.
36 ____. 1994a. *Das Recht der Gesellschaft.* Frankfurt am Main: Suhrkamp Verlag.
37 ____. 1994b. *Soziologische Aufklärung 4: Beiträge zur funktionalen Differenzierung der Gesellschaft.* Opladen: Westdeutscher Verlag.
38 ____. 1995 [1984]. *Social Systems*, (translated by Dirk Baecker and Eva Knodt). Stanford: Stanford University Press. Originally published in German as *Soziale Systeme: Grundriß einer allgemeinen Theorie* (Frankfurt am Main: Suhrkamp Verlag, 1984).
39 ____. 1995. *Die Kunst der Gesellschaft.* Frankfurt am Main: Suhrkamp Verlag.
40 ____. 1997. *Die Gesellschaft der Gesellschaft.* Frankfurt am Main: Suhrkamp Verlag.
41 ____. 2000. *Die Politik der Gesellschaft.* Frankfurt am Main: Suhrkamp Verlag.
42 Maruyama Masao. 1974. *Studies in the Intellectual History of Tokugawa Japan* (trans. Mikiso Hane). Tokyo: University of Tokyo Press.
43 Maturana, Humberto. 1981. "Autopoiesis." In Milan Zeleny (ed.), *Autopoiesis: A Theory of Living Organization.* New York: North Hollan
44 McLuhan, Marshall. 1964. *Understanding Media – the Extensions of Man.* New York: New American Library.
45 Miyoshi, Masao and Harootunian, David. 1988. *Postmodernism and Japan.* Durham, N.C.: Duke University Press.
46 Ooms, Herman. 2000. "Forms and Norms in Edo Arts and Society." In Edo – *Art in Japan.* New Haven and London: Yale University Press.
47 Ong, Walter. 1982. *Orality and Literacy – The Technologizing of the Word.* London/New York: Methuen.
48 Parsons, Talcott. 1967. "On the Concept of Political Power". p. 297-354 in *Sociological Theory and Modern Society.* New York & London: The Free Press.
49 Samuels, Richard J. 1994. *Rich Nation, Strong Army.* Ithaca and London: Cornell University Press.
50 Schiltz, Michael/Verschraegen, Gert. Forthcoming. "In Search of an Autologic Epistemology: Niklas Luhmann Indebtedness to the *Laws of Form.*"
51 Screech, Timon. "Dodoens and Popular Culture in eighteenth century Japan" (Paper prepared for the Conference 'Translation of Culture, Culture of Translation,' Katholieke Universiteit Leuven, Belgium, 12-15 October 1998).
52 Singh, Simon. 1997. *Fermat's Last Theorem: The Story of a Riddle that Confounded the World's Greatest Minds for 358 Years.* London: Fourth Estate.
53 Spencer-Brown, George. 1994 [1969]. *Laws of Form.* Portland: Cognizer Co.
54 Williams, David. 1994. *Japan and the Enemies of Open Political Science.* London: Routledge.
55 Willke, Helmut. 1996 [1992]. *Ironie des Staates – Grundlinien einer Staatstheorie polyzentrischer Gesellschaft.* Frankfurt am Main: Suhrkamp Verlag.
56 Wittgenstein, Ludwig. 1989. *Logisch-philosophische Abhandlung: Tractatus logico-philosophicus*, (critical edition by Brian McGuinness and Joachim Schulte). Frankfurt am Main: Suhrkamp Verlag.

NOTES

1 We refer to the numerous contributions to the debate about the 'paradigm in Japanese Studies', published for instance in the Japan Foundation Newsletter; also the concluding sessions of many Japanese Studies-related conferences etc. The more impressive, if controversial, fruits of reflexive literature are doubtlessly: David Williams (1994) and Miyoshi & Harootunian (1988).

2 We refer here to Derrida 1972a.

3 See particularly: Luhmann 1990: 82-85.

4 See: Luhmann 1990: 201; Barnes 1977: 10.

5 That this condensation jeopardizes the communicability of research results, has been the subject of some workshops and has been thematized within systems theory as well. It is a question of *Gesprächsfähigkeit* (or even: *Anschlußfähigkeit*, 'connectivity') of the theory. See Fuchs 1998.

6 With a strong reference to Luhmann 1971.

7 See Eva Knodt in the 'Introduction' to the English translation of *Soziale Systeme: Grundriß einer allgemeinen Theorie*: "*Social Systems* does not present a sociological analysis of modern society or a theory of society (*Gesellschaftstheorie*) but elaborates the general conceptual framework for such a theory. It supplies the instruments for observing a variety of social systems – societies, organisations, and interactions – not primarily such observations themselves. The distinction is far from trivial. In positing a difference between "what" questions and "how" questions, the theory of social systems situates itself within the "de-ontologized" realm of "second-order observations," a level of abstraction where, to speak in Kantian terms, questions concerning conditions of possibility arise" (Luhmann 1984: XVII).

8 'Communication' is therefore discussed in contrast to 'action' (Luhmann 1984: 135-175). Note furthermore that the addition of 'understanding' as the third selection actually seems to reverse the temporal process of communication: "Communication is made possible, so to speak, from behind, contrary to the temporal course of the process" (Luhmann 1984: 143).

9 The term 'attractors' is taken from non-linear systems theory. On the way the term is received in the social sciences, see Kiel and Elliott (1996: esp. 19-116). An application of the concept in strategic studies can be found in Alberts and Czerwinsky (1997).

10 I thank Jan van Bavel and Rudi Laermans for drawing my attention to this point; the discussion was conducted in the context of a Ph.D. seminar on Luhmann's theory of evolution.

11 A very good illustration of this is provided by Wittgenstein's remarks on logical expressions: "Die Sätze der Logik sind Tautologien" (Tractatus: 6.1 and further). Mathematics gains its clarity precisely by its tautological nature. Numbers, for instance, cannot be said to refer to something that is 'there' in the environment. They acquire their meaning in relation to themselves: self-referentially. In spite (or rather: *because*) of this, their range of applicability is vast.

12 Again, we have to point to the importance of Spencer-Brown's calculus of indications for Luhmannian theory design. In his calculus, forms are understood as distinctions, indicating one side of the two sides separated by the distinction. On the other hand, there is Luhmann's indebtedness to Gotthard Günther. See, in the context of differentiation, "Life as Polycontexturality" (Günther 1976-1979 II: 283-306).

13 The system/environment paradigm thus replaces the 'old-European' distinction between the parts and the whole.

14 On a theoretical treatment of the systems theoretical concept of (binary) codes (*Kodierung*), see: Luhmann 1975: 170-192 and 1994: 13-31.

15 The meaning of a social 'event' can therefore only be judged in relation to a system. Sticking to the topic of this paper, a translation will be viewed by science in terms of the scientific truth it embodies, by the economy in terms of the money that can be made from it ad so on. When we want to talk of a social event, it will thus always be necessary to do so in reference to an observing system.

16 In the twentieth century, one may add electronic media (the computer, ...) to the list (Luhmann 1997: 302-311). Note that they pose specific problems for and have different effects on the nature of communication.

17 Luhmann devotes an entire chapter to this topic (Luhmann 1997: 249-290). Probably most fundamental is the fact that, the introduction of written communication has the ramification of differentiating the different dimensions of meaning. *Temporally* speaking, writing has several 'presents' at its disposal. *Socially*, the increasing insecurity of acceptance must be stressed. And third, as the physical presence of speaker and listener is less relevant, the *factual* dimension emerges as the one drawing most attention.

18 See in this context: Kasaya's essay in the present volume.

19 One may be tempted to speak here of 'Western science', but why would we? With its introduction into 'the East', science loses its regionally 'Western' characteristics. This does not mean that it is impossible to observe regional differences, e.g. in the organisation of universities etc. But such regional differences do not have primacy over the 'core' of scientific communication that is the difference between the – abstractly defined – code values of 'true' and 'false'. Regional differences do moreover get a place in our theory, namely as conditioning (*Konditionierung*); they have the function of enabling the global introduction and dissemination of the evolutionary highly improbable achievement that is functional differentiation (cf. infra).

20 This is not to say that other ways of differentiation have become outdated: segmentary differentiation, for instance, may connect to the primacy of functional differentiation. The system 'science' is differentiated into segments, called disciplines (e.g. physics, biology, etc.), the political system into segments called states and "authorities". There seems to exist discussion concerning the latter point, however.

21 For a treatment of Oda Nobunaga 織田信長 as a 'modern' ruler (in the sense of Machiavelli): see Lamers 1997.

22 See in this context: "Distinction is perfect continence" (Spencer-Brown 1969: 1)

23 On the importance of the distinction between societal *structure* and societal *semantics*, see especially: "I start the analysis of the modernity of modern society with the distinction between societal structure and semantics. My preference for this start – a preference that cannot be justified at the start – has to do with the confusing characteristic of this distinction, namely the fact that contains itself. In itself it is a semantic distinction" (Luhmann 1992: 11).

24 On the workings and functions of this semantics in Japan, see Gluck 1985.

25 See, for instance: Douglas Hofstadter (1997) and Jacques Derrida (1988).

26 See in this connection, with strong reference to de Saussure, Luhmann 1993b: 45-70.

27 Derrida 1972b: 'La pharmacie de Platon'.

VI

LINGUISTICS AND TRANSLATION IN PRE-MODERN JAPAN AND CHINA: A COMPARISON

W.F. Vande Walle
Department of Oriental and Slavonic Studies, K.U.Leuven

1 The Culture of Translation

Translation is often conceived of as a peripheral phenomenon, but its influence is incisive and long lasting. Admittedly, in some cultures translations have a higher status than in others, but this does not detract from their real importance even in cultures that claim a high level of self-sufficiency. The Japanese have coined the term *honyaku bunka* (翻訳文化), meaning "culture of translation", which suggests that there is a whole set of values, rules, patterns and attitudes involved in it. I will try to prove this point by comparing the role of translation in traditional Chinese and Japanese culture, pointing out the differences and the similarities between the two.

Chinese culture was deeply influenced by Indian culture, even to the point that the famous scholar Hu Shi (胡適) has deplored its Indianization. The translation of the Buddhist scriptures, and the concomitant transmission of Indian culture, which started in the first century A.D. and went on for a millennium, constituted a major cultural enterprise with wide ramifications. In a similar wave, a conflation of Indian and Chinese culture was transmitted to Japan in a process that was even more incisive and long-lasting. The second major encounter with another civilisation, again involving both China and Japan, was the encounter with the West, a process that started in the sixteenth century and has been going on ever since. If religion played an essential role in both encounters, the exchange was by no means limited to that category, but it equally involved a broad range of lore, proto-scientific and scientific knowledge. Needless to say, at the core of this movement of transmission was translation. Tremendous intellectual capacities were marshalled to devise equivalents for all the new concepts that were introduced. What interests us most here in view of the subject of this collection of essays is the way in which scientific notions were translated in the course of the second encounter. Both Chinese and Japanese evidently turned to the medium of Classical Chinese. Nakayama Shigeru has proposed a periodization for the translation of Western scientific terms into Chinese, which is useful for our argument. The phases he distinguishes are:

1 The period of the Jesuit translation: seventeenth to eighteenth century.
2 The period of Dutch Learning in Japan: late eighteenth and early nineteenth century.
3 Protestant and Western-business schools: latter half of the nineteenth century.
4 Efforts at standardising by the Meiji scientists: 1880 and 1890.
5 Translation by Chinese students residing in Japan from Japanese textbooks: 1895-1918.
6 Period from the First World War onwards.[1]

Both Chinese and Japanese were involved in the same process. It has to be noted here that, just as had been the case with the Buddhist scriptures, the impact that the translations made during the seventeenth and eighteenth centuries was initially very limited, because of their restrained circulation, but there was an effect of retardation at play, and eventually the full impact made itself felt and it was pervasive and long-lasting.

On the whole, it is fair to say that the translation effort in China was mostly done by Westerners, who were assisted by Chinese. In Japan it was mostly the Japanese who bore the brunt, although in ideal circumstances they received the assistance of Westerners. This is probably related to the phenomenon of diglossia, which played an important role in Japan, while none in China, although admittedly, in Japan too it remained the privilege of the happy few, as was the case with the coexistence of Latin and the vernaculars in Europe. Diglossia, however modest in depth or scale, had at least the merit of fostering a modicum of awareness of the linguistic phenomenon. If the Japanese in spite of this paid less attention to the grammatical issue, it may have had to do with the educational tradition that set great stock by rote learning, as well as with the alleged absence of grammar in classical Chinese, which normally would have had to serve as model. In the European tradition the "rational" grammatical structure of Greek and Latin served as the model for the study of any other vernacular language. In China the only period that came anywhere near the state of diglossia in the limited definition as suggested above, was during the time the Buddhist scriptures were translated. Even then there were only a handful of monks who studied Sanskrit or another Central Asian or Indian language. But even so, this was never paralleled in later times before the nineteenth century. It is here that we hit upon the basic difference between the first and the second wave of translation in China. After prolonged contacts, the transmission of Buddhism had become "Chinese-driven", which does not mean that there was no place for foreign Buddhist missionaries, who often continued playing a vital role. The transmission of Christianity and Western science never reached this level of acceptance, with the exception perhaps of astronomy. In order to achieve this level of acceptance, this particular mathematical discipline had to be dissociated from the ideological basket it was presented in by the missionaries, because they tried to present all they were transmitting as one package.

It would also have been natural for them to try and promote the study of Latin among the Chinese, since this was both the language of Christianity and of Western science. Yet, apart from a few half-hearted attempts, no serious efforts were made to promote the study of Latin, at least not among the Jesuits who were the most successful missionary order. This appears to have been a deliberate choice. The Flemish Jesuit François de Rougemont (1624-1676), who was active in China during the sixties and seventies of the seventeenth century, was very much in favour of using Chinese instead of Latin for a number of reasons. The Chinese authorities were suspicious of foreigners using a language that they could not understand. If the Chinese clergy were to use Latin, it might make them suspect in the eyes of the Chinese authorities who could think them to be spies. Also, the mastery of Latin might lead the Chinese to reading heretical books.[2] Moreover, with regard to Latin, de Rougemont considered it next to impossible to teach the literati of mature age Latin. The Chinese language lacking certain sounds that are used in Latin, it is hard for the Chinese to pronounce the words correctly, and this might also affect the validity of the sacraments.[3] Moreover, there were good social reasons for choosing Chinese rather than Latin, because of the low esteem of anything foreign in the eyes of the literate classes.

Instead of using Latin, with a view to spreading the Gospel, they might have opted for a movement to romanise Chinese, but they did not make a sustained effort to do so. Far greater efforts appear to have been made in trying to romanise Japanese. The publication of romanised versions of many varied texts, translated ones as well as original Japanese ones, testifies to a conscious effort to rely on *rômaji*. The Jesuits never attempted anything similar in China. One reason must have been the difficulty of romanising Chinese because of its great number of homonyms, but the main reason must be looked for in the bewildering effect romanised Chinese would have had on the Chinese literati, as well as the unassailable prestige of the Chinese characters. Yet another reason may have been the lack of a printing press using movable type, although the absence of such a device is probably more the result than the cause. When the situation deteriorated markedly for the Christians in Japan, the press the Jesuits had set up in Nagasaki was sent to Macao in 1611[4], yet some time after 1620 it was sold and shipped to Manila. Apparently it occurred to no one that it could be put to use for printing romanised texts.

At any rate, notably in the field of language study, the Jesuit mission in China failed to print books on linguistics, apart from a few exceptions, such as Nicolas Trigault's (Chinese name Jin Nige, 1577-1622) 金尼閣 romanised character dictionary *Xiru er-mu ci* ('An Audio-visual Aid 西儒耳目資 for Western Scholars', Hangzhou, 1626).[5] In Japan prospects of mass consumption of the printed books were certainly bigger. There were well-established colleges and seminaries, whereas in China, the Jesuits never had such strong institutional foothold. Another difference seems to be that in Japan the Jesuits made a marked effort to use performing arts as a means to instruct as well as to study the language. This is clearly shown in the choice of texts they translated. In China they were much more concerned with doctrinal issues.

The Jesuits in China have left a considerable number of writings in Chinese, notably Matteo Ricci, Adam Schall and Ferdinand Verbiest were involved in the composition of numerous books and treatises. It is clear that most if not all these Chinese works went through a process of editing by native speakers. In China during the seventeenth century translation was still done in much the same way as it had been done during the era of Buddhist translation: by the method of *koushou* 口授筆受 *bishou*. While the Westerner was telling what the content of the text was, a Chinese held the brush and committed it to the paper in a proper phraseology.[6] In this relationship the Westerner and the Chinese are very dependent upon one another. As soon as Western presence decreased or ceased, translations were no longer made. In Japan the situation was different in the sense that virtually all translations were done during the period of seclusion. There was no room for a similar reliance upon one another. This made for a higher degree of autonomy on the side of the Japanese translators and interpreters and for a higher level of deficiency as well.

During the seventeenth century a few hundred books must have been translated into Chinese. It is however hard to make a clear-cut distinction between what is really a translation and what is a free adaptation or compilation of extracts from several source books. According to a preliminary inventory, about 600 writings in Chinese were produced by missionaries and Chinese converts during the seventeenth century. About 120 of these deal with the West and Western sciences, while no less than 470 texts are related to religious and moral issues. Out of these 470 some 330 titles are attributable to Western missionaries.[7] Which of these are to be labelled "translation" and which "free adaptation" is a moot point. At any rate they did not have a pre-existing framework to refer to in the process of translation or adaptation. Yet by trial and error they managed to build up a specific translator's vocabulary of considerable amplitude. Incidentally, at the end of the Edo period the library of the Bakufu contained around 3500 volumes,[8] which gives an idea of the largest collection of foreign titles that was available in pre-Meiji Japan.

II A poor level of grammatical knowledge

Buddhist scriptures were not normally translated in Japan, but read in the original language. Yet as time passed, there was a growing indigenisation, manifesting itself in paraphrased versions, compilations and original essays. Especially during the Edo period, the output of Buddhist writings composed by Japanese was impressive. At the same time scholars devoted much attention to the translation of Western texts with practical content. The misreadings and mistranslations that interfered in negotiating the language gap, are part of the translation process. Westerners describing the Japanese language tended to emphasise the grammar, while Japanese studying Portuguese or Dutch were more concentrated on vocabulary. Indigenization of the signified was a primordial concern for them, while the Westerners were more interested in the sign, because they were not interested in indigenizing Oriental contents. Such a trend only started manifesting itself during the nineteenth century.

During the period of *Sakoku*, the Dutch were the only Western 鎖国 people who were theoretically in a position to learn Japanese. However, only a small percentage actually did. For one thing, the Bakufu made every effort to discourage them from doing so. Carl Peter Thunberg (1743-1828), who stayed in Japan during the years 1775-1776, deplored the lack of interest of the Dutch in the study of Japanese and compared them unfavourably with the Portuguese.[9] However, there were notable exceptions, such as, for instance, Hendrik Doeff, who could read and write Japanese,[10] but, if he could, it was largely due to coincidence. The Napoleonic Wars in Europe kept him a virtual prisoner for several years on the tiny island of Deshima. Since in prin- 出島 ciple the Dutch were not supposed to learn Japanese, it was incumbent upon the Japanese to make the efforts. The memoirs and reports of the Dutch residents in Deshima give us an idea about the proficiency of the professional interpreters (*tsûji*). During the seventeenth century at 通詞 least, they concentrated mainly on the spoken language, and had a very limited range of vocabulary. It was not until the eighteenth century that members of the intellectual and social elite started the study of Dutch as an academic pursuit. Aoki Kon'yô and Noro Genjô began 青木昆陽, 野呂元丈 studying the language at the behest of Shogun Yoshimune. It is not 吉宗 clear how well Kon'yô knew Dutch: allegedly somewhere between 400 and 700 words. Yoshimune's interest in Dutch started when he browsed through a few books (or one book) that were in the Bakufu library. He was surprised by the accurateness of the illustrations. Shogunal interest at once put the study of Dutch on another social footing. However, Aoki and Noro had no books and no teacher. The only possibility for them to learn the language was to seek contact with the Dutch, when they were in Edo for their yearly audience with the Shogun. On these occasions they had to rely on the *tsûji* who accompanied the Dutch. Because the intellectual *Rangakusha* wrote their own 蘭学者 history, they have tended to downplay the scholastic merits of the professional interpreters, and overrated their own learning and knowledge. Ôtsuki Gentaku in his *Rangaku kaitei* (1788) divides the students 蘭学階梯 of Dutch into two categories: the interpreters who simply speak Dutch and the *Rangakusha* of Edo, who not only learn the language but also translate and study Western learning. While, in many instances, this general assessment may have been correct, there are notable exceptions. A few of the interpreters have indeed distinguished themselves,

今村源右衛門 such as Imamura Gen'emon, the assistant of Engelbert Kaempfer, and it cannot be denied that *Rangaku* was initiated by the Nagasaki interpreters and that they continued to exert their influence on the *Rangaku* of Edo until the end of the Edo period.[11]

Yet, in general the standard of most interpreters was low. Kaempfer himself complains bitterly about them. Although these interpreters receive yearly allowances, they do a poor job. In his assessment, they simply seemed to make strings of foreign words, which they connected according to Japanese syntax, the result of which was so poor that one would actually need another interpreter to translate their warped translations.[12] This complaint is not an isolated case. We find it regularly in the *Dagh-registers* (*Deshima diaries*). The poor standard is due to two factors: first, although during the seventeenth century there were quite a number of interpreters, they had been trained in Portuguese, rather than in Dutch; secondly, since they had to assist in the commercial transactions with the Dutch, vocabulary was more important than syntax.

In a generalising way it is often said that Dutch-Japanese relations lasted about 250 years, and for convenience's sake, the starting date is usually taken as 1639. However, the use of Portuguese as the vehicle of communication persisted for a considerable time after that date. A 井上政重 cogent proof of this is the case of Inoue Masashige (1585-1661), the notorious Bakufu official known for his ruthless persecution of Christians. Besides his anti-Christian attitude, he also had a very keen interest in Western knowledge, especially medicine. The diaries of the Dutch 甲比丹 *Opperhoofden* ('overseers', Jap.: *kapitan*) relate many instances testifying to the frequent contacts he had with them. He placed orders with the Dutch for books on anatomy, on medical drugs and scientific instruments, and often addressed queries of a scientific nature to them. He seems to have had (a) private interpreter(s), not of the Dutch language, however, but of Portuguese. In a diary entry of 17th December 1652, the Dutch *Opperhoofd* relates that Inoue inquired with the Dutch whether anyone among them could translate the *Cruijdeboeck* of Dodoens into Portuguese.[13] They replied that there was no one who 安藤右京 could. In the same year Andô Ukyô ordered a book on anatomy "illustrated and in Portuguese", as well as one on botany equally "illustrated and in Portuguese".[14] Inoue Masashige owned a copy of the *Vocabulario da Lingoa de Iapam*, and after he lost his copy in a fire, his successor Hôjô Ujinaga asked the Dutch captain for another copy, in vain.[15]

When in 1673 the English vessel *Return* sailed to Nagasaki, its captain Simon Delboe negotiated with the Nagasaki magistrates (*Nagasaki bugyô*) 長崎奉行 in Dutch and Portuguese. In his diary he recorded that all questions were put to him in Portuguese, were answered in Portuguese or Spanish, and then put into Dutch.[16] Evidently both languages were used on a more or less equal footing, but the text does seem to suggest that the interpreters were more confident in Portuguese. In 1695 Imamura Gen'emon took an examination in both Dutch and Portuguese,[17] while even in the second half of the eighteenth century, one of the senior interpreters that Thunberg met, still cherished a copy of the *Dictionarium Latino Lusitanicum ac Iaponicum*, also known by the eponym *Calepino* after its original compiler Ambrogio Calepino[18]. It was probably knowledge of Portuguese that served him best when he had to assist Arai Hakuseki 新井白石 with the questioning of the Italian missionary Sidotti. We also have to take into account that the Dutch East-India Company (Vereenigde Oostindische Compagnie, V.O.C), one of the first truly multinational companies, did not actively promote the use of its own language in its transactions, or if it did, not nearly as aggressively as its colonial competitors, the Portuguese and the Spaniards. In 1674, at the time when the company was flourishing as never before and had ousted the Spaniards and the Portuguese from the Indonesian archipelago, Joan Maetsuyker (1606-1678), the governor-general in Batavia, lamented that, although he had promoted the use of Dutch throughout, to his dismay Portuguese had continued to prevail, not in the least due to the "stupidity of the Dutch themselves, who take so much pride in being able to speak a foreign language".

The study of Dutch among the *tsûji* was far less methodical than had been the study of Portuguese, Latin, and Japanese among the Iberian missionaries. The knowledge that they had accumulated was not passed on to the Dutch, a fact easily explained by the rupture caused by the expulsion and persecution of the Christians. As a result, the general standard of Dutch among the *tsûji* was low. Emphasis was much more on lists of words, and less on proper sentences. Kaempfer wrote that he taught grammar to Imamura Gen'emon, a rather puzzling statement, unless we take into account that most *tsûji* contented themselves with memorising vocabulary.

In his *Rangaku jishi* (or: *Rangaku kotohajime*) Sugita Genpaku 蘭学事始 杉田玄白 writes that it took hundred years before Dutch writing and reading were practised by the Nagasaki *tsûji*. From the closure of the country to the middle of the eighteenth century the Nagasaki *tsûji* used *katakana* to transcribe the Dutch words they had picked up aurally and managed to remember. This contention is generally contested by scholars now. Although, in a general sense, Sugita's statement cannot have been very far from the truth, there are testimonies about the training of *tsûji*, which included reading and writing, taught by Dutch at Deshima. However, that the Dutch that was taught cannot have been of a very high level, hardly needs to be argued. The great majority of the Dutch staff consisted of men of modest educational background themselves. At any rate, the study seemed to be concentrated more on vocabulary than on grammar and syntax. As government officials in charge of controlling the goods that were being unloaded and put up for sale, the *tsûji* before anything else had to be familiar with nomenclature. There are quite a number of vocabulary lists, preserved mostly as manuscripts in libraries. As a rule, the words are transcribed in *katakana*. However, in recent years, collections of conversation sentences in Dutch and Japanese have been found, so that Genpaku's statement has at least to be qualified.[19] Apart from the technical problem of reading and writing *rômaji*, it is easy to assume that there was a kind of prejudice against it, since for the Japanese it was impossible to distinguish between Portuguese and Dutch, and since Portuguese was indelibly associated with things Christian.

III The Nanban or Iberian Connection

The conditions for transmission of Western science appear to have been much better during the Iberian age than during the Dutch period. This can be demonstrated by the extent to which a systematic presentation of the tree of scientific disciplines penetrated Japan and circulated among Japanese scholars. If we assume that the *artes liberales* were indeed taught by the Jesuits in the *seminarios*, we may ask what kind of teaching material was used? Yoshida Tadashi surmises that they could have used Duarte De Sande's *Nihon ken'ô shisetsu taiwa roku*. In this book there is a survey of the categories of science. Naturally, it mentions the seven *artes liberales*, which comprise the three artes related to language and the four mathematical sciences (geometry, calculus, music and astronomy). He assumes that arithmetic, astronomy, medicine, ethics etc. will have been fairly understandable, since there were similar disciplines in traditional learning, although their methodology was different. As to the language-related *artes* however, he sees no existing equivalent. They could only be acquired through education.

吉田忠志

日本遣欧使節対話録

Very few persons if any, will have had knowledge of them.[20] This lack of grammatical background will prove to be a handicap for *Rangaku* throughout its history. The difference between the approach displayed in Nanban writings and in *Rangaku* writings no doubt has also to do with the fact that in the first case it were missionaries aided by Japanese assistants who tried their hands at translating and transliterating Western languages into Japanese, whereas in the latter case it were Japanese, sporadically assisted by Dutchmen.

Giulio Aleni's (1582-1649) *Xixue fan* 西学凡 (Jap.: *Seigaku han*) was a forbidden book. Notwithstanding the prohibition, quite a few scholars had copies of the book in their library. The *Xixue fan* was incorporated in the compilation *Tianxue chuhan* 天学初函, published by Li Zhizao in 1629.[21] The series was brought to Japan in 1771 and in view of its being blacklisted it was investigated by the censor (*shomotsu aratame yaku* 書物改役) Mukai Kanemi. The latter wrote a report about the collection, summarising the content of eight books contained in the series. This unpublished report is entitled *Tengaku shokan daiisho* 天学初函大意書 ('*A Digest of the Tianxue chuhan*') and has recently been published by Ôba Osamu 大庭修. The book that gets the longest summary was the *Xixue fan*. For the greater part the summary is an extensive *yomikudashi* 訓み下し version of the original, which suggests that the censor went out of his way to transmit the contents as faithfully as possible.[22] This shows how the prohibition on books was bypassed. Nevertheless, it is clear that the work had only a limited circulation. Far more widespread was another writing by Aleni, namely the *Zhifang waiji* 職方外紀 (1623), which is much more concise than the *Xixue fan*, but does use similar terminology.

During the Edo period scientific knowledge came into Japan through three channels. There was the Iberian heritage and the Dutch, and in between the Chinese, which itself may be divided into two tributaries, the traditional Chinese *Honzôgaku* 本草学 and the Chinese translations and adaptations of Western works, which again mostly derived from the Iberian or the Renaissance tradition. It is interesting to note that the *Rangakusha* tried to link Dutch learning to the Chinese versions of Iberian learning. Yoshida has demonstrated that the notions the Japanese scholars of the Edo period had about the classification of science in the West were erratic and dim. What ideas they had, they had received through Iberian learning, in its Chinese guise. Through *Rangaku* they hardly received a systematic classification of Western science. The most systematic classification appears to be the one explained by Aleni in

幾何原本

Xixue fan. This in its turn appears to be largely borrowed from Ricci's preface to his translation of Euclid into Chinese (*Jihe yuanben*, Beijing, 1607).[23]

iv Motoki Ryôei

長崎通詞由緒書
星術本原太陽窮理了解新制天地二球用法記

That some of the Nagasaki interpreters were scholars of considerable standing is demonstrated by the case of Motoki Ryôei (1735-1794). He was born in a family of hereditary interpreters. His was the third generation. According to the manuscript *Nagasaki tsûji yuishosho*, in the eleventh month of the year *Kansei* 3 (1791)[24], Motoki Ryôei received the order to translate a Dutch book. He started work on the translation in the autumn of 1791 and completed the manuscript in the spring of 1792[25]. He titled it *Seijutsu hongen taiyô kyûri ryôkai shinsei tenchi nikyû yôhô no ki*. This was the first more or less systematic presentation of the Copernican system (the heliocentric system) in Japan. It did not only mark the beginning of modern astronomy in Japan, but it is also considered an important contribution to the study of Dutch and therefore to the development of *Rangaku*[26]. It is a rather unusual conflation of a treatise on astronomy and one on comparative linguistics, in which he deals with the major phonological and grammatical differences between Japanese and Dutch.

天地二球用法
正訳
義訳、仮借、略文
語路同じからず
松村君紀
漢訳の名義
字句

The first time that Ryôei touched upon the Copernican system was in *Tenchi nikyû yôhô*. This was his translation of a Dutch book, which according to him bore the title *Onderwijs van de hemel en aardse globen*. He says that it had been written by Willem Johan Blaeu and published by Johan Blaeu in Amsterdam in 1666. In his translation, compiled in the year *An'ei* 3 (1774), Ryôei writes that in translating this book "he did not rely on the stylistic rules that govern Chinese or Japanese, but stuck to the meaning of the Dutch original, in a mixture of literal translation (*seiyaku*), translation of the intended meaning (*giyaku*), transliteration (*kasha*) and abbreviation (*ryakubun*). Any other method would fail to grasp the meaning of that language, for Dutch and Japanese have different syntaxes (*goro onajikarazu*)". He adds that he has consulted his friend Matsumura Kiminori about the Chinese translations (*kanyaku no myôgi*) of the Dutch terms, and also asked him to do the revision of the expressions (*jiku*).[27] This passage, dating from 1774, the year in which *Kaitai shinsho* was published, already shows a sophisticated approach to the problem of translation.

In the preface (*hanrei*) to *Kaitai shinsho* we find similar categories. Here the author distinguishes between *honyaku, giyaku*, and *chokuyaku*. *Honyaku* is substituting a Japanese equivalent for a Dutch word, e.g. *hone* for *beenderen*. *Giyaku* are neologisms, such as *nankotsu* as a translation for *kraakbeen*. It is actually a literal translation from the Dutch. *Chokuyaku* then is a transliteration of the original word, since no equivalent can be found in the existing lexicography.[28] *Honyaku* refers to the case where there is a Chinese equivalent (*kango*) available. *Giyaku* refers to neologisms, in other words Japanese-made *kango*, such e.g. *kôsei* and *nankotsu*. *Chokuyaku* is what is usually called *onyaku*, in other words transcription or rather transliteration and what Motoki calls *kasha*.

翻訳
義訳、軟骨
直訳
恒星
音訳

Seijutsu hongen taiyô kyûri ryôkai shinsei tenchi nikyû yôhô no ki contains a reference to a *filosofische onderwijzer* (philosophical teacher), which Motoki translates as *jukyô* or *chigaku*. According to Mie Hirone, this is the first instance where the Western term 'philosophy', later translated by Nishi Amane as *'tetsugaku'*, is being problematized. Other translation words (*yakugo*) that are featured in this text include *kôsei*, *wakusei*, *kasei*, *kinsei*, *mokusei*, *dosei*, the seven days of the week etc.[29] According to Mie Hirone, who edited and published the text in *Nihon tetsugaku zensho dai 8 kan-Dai 2 bu: Shizen tetsugaku: Tenmon-butsurigakka no shizenkan*, the manuscript existed in two versions, one preserved at the time of editing in the Town Hall of Nagasaki and one in *Naikaku bunko*. The text he edited is a collation of the two versions. The *Naikaku bunko* manuscript lacks the second and final *maki* (*gekan*). It is the second *maki* that contains the *Wage reigen* and the drawings. Moreover, the *Naikaku bunko* manuscript has omitted most of the discussion about astronomical terms and translations of Dutch terms.[30]

儒教、智学
恒星、惑星、火星
金星、木星、土星
内閣文庫
下巻、和解例言

I consulted the manuscript in *Naikaku bunko* a few years ago. I found it in a box (*chitsu*) that contained a copy of Ferdinand Verbiest's astronomical treatise in Chinese titled *Xinzhi lingtai yixiang-zhi*.[31] Someone had supplemented Motoki's text to Verbiest's treatise in the assumption that it could serve as a Japanese explanation of the theory expounded and applied in Verbiest's work, and had therefore given it the title *Wage ikkan*. This *wage* starts by stating that the Hollanders call this book "Gronden der Starrenkunde gelegt in het Zonne stelsel bevatlijk gemaakt in een Beschrijving van 't Maaksel en gebruik der Nieuwe Hemel en Aardgloben." This Dutch clause at once makes the

新製霊台儀象志
和解一巻

meaning of the Japanese title clear, for the word order of the Japanese title closely follows that of the Dutch title. The manuscript then goes on to explain the Dutch alphabet and the pronunciation of the letters in Dutch. At the end it bears the date *Kansei* 12 (1800), and is signed 魯鈍斎 by Rotonsai.

Now, this passage appears to be identical with what we read in the beginning of the *Wage reigen* in the second maki of the emendated version published in the *Nihon tetsugaku zensho dai 8 kan-Dai 2 bu: Shizen tetsugaku: Tenmon-butsurigakka no shizenkan*. The body of the text is a detailed treatise on the phonological characteristics of Dutch. Ryôei is almost overwhelmed by the phonological differences between Dutch and Japanese. He says that for all the devices that one can use to represent Dutch sounds, such as katakana, supplemented with ツ *dakuon*, bullets, combining two kana signs, inserting a *tsu* to represent the *sokuon*, or adding others symbols next to the katakana, they still are inadequate for representing the sounds faithfully. Therefore he 石崎次郎左衛門 has consulted an interpreter of Chinese by the name of Ishizaki Jirôza- 唐音 emon, from whom he has learned the Chinese sounds (*Tôon*). He has therefore adopted the practice of transliterating the Dutch letters and sounds into Chinese characters. To us this may seem remarkable. We would be inclined to think that transliteration into the Japanese syllabary would be much more convenient, yet he allegedly prefers awkward Chinese characters – or rather their awkward sounds – to assure greater accuracy. Whether the prestige of Chinese is enough to explain this choice is hard to tell. It may also be inspired by a desire to distance himself from a practice which might have reminded the authorities of Christianity, for during the period of *Nanban* culture, it was common practice to transcribe foreign words into hiragana. Another reason why *Rangakusha* may have preferred to transliterate Dutch into *Tôon*, was perhaps that some phonological features of Dutch were easy to assimilate to *sokuon*, *batsuon* and *yôon*, which were and are characteristic of Japanese words derived from Chinese (*kango*).[32] In addition we may also point out that among some segments of the Edo period intelligent- 唐話 sia there was a strong interest in contemporary spoken Chinese (*Tôwa*).

Motoki ventures into some sweeping remarks about Dutch phonology and thereby seems to subvert the relationship between the source language and the target language. One would think that his book is intended as a treatise about the Dutch language for a Japanese readership, yet he strikes a tone as if the Dutch were the ones trying to study Japanese. Thus, he writes that because the Dutch pronounce *u* as *yu* ([y] and [y.]), there are many corruptions (*namari*) in two letter combinations (*niji rengô hansetsu kinô*). However, this is not the issue here, for he is writing about Dutch for Japanese trying to learn that language. He notably says that the Dutch pronounce the Japanese vowel series "*a i u e o*" as "*a e i o yu*". The question here is: does he mean to assist Dutch-speakers trying to read the Japanese syllabary or rather Dutch-speakers reading transcribed Japanese words? If the latter is the case, we need to know what system of transliteration was in use among the Dutch. Yet in my view neither case is intended here. It is hard to believe that Motoki actually means to say that the Dutch corrupt many sounds. He simply seems to view the different order the Western vowels are listed in (i.e. *a e i o u*) as 'wrong', in the sense that it deviates from the Japanese order *a i u e o*. The observation that the Dutch pronounce the u as yu (i.e. [y] or [y.]) is of course correct. As *a e i o u* is incidentally the customary order of the vowels in the Western alphabet, he most probably simply means to say that whereas the Japanese order is *a i u e o* it is *a e i o u* in the Western alphabet.

ウ
ユ、訛り
二字連合反切帰納

Motoki presents a table of all Dutch sounds; or rather syllables, although not in the technical sense the word 'syllable' has in modern Dutch phonology. He draws up an extensive table with all possible letter combinations, arranged in the order of the Dutch alphabet. Starting with *a e i o u*, continuing with *aa ee ii oo uu aab eeb iib oob uub*, running down the gamut of all the other possible consonants, then going on with all possible clusters that start with a *b* such as *ba be bi bo bu* and *baa bee bie boo buu*, etc. For each letter or letter combination he lists the pronunciation, the *kanagaeshi*[33] (actually only for the clusters starting with the five vowels, for the remainder he has probably forgotten to do or never got that far), the pronunciation (in katakana) and the corresponding Chinese character (*Tôon kasha moji*).

仮名返し
唐音仮借文字

The organisation of the table is analogous to the phonological table found in the treatise on Chinese phonology compiled by Nicolas Trigault, which we mentioned earlier. Trigault, while adopting the traditional Chinese rime-table format,[34] made the first systematic analysis of Chinese phonology in the Western alphabet. He arranged all Chinese morphemes according to their initial sound. Under each initial sound he arranged all morphemes that consist of the initial sound plus one letter, then of the initial sound plus two letters and finally of the initial sound plus three letters. He repeated the same procedure for all vowels and then for all initial consonants that exist in Chinese (in Trigault's transcription: c, ch, k, p, t, j, v, f, g, l, m, n, s, x, h, the remaining letters not being found). The maximum number of letters a combination can consist of is four. Each possible combination can theoretically appear in one of the five different tones of the Southern Mandarin 官話 (*Guanhua*).

Motoki's table for Dutch syllables is analogous and here too the maximum number of letters combined is four. The order of the initial vowels is also similar, but not that of the initial consonants, since Trigault uses a system that is apparently based on an Iberian alphabetical arrangement. In spite of the similarities, it is doubtful that Motoki had direct recourse to Trigault's book. The characters selected by Motoki to represent single consonants all differ from those used by Trigault. Moreover, while the Jesuit married the sophisticated rime-table system with the Western alphabet, Motoki constructed the Dutch sounds as combinations of initials and finals in a fashion that echoes the way traditional Chinese phonology glossed morphemes-a system known as 反切 *fanqie*, a system that preceded but was subsumed in the rime-table system. Significantly, he calls his procedure 連合反切帰納 *rengô hansetsu kinô* (induction by conflating and cutting in half). That also explains why the majority of all Dutch 'syllables' is represented by combinations of two Chinese characters, although there are exceptions, notably in the sounds that contain a *u*, which are often represented by three-character combinations. It also explains why the end consonants are just as consistently represented by the same character as the initials are. For instance, the consonants in end positions are represented thus: *k* as 郭, *l* as 郭, *r* as 耳, *s* as 数, *t* as 鐸 and *d* as 啜.

Furthermore, Motoki describes for each letter of the Dutch alphabet its pronunciation and the peculiarities of the combinations with the five vowels and the consonants. Motoki obviously had a grounding in Chinese phonology. That is clear from the terminology he uses, such as *hansetsu* 反切, *on* 音, *in* 韻 etc. What interests us here in particular is the choice of the Chinese characters that have to serve to transliterate the Dutch words and the way their pronunciation is represented in katakana to their right hand side. Since he consulted a Nagasaki interpreter of Chinese and the pronunciation of the characters is claimed to be *Tôon*, the sound system must be akin to the pronunciation of the Jiangnan area in China, because this is the area whence most monks and merchants that came to Nagasaki, hailed from.

What did this pronunciation sound like? In all likelihood it was rather close to the kind of pronunciation that was used by the Jesuit missionaries who were active in Southern China at the end of the Ming and the beginning of the Qing periods, such as Nicolas Trigault. Since recently some research has been done on the phonology of this so-called southern *Guanhua*, it may be instructive for our purpose to see how the characters used by Motoki Ryôei were actually pronounced in that vernacular. Let us go back to his phonological table. The end *k* in sounds like *aak* is transliterated as 郭 and the end *t* in sounds like *aat* as 鐸. According to Trigault's *Xiru er-mu ci*, 郭 was an aspirated sound in Chinese (in Trigault's system represented as *kuo* in the fifth tone). Since the end *g* is represented by an unaspirated sound, i.e. *ku* in the third 古
tone in Trigault's system, one surmises that an attempt has been made to distinguish between voiceless and voiced plosives, but this is by no means a systematic procedure. 鐸 representing *t* is pronounced *to* in the fifth tone in Trigault's transcription and is not aspirated. The end *d* is represented by the character 啜, which in Trigault's transcription equally sounds as to and has the fifth tone. So, although the aspiration is sometimes used *to* approximate plosive sounds, just as often it is not. The plosive *p* for instance is transliterated as *fu*, which is in the third 甫
tone, whereas *b* is written as *pu*, which has the fourth tone in Trigault's 捕
system.

From what precedes it may be inferred that, although there is a considerable resemblance, it is far from perfect. Consequently, Motoki may have been using a system that was not inherently consistent, or else one that was based on the phonology of a Chinese vernacular, that was closely related to, yet different from the one featured in Trigault's book. There are noteworthy differences. For *a* and *aa* Motoki uses the characters 曷 and 阿, which in Trigault's table are pronounced *o* in the fifth tone and *o* in the first tone respectively. The character 耳 is used by Motoki to represent the sound *r*, although rather surprisingly Trigault gives the pronunciation *iu* in the third tone. *Ne* is represented by 捏, transliterated by Trigault as *hûm* (the circumflex indicates the second tone), which is probably a mistake.

By way of conclusion, we may say that Motoki's system was not directly based on Trigault's, but that the sound values were close to the language described by the Jesuit, and therefore closely related to the Southern *Guanhua* of the seventeenth-century. Further research will have to establish the exact degree of congruence between the two kinds of pronunciation. In addition, the choice of characters must have been suggested by the interpreter of Chinese Ishizaki Jirôzaemon. He must have based himself on some Chinese writing, which remains to be identified. We have at any rate other evidence about the use of the Southern *Guanhua* pronunciation. In the preface to Ôtsuki Gentaku's *Chôtei kaitai shinsho*, we read: "in transliterating we always use the pronunciation of Hangzhou, but in an approximative way. For place names we adopt already existing Chinese translations. Even if there is no appropriate translation yet, we adopt it as it stands, without altering it. If there is no Chinese translation yet, we refer to examples and adopt a character with (the appropriate sound)".[35]

山村才助 外紀西語考 落日加 費西加 黙達費西加 費録所斐亜 瑪得瑪第加

What did *Rangakusha* transcribing Dutch names refer to? Since Motoki Ryôei's book is in manuscript form, the question remains how widespread it was. Copying manuscripts was common practice enough, but still had its limits. Yamamura Saisuke is the author of *Gaiki seigo-kô*, a kind of glossary of place names and other proper nouns. Here we find a list of Chinese characters, juxtaposed to their Latin and Dutch equivalents. Thus we find *Luo ri jia* – Ladica (sic) – Redenkonst, referring to logic; *Fei xi jia* – Phisica (sic) – Natuurkonst; *Me da fei xi jia* – Metaphisica – Overnatuurkonst; *Fei lu suo fei ya* – Philosophia – Wijsbegeerte der Waerde; *Ma de ma di jia* – Mathematica – Wiskonst. It is not hard to surmise that Yamamura Saisuke must have tried to identify the terminology he found in Aleni's *Zhifang waiji* (Jap.: *Shokuhô gaiki*) with Dutch terminology.[36]

窮理、易經 川本幸民 気海観瀾広義 青地林宗 広瀬元恭、理学提要

During the Edo period the term *natuurkunde* was usually translated into the compound *kyûri*, a term derived from the *Yijing* (Jap.: Eki-kyô, 'Book of Changes'). In Kawamoto Kômin's *Kikai kanran kôgi* (completed 1850), a commentary on Aochi Rinsô's *Kikai kanran* (1827), we find the definition: *Fei xi jia to wa butsuri o kiwamuru no gaku nari* 費西加者 窮物理之学也. Hirose Genkyô's *Rigaku teiyô* (1856) transcribes the Dutch word 'natuurkunde' in Chinese characters as 納都烏爾君垤 (present-day pronunciation transcribed in *pinyin* as: *na du wu er jun zhi*) and defines the term as *Kyûbutsurigaku*.[37] This transliteration only makes sense when read in contemporaneous Chinese. Likewise 独度涅烏斯 can only represent 'Dodonæus' when pronounced in Chinese.

This shows how until the middle of the nineteenth century traditional Neo-Confucian terminology was still maintained, while it also demonstrates how the Dutch term was transcribed in Chinese characters on the basis of their Chinese pronunciation and not their Sino-Japanese reading. Again this points to the latent persistence of Iberian learning in its Chinese guise. Udagawa Genshin says this explicitly in his preface (*hanrei*) to *Zôho chôtei naika senyô* (1822): "Whenever a term is first encountered in a Dutch book that has already been translated by a Chinese, we follow (that translation). [...] When there is a transliteration made by a Chinese [...], we always adopt that one."[38] It must be noted however that there are also some remnants of Buddhist transcriptions, but we cannot go into that matter here.

宇田川玄真
增補重訂内科撰要

In Udagawa Yôan's *Shokugaku dokugo* (*Itô bunko* in the Diet Library) we find the term 'zoology' transcribed as *Suo luo yi ya*, which is clearly closer to the original in the Chinese pronunciation; 'botanica' is transcribed as *Pu duo e he*, which arguably is closer to the Sino-Japanese reading *botayakuka*, although the third character is anomalous; 'mineralogy' is transcribed as *Mi nie la luo yi ya*, which again is closer in Chinese pronunciation than in Sino-Japanese. In *Shokubutsu keigen* (1833) he discerns three categories of science which he calls *benbutsu*, transcribed as *Fei si duo li*, which in Sino-Japanese would read *hisutari*, for Dutch 'historie'; *kyûri*, transcribed as *Fei xi jia*, meaning 'physica'; and the third category is *She mi*, i.e. the Dutch 'chemie'.[39]

宇田川榕庵、植学独語
索羅義亜
菩多厄訶
密涅剌羅義亜
植物啓原
弁物
斐斯多里
究理、費西加
舎密

The Chinese versions of Iberian learning offered a basic framework for the *Rangakusha* and continued to exert a considerable influence on Western learning throughout the Edo period. Whether this is to be attributed to a deeply engrained strain of atavism in Japanese culture or simply to the fact that Iberian learning had been much more systematically introduced in the East than Dutch learning was, or both, is hard to tell. At any rate, most translators referred to the background of diglossia that prevailed in learned society. When translating Western books, they translated them first into *kanbun*. Only subsequently did they translate them into Japanese (*wage*). The *wage* versions were in a way the Japanese equivalents of the Dutch versions of Latin books that the Dutch East India Company imported into Japan. It must be pointed out that despite a departure from the Chinese model, Chinese learning still commanded unsurpassed prestige during most of the Edo period.

漢文
和解

IV Translation of Culture

Both China and Japan were informed by a xenophobia whose rationale was ideologically founded on Confucianism. Yet the way the foreign element was controlled was different. The Japanese did it by identifying the Christian element as dangerous. Therefore they ritualised the expulsion of anything Christian, but the elements of practical learning were acceptable and were studied. The Chinese did it by confining it to the field of entertainment. Science did not have any practical function, except in astronomy. But this was so limited an area that it was containable. Incidentally, even in the West much of the new science was initially gentlemen's entertainment.

The completely different perspective as well as the poor linguistic level explains why a work like *Kaitai shinsho* is anything but a literal or close translation of Johannes Adam Kulmus' original. Kulmus wanted to reconcile the revelation of the Bible with the new knowledge of anatomy. He amply drew on Bible texts, to demonstrate for instance that King Solomon was already aware of the facts of the circulation of the blood and the chyle system.[40] This runs parallel to the attempts of Athanasius Kircher to reduce all civilisation to the one described in the Bible, and the Chinese efforts to characterise the advent of Christianity as the return of a teaching that had actually originated in China. In translating the *Ontleedkundige tafelen* the Japanese translators did not translate the hard to negotiate (culturally difficult) elements, such as the casuistic footnotes, historical passages and the discussions of Bible texts, but they limited themselves to the texts that accompanied the plates.[41] The title page featuring Adam and Eve must have had a different connotation for the Japanese than it was meant to have in the West. In the seventeenth century nude male and female figures on the title page of atlases connoted the Fall of man, but this message was evidently lost on the Japanese reader.

Yet, *Rangaku* made tremendous strides in terms of linguistic proficiency and insight in the beginning of the nineteenth century. The scholarly interpreters Yoshio Kôgyû, Motoki Ryôei and Nakano Ryûho had all been active in Nagasaki. When Baba Sajûrô was summoned by the Bakufu to Edo in 1808 (*Bunka 5*), it meant that *Rangaku* transformed itself from a private avocation to an official position. Baba was ordered to make a world map, but in 1811 he was appointed "official translator" (*Oranda shoseki wage goyô*, commonly called *honyaku kyoku*).

吉雄耕牛
中野柳圃
和蘭書籍和解御用

In this capacity he was commissioned with the translation of Noël Chomel's encyclopaedia, with the collaboration of Ôtsuki Gentaku, Udagawa Genshin, Udagawa Yôan, and others. Baba came to Edo at the age of 22 and spent most of his time in that city until he died at the young age of 36. In that period he managed to write a few Dutch grammars including *Oranda bunpan tekiyô* (1813), and *Rangaku teikô* (1814), as well as translating a few books of Russian grammar. Edo became the undisputed centre of the systematic linguistic study of Dutch and Western learning, but the foundations had been laid by scholarly interpreters from Nagasaki. No lineage was to make a bigger contribution than the Udagawa family. The first great scholar in that lineage was Udagawa Genzui (1755-1797). About him the *Rangaku kotohajime* (*ge*) says the following:

翻訳局 和蘭文範摘要 蘭学梯航 宇田川玄随

> "The lord of Tsuyama had as his physician in ordinary a certain man named Udagawa Genzui. He was an expert in Chinese studies, was widely read and had a formidable memory. He also learned several languages. Since he was so gifted and strong-willed, he made tremendous progress in his study so that he was able to translate a book, which consisted of 18 *maki* and was entitled *Naika senyô*. Although this is a compact book, it constitutes the first new translation of internal medicine in Japan. It is a pity that he died when he was just over forty years of age."[42]

The book that is meant here is the *Seisetsu naika senyô* (1793). He was also the author of *Seiyô igen*, a glossary of medical terms listing the Dutch and the Japanese equivalents in kanji either in transliteration or translation. His *Ranyaku benbô* is a kind of grammar. Udagawa Genshin was the posthumously adopted son of Udagawa Genzui. He studied Chinese studies and Chinese medicine with Genzui and Dutch with Gentaku and Katsuragawa Hoshû. In 1813 (*Bunka* 10) he was appointed "official translator". He was the author of three important works: *Seisetsu ihan teikô shakugi* (3 *maki*, 1805), *Ensei ihô meibutsu-kô* (36 *maki*, before 1822) and *Oranda yakukyô* (18 *maki*, preface dated 1828). *Seisetsu ihan teikô shakugi* is actually the concise version of *Seisetsu ihan* (30 *maki*, only 4 *maki* preserved), a kind of vade-mecum to read *Seisetsu ihan*. It is the record of lectures given by Genshin and noted down by a few of his disciples. It is a compilation of high scientific achievement, and represents for internal medicine what *Kaitai shinsho* represents for surgery. It has a supplement of anatomical charts, printed from copperplates, the first of their kind in Japan.

西説内科撰要 西洋医言 蘭訳弁髦 桂川甫周 西説医範堤鋼釈義 遠西医方名物考 和蘭薬鏡 西説医範 解体新書

国字体 The text is written in so-called *kokujitai*, by which is meant *kanji-kanamajiri*, and the author goes to great length to explain technical terminology in common terms and to add the reading to difficult Chinese characters. Here we find a felicitous combination of a sophisticated linguistic grasp and a high-level of medical expertise.

It is interesting to note that Western medicine constituted the most important discipline of Western science that was introduced into Japan, while in China it was astronomy. The emphasis on medicine also constitutes a major departure from Iberian learning. It must be noted in this context that for the Jesuits the spread of Western medical science did not enjoy a high priority. The *Constitutions* of the Society did not consider medicine a central concern, as a result of which there were indeed very few trained physicians among the Jesuits. In China they did not even found any hospitals, only in Macao, and the Franciscans too had just an infirmary at Canton. The same holds true for Japan, with the notable exception of Luis de Almeida (1525?-1583), who established a foundling-asylum at Funai in 1556. The following year he expanded his institution with a hospital for the treatment of leprosy and venereal diseases. The hospital also included a pharmacy that was supplied with herbs from Macao. Almeida also taught medicine. Almeida's superiors had but little appreciation for medical practice by their confreres. In his second *Summario* (1583) Valignano stipulated that the Jesuit hospitals would "receive only samurai and nobles as patients. Lepers and sufferers from venereal diseases were on no account to be admitted [...]. If they admitted low-class patients, it would give the hospital and the missionaries a bad name."[43] In the *Obediencias* the standpoint voiced earlier by Valignano is confirmed: "Nobody of the Companhia will be allowed to learn medicine or surgery, nor to practice anything of these two arts which he may already know, nor to have books which treat of these subjects, and the same prohibitions apply to the *dôjuku*."[44] If in the end medicine figured so prominently in Western Studies in Japan, it was as a result of both the official ban on writings with a (Christian) ideological slant and the deliberate practical orientation of the *Rangakusha*.

The practice of literal translation appears to be a development of the nineteenth and twentieth centuries. Before, most often translations were adaptations involving omissions as well as additions. When these translations were for practical use, they tended even more to be free adaptations. It is only when the interest is purely academic that translations need to be literal. There is no need for us to adapt and rework e.g. *Genji monogatari*, although even in this case, the first translation by Arthur Waley was more of a free reworking than a literal translation. In the translation process there is a filter at work, a reference to a well-established framework. This makes for a very gradual and smooth transition and I wonder whether the dramatic paradigm shift does work here. Possibly the concepts *episteme* or *metabletics* are more appropriate here. In his article "The Leiden Anatomical Theatre and its Moral Lesson"[45] Harmen Beukers reminds us of how shocking a happening the first public dissections must have been. As a historical phenomenon their public acceptance presupposes general changes in cultural patterns on a broad level. Do we attribute these changes to an *episteme*, like Foucault would,[46] or to changes in man's subjective experience of reality, as J.H. van den Berg claims?[47] Both agree on the idea that science develops in a discontinuous manner. The history of science during the Edo period as seen through the activity of translation seems to bear this out in dramatic fashion.

In the already cited article Beukers explains how in the early fourteenth century, when Mundinus performed the first dissections, the professor read and explained the anatomy text, while a *prosector* did the actual dissection.[48] It was not what we would call a hands-on approach. The dissections were meant as demonstrations or references for Galen's texts which were considered logical, conclusive. That the theories of Galen went largely unchallenged is to be attributed to the fact that the fourteenth – century anatomist, despite the appearances, did not really look into the body, whereas Andreas Vesalius did so two centuries later. With a delay of yet another two centuries the same thing happened in Japan, when Sugita Genpaku and his colleagues really looked into the body of a dissected person. Whether this is to be attributed to a metabletic process, by which the contemporaries of Sugita went through a shift of vision, shedding their traditional closed body vision and adopting the opened body vision of modern Western anatomy, I cannot prove. What is striking at any rate is the power of the image, the visual medium. There clearly is a link, I do not know whether it is causal or not, between the progress of representational techniques and the change in vision.

Although a lengthy treatment of this subject falls outside the purview of this essay, we only have to refer to the popularity Western pictures, images and icons had during the Nanban period (*Nanban jidai*). Flemish prints were widely copied and reproduced. The same holds for Chi- 南蛮時代

nese representations of Western art, as well as for the impact Verbiest's astronomical instruments for the Chinese emperor Kangxi 康熙 had: they were not meant for use, but simply to impress. It were also the drawings of the *Ontleedkundige tafelen*, of Jan Palfijn and of Dodoens that made a major impact. Giacomo Niva (1579-1638), who had been born in Japan from a Chinese father and a Japanese mother, studied Western painting under the Jesuit painter Giovanni (Niccolo) Cola (1560-1626). He arrived in Beijing in 1602 as a brother coadjutor and was accepted into the Society in 1606. "As a painter he fulfilled a very important role in the Mission. The Jesuits were very much aware of the importance and power of visual material as an aid in converting and a complement to the written word."[49] The same power would be instrumental in introducing Western scientific thought into Japan in the eighteenth century.

NOTES

1 Shigeru Nakayama, "Translation of Modern Scientific Terms into Chinese Characters-the Chinese and Japanese Behavior in Comparison", in *Science and Technology in Modern China*, ed. Tsui-hua Yang & Yi-long Huang (Taipei: Institute of Modern History, Academia Sinica & Institute of History, National Tsing-hua University, 1990) [Yang Cuihua & Huang Yilong zhubian, *Jindai zhongguo keji-shi lunji* (Taibei: Zhongyang yanjiuyuan jindaishi yanjiusuo & Guoli qinghua daxue lishi yanjiusuo, 1990)], 295-305.

2 Albert Chan, "Towards a Chinese Church: the Contribution of Philippe Couplet S.J. (1622-1693)", in J. Heyndrickx C.I.C.M., ed., *Philippe Couplet, S.J. The Man who Brought China to Europe* (Monumenta Serica Monograph Series, XXII) (St. Augustin-Nettetal, 1990), 65.

3 Ibid., 65.

4 Henri Bernard S.J., "Traductions chinoises d'ouvrages européens au Japon durant la période de fermeture (1614-1853)", *Monumenta Nipponica* (1940), vol. 3, 42, quoting an article by Schilling, "Vorgeschichte des Typendrucks auf den Philippinen", in Gutenberg Jahrbuch 1937, 211-212.

5 Paul Fu-mien Yang S.J., "The Portuguese-Chinese Dictionary of Matteo Ricci: A Historical and Linguistic Introduction", in *The Proceedings of the Second International Conference on Sinology. Section on Linguistics and Paleography* (Taipei: Academia Sinica, 1989), 202.

6 Peter Engelfriet, "Euclid in China", (Ph.D. Dissertation, Leiden, 1996), 107.

7 Information supplied by Ad. Dudink and N. Standaert.

8 Marius Jansen, "New Materials for the Intellectual History of Nineteenth-Century Japan", *Harvard Journal of Asiatic Studies* 20 (1959): 567 *f*.

9 Frits Vos, "Mihatenu yume-An Unfinished Dream: Japanese Studies until 1940", in *Leiden Oriental Connections* 1850-1940, ed. Willem Otterspeer (Leiden, New York, Kobenhavn, Köln: E.J. Brill, 1989), 359.

10 Ibid., 360-1.

11 Sugimoto Tsutomu 杉本つとむ, *Kindai Nihongo no seiritsu to hatten* 「近代日本語の成立と発展」 (Sugimoto Tsutomu chosaku senshû II 「杉本つとむ著作選集 2」), (Tokyo: Yasaka shobô 八坂書房, 1998), 266.

12 Engelbert Kaempher, *History of Japan, together with a Description of the Kingdom of Siam*, 1690-1692, 3 vols (Glasgow, 1906), vol. 2, 101.

13 Sugimoto Tsutomu 杉本つとむ, *Kaitai shinsho no jidai: Edo no honyaku bunka o saguru* 「解体新書の時代—江戸の翻訳文化を探る」 (Tokyo: Waseda daigaku shuppanbu, 1987), 31.

14 Ibid., 29.

15 Nagazumi Yôko, "Foreign Intelligence and Its Interpreters", in *Engelbert Kaempher – Werk und Wirkung – Vorträge der Symposien in Lemgo und in Tokyo*, herausgegeben von Detlef Haberland (Stuttgart: Franz Steiner Verlag, 1993), 34.

16 Ibid., 33 ff.

17 Paul van der Velde, "Die Achse, um die sich alles dreht. Imamura Gen'emon Eisie (1671-1736) Dolmetscher under ebenbürtiger "Diener" Kaempfers", in *Engelbert Kaempher – Werk und Wirkung –*, 177.
18 See Blomberg's essay in this volume.
19 Sugimoto, *Kaitai shinsho no jidai*, 78 ff.
20 Yoshida Tadashi, "Edo jidai ni okeru seiyô gakumon bunrui no ninshiki",「江戸時代における西洋学問分類の認識」, in *Reports of the Research Institute for Japanese Culture Nihon bunka kenkyûsho kenkyû hôkoku* 28 (1992): 52.
21 Engelfriet, op.cit., 68; Yoshida, op.cit., has the date 1623.
22 Yoshida, op.cit., 56.
23 Ibid., 55.
24 Mie Hirone, "Seijutsu hongen taiyô kyûri ryôkai shinsei tenchi nikyû yôhô ki kaisetsu" in Inoue Tetsujirô, Koyanagi Shikita, Takakusu Junjirô, Fujikawa Yû (kanshû), Mie Hirone (hensan) 「井上哲次郎・小柳司気太・高楠順次郎・富士川游監修・三枝博音編纂」 *Nihon tetsugaku zensho dai 8 kan-Dai 2 bu: Shizen tetsugaku: Tenmon-butsurigakka no shizenkan*「日本哲学全書第八巻第二部自然哲学天文・物理学家の自然観」、Daiichi shobô 第一書房, 昭和 11, 207.
25 Sugimoto dates it to *Kansei* 5, which corresponds with 1793; see Sugimoto Tsutomu 杉本つとむ, *Kindai Nihongo no seiritsu to hatten*「近代日本語の成立と発展」(Sugimoto Tsutomu chosaku senshû II 「杉本つとむ著作選集 2」), (Tokyo: Yasaka shobô 八坂書房, 1998), 267.
26 Mie Hirone, "Seijutsu hongen taiyô kyûri ryôkai shinsei tenchi nikyû yôhô ki kaisetsu", 209.
27 Ibid., 210.
28 Sugimoto Tsutomu 杉本つとむ, *Kokugogaku to Rangogaku*「国語学と蘭語学」(Tokyo: Musashino shoin 武蔵野書院, 1991), 378; Sugimoto Tsutomu, *Nihongo no rekishi*「日本語の歴史」(Sugimoto Tsutomu chosaku senshû I「杉本つとむ著作選集 1」), (Yasaka shobô, 1998), 255.
29 Sugimoto Tsutomu, *Kindai Nihongo no seiritsu to hatten*, 269.
30 Mie Hirone, "Seijutsu hongen taiyô kyûri ryôkai shinsei tenchi nikyû yôhô ki kaisetsu", 211.
31 漢 7953冊数 17函号 305/213.
32 Sugimoto Tsutomu, *Nihongo no rekishi*, 73.
33 This is a system to represent the pronunciation of Japanese syllables inspired on the *Chinese hansetsu* (cutting in half) system. In Chinese the pronunciation of a character is represented by two other characters: the initial of the first of these two is identical to the initial of the character whose pronunciation is being represented, the final of the second character is identical to the final of the character whose pronunciation is being represented. Applied to Japanese the first kana indicates the initial consonant or vowel, the second kana indicates the end vowel.
34 See David Prager Branner, "The rime-table system of formal Chinese phonology", in Sylvain Auroux e.a. (ed), *Geschichte der Sprachwissenschaften. History of the Language Sciences. Histoire des sciences du langage. Ein internationales Handbuch zur Entwicklung der Sprachforschung von den Anfangen bis zur Gegenwart. An International Handbook on the Evolution of the Study of Languages from the Beginnings to the Present. Manuel international d'histoire des etudes linguistiques des origines a nos jours* (Berlin-New York: Walter de Gruyter, 2000), 46 ff.
35 Sugimoto Tsutomu, *Kindai Nihongo no seiritsu to hatten*, 351.
36 Ibid., 56-57.
37 Ibid., 57.
38 Sugimoto Tsutomu, *Nihongo no rekishi*, 256.
39 Ibid., 58.
40 A.M. Luyendijk-Elshout, "'Ontleedinge' (Anatomy) as Underlying Principle of Western Medicine in Japan", in H. Beukers, A.M. Luyendijk-Elshout, M.E. van Opstall and F. Vos, eds., *Red-Hair Medicine: Dutch-Japanese Medical Relations* (Amsterdam: Atlanta, G.A., 1991) (Publications of the Netherlands Association for Japanese Studies no. 5), 30 quoting Kulmus, *Ontleedkundige tafelen*, 197.
41 Ibid., 31.
42 Sugimoto Tsutomu, *Kindai Nihongo no seiritsu to hatten*, 272.
43 C.R. Boxer, *The Christian Century in Japan* 1549-1650 (Berkeley, Los Angeles and London, 1951), 203-204.
44 Ibid
45 *Reports of the Research Institute for Japanese Culture Nihon bunka kenkyûsho kenkyû hôkoku* 31 (1995), (1)-(2).
46 Michel Foucault, *Les mots et les choses* (Paris, 1966).
47 J.H. van den Berg, *Het menselijk lichaam* (Nijkerk, 1956-1961),vols. I and II.
48 "The Leiden Anatomical Theatre and its Moral Lesson", in *Reports of the Research Institute for Japanese Culture Nihon bunka kenkyûsho kenkyû hôkoku* 31 (1995):(4).
49 Engelfriet, op.cit., 89.

VII

REINTERPRETATION OF THE WESTERN LINEAR PERSPECTIVE IN EIGHTEENTH- AND NINETEENTH-CENTURY JAPAN: A CASE OF CULTURAL TRANSLATION

Shigemi Inaga
International Research Center for Japanese Studies, Kyoto

Can the canon of one culture be translated into that of another without violating the integrity of the criteria that govern the latter? Here is a vital question pertaining to cultural translation. Let me take up the case of Western linear perspective and its reinterpretation in Tokugawa Japan as a case study in the cultural translation of a canon in non-verbal communications.[1]
In a sense the linear perspective was more than a technique, it was a culture. For a long time, it represented, as a metaphor or a metonymy, Western science as a whole and incarnated the progress of modern knowledge. By virtue of its monopoly of the pictorial plane in Western academic education of the fine arts, its exclusive dominance as the unique grammar of architectural drawing, its rational determination and reduction of the three-dimensional space into two-dimensionality, its panopticonlike *régime du regard*, the linear perspective has occupied a position comparable to that of a kind of secularised monotheistic theology. Symbol of advanced technology, the linear perspective was believed to be universally valid and served not only as the measurement for the stage of psychological development of an individual, but also, in a larger context, as a sign of mental evolution and enlightenment of the races and mankind. Thus the linear perspective spread hand in hand with European expansion into the world.[2]

However, this absolute criterion in spatial construction was gradually to lose its supremacy over Western painting in the second half of the nineteenth century. Curiously enough, this period coincided with the vogue of Japanese art in Europe. It has been asserted that Japanese art served as a catalyst to Western artists by suggesting to them a new type of pictorial plane free from the restraint of the academic linear perspective. However, this assertion overlooks the fact that the linear perspective had been introduced in Japan in the second half of the eighteenth century and was widely applied by *ukiyo-e* print designers.[3] Here the problem of cultural translation can be raised. How did the Japanese craftsmen of the Tokugawa period (1600-1868) "translate" Western linear perspective? To which extent was this translation meaningful for the reconstruction of the pictorial plane and visual culture in Tokugawa Japan as well as in the late nineteenth-century West? What was the contribution of this "translation" to the transformation of the space conception and configuration in the context of the cultural exchange between East and West, as well as in terms of the formation of "Modernist aesthetics"?

浮世絵

I

To answer these questions, let us briefly analyse the process of adaptation of the linear perspective in Tokugawa Japan. Okumura Masanobu (1686-1764) appears to have claimed for himself the title of initiator of the linear perspective (*uki-e kongen*). *Enjoying the Cool at Ryôgoku-Bridge* (*Ryôgokubashi yûsuzumi*, ca. 1745) [ILL. 1], one of his typical pieces, enables us to see Okumura's way of "understanding" and applying a new technique from abroad. While faithfully imitating the linear perspective in frontal projection to depict the interior space in the foreground, Masanobu could not apply, or rather did not feel it necessary to apply the same principle to the exterior landscape, lying in the background. The popular quarter of Ryôgoku-Bridge is depicted from a bird's-eye view, the traditional and virtually only possible viewpoint known to Japanese painters since the eleventh-century *Monogatari-emaki* or sixteenth-century *Rakuchû-rakugai-zu byôbu* screens. As one can also see in *View of the Port of Nagasaki* [ILL. 2], attributed to Maruyama Ôkyo (1733-1795), the bird's-eye view was the only way of depicting panoramic scenery available to the Oriental artist.

奥村政信
浮絵根源
両国橋夕涼
物語絵巻
洛中洛外図屏風
円山応挙

Looking at *Enjoying the Cool at Ryôgoku-Bridge*, modern viewers will immediately detect a lack of unity between foreground and background. The bold superposition of the newly adapted Western technique upon the conventional pictorial space gives us the impression that the hall in the foreground is floating in the air. Incidentally, *uki-e* meant "floating pictures". Modern mentality would find in this arbitrary composition evidence of a lack of understanding of, or a neglect of the principle of the linear perspective.

However, what characterises Okumura's way of translating precisely

ILL 1

ILL 2

consists in the arbitrariness of this fanciful combination. Or more accurately, Okumura and his public probably did not find any inconvenience or arbitrariness in this combination. As is well known, in the *Rakuchû-rakugai-zu byôbu* screens of the sixteenth century as well as in the *View of the Port of Nagasaki*, the artist's (and the viewer's) point of view shifted easily and moved around on the pictorial plane. While the general scenes of the quarters of downtown Kyoto or the artificial islet of Deshima in the Nagasaki Bay were depicted from a bird's-eye view, the details of the human figures on the streets as well as the ships that are minutely depicted in the Nagasaki Bay appear to be taken from a horizontal point of view. In addition, this constantly shifting viewpoint also accounts for another feature, i.e. the lack of a fixed horizon line in these Japanese bird's-eye view panoramas. 出島

Capriciously shifting viewpoints and the lack of horizon, both of these "Japanese" characteristics still remain intact in Masanobu's Western-style *trompe l'oeil*, and would look inconsistent and confusing to our modern eye. Still, these features may have contributed to accentuating the unexpected illusion of a receding background and may have aroused excitement among the contemporary public of Masanobu. Let us note that versatile adaptability in cultural translation was already at work in this experimental tentative of Masanobu. The superimposition of two pictorial layers enabled the combination of heterogeneous elements without taking into account the incompatibility of the linear perspective with such an offhanded combination. Instead of a struggle for supremacy or parasitic relation between Oriental conventions and Western criteria, a sort of symbiosis (or "cohabitation") is realised, which we can compare, at least in a metaphorical sense, to the phenomenon named *sumiwake* by Japanese eco-biologists like Imanishi Kinji and Miyaji Denzaburô.

Theoretical-minded scholars will wonder what the origin of this Japanese *trompe l'oeil* was. A pioneering work in this field, "The Origin and Development of Japanese Landscape Prints," an unpublished Ph.D. thesis by Julian Lee (Washington, 1977), points to a Latin treatise by Andrea Pozzo, *Perspectiva pictorum et architectorum* (1693-96), translated into Chinese for the first time in 1729 and again in 1735 with illustrations. As Tokugawa Yoshimune had been promoting the importation of foreign scientific books since 1720, the Chinese translation of this work could theoretically have been accessible. However, no evidence has been found as to its diffusion and influence in Japan. Practi- 徳川吉宗

蘇州

cally speaking, some examples of Westernised Suzhou prints, of which copies are known, were enough for Okumura Masanobu and other print designers to conceive and realise the imitations of linear perspective in their *uki-e* prints.[4]

岸文和 歌舞伎 平安時代

It is noteworthy that this first generation of *uki-e* producers was not so much interested in the principle as in the effect of the linear perspective. As Kishi Fumikazu has recently pointed out, the early *trompe l'oeil* prints depicting the interior of Kabuki theatres did not necessarily search for an accurate application of the linear perspective. Instead of becoming progressively faithful to its principle, they rather manifested increasing deviation from it [ILL. 3,4]. The unique vanishing point was sacrificed for the sake of the visibility of the famous characters and scenes on the Kabuki theatre stage, where vanishing lines suddenly disappear and are replaced by parallel lateral lines, just as had been the case in Heian period (*Heian jidai*, 794-1185) scrolls. Even the framing technique was introduced in the editing process. While reutilising again and again the same wood block for the theatre interior, the part of the block representing the stage would be cut out and replaced by a new block that was inserted into the old one, in an effort to catch up with the latest popular performance and favourite actors. Clearly, it was not the accuracy of linear perspective *per se* but the rapid printing with its possibilities of wide diffusion that was the primary concern of the Japanese editors (then and now).[5]

II

眼鏡絵 吉良上野介

While the Edo theatre *uki-e* declined in the 1750's, from the end of that decade on Maruyama Ôkyo executed several *megane-e*, i.e. prints destined for optical view, which was then in vogue all over the world.[6] The "naturalness" with which Ôkyo could combine architecture in linear perspective with natural sceneries is "remarkable," given the fact that such kind of integration was out of the question for the former *uki-e* prints. Still, one work, *The Vendetta of the 47 Samurai* [ILL. 5] betrays that Ôkyo's understanding was no "better" than that of his predecessors. There are unnatural double vanishing points in the frontal projection of the interior of the mansion of Kira Kôzukenosuke. As a result, the elements situated in between the double vanishing lines confuse us. If we try to restore the building into a three-dimensional model, or simply to plot it on a floor plan, this part would remain indeterminable and impossible to explain.

ILL 3

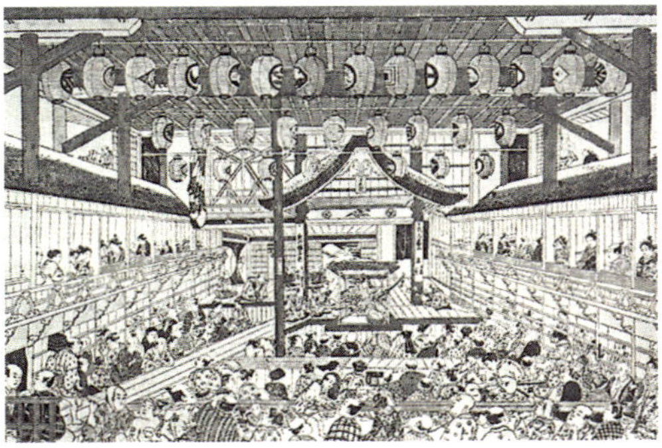

ILL 4

ILL 5

Once again, what is at stake here is not the lack of coherence in geometrical operation, but the lack of consciousness to feel this incoherence as incoherent. Ôkyo was not so much interested in the objective rendition of the space in linear perspective as in the dramatic effect created by the illusion of depth. Rather than the homogeneous space that is presented to the fixed monocular viewpoint, it was the aesthetic singularity and supernaturalness of the bloody and atrocious scene of revenge that was pursued by the artist at the price of revealing his lack of respect for the principle of the linear perspective.

歌川豊春
明和
浮絵紅毛フランカイ湊万里鐘響図

Shortly after Ôkyo's *vue d'optique*, Utagawa Toyoharu (1735-1814) renovated the *trompe l'oeil* in the Edo of the *Meiwa* period (1764-1772). *View of a Western Port* (*Uki-e Kômô Furankai-minato Banri dôkyô no zu*) [ILL. 6] has already been identified as an imitation of a copper etching of the Canal Grande in Venice by Antonio Vicentini.[7] Harmonious in general, the details betray Toyoharu's carefree attitude toward the original. The stairs on the quay at the right side, depicted either in parallel or even in diverging lines, do not follow the supposed vanishing lines, which remain difficult to locate. The vanishing point is also difficult to locate, and anyway would not converge on the horizon. This lack of consistency clearly reveals that the notion of horizon line was lacking when Toyoharu copied the original. Needless to say, the horizon line, as the viewer's standpoint projected to the infinite distance, is the base line for the linear perspective construction as a whole. What interests us, in the context of cultural translation, however, is not Toyoharu's evident "incomprehension" of this principle but rather the fact that Toyoharu, either consciously or unconsciously elected to ignore such principle in his adaptation of the linear perspective.

浮絵洛陽四条河原夕涼図

Later, Toyoharu did take notice of the horizon as the base line of the whole composition *Enjoying the evening Cool at the Shijôgawara in Kyoto* (*Uki-e Rakuyô Shijôgawara yûsuzumi-zu*) [ILL. 7] being an example. Still, the viewpoint is raised as if to reconcile it with the bird's-eye view tradition. What are intriguing, however, in the generally seamless composition, are the lines of people walking across the foreground. As the Kamogawa River is flowing from the rear to the fore, a bank on which people are strolling seems to dam up the stream. Why did the artist make such an absurd and incomprehensible addition to the foreground? Toyoharu seems not to have been satisfied with leaving the foreground empty and filled it up with the mentioned motif. Although not reasonable from a realistic point of view, these additions were

ILL 6

ILL 7

made so as to satisfy his aesthetic sensibilities. As we shall see, this apparently superfluous addition will account for the "Japanization" of the linear perspective in the following period.

III

佐竹義敦、大名 秋田、曙山 平賀源内

The first attempt to formulate the rules of Western linear perspective was made by Satake Yoshiatsu (1748-1785), *daimyô* of Akita, also known by his artistic sobriquet Shozan, in a pair of short essays completed in 1778. Probably initiated by Hiraga Gennai (1728-1779), a polymath and Westernist, who had visited Akita in 1773, lord Satake began his study of Western painting. Among their main references that have been identified figure the so-called *Groote Schilderboek* by Gerard de Lairesse (1707), a pupil of Rembrandt, and *Naeukeurige beschryving van de natuur der viervoetige dieren, vissen en bloedloze waterdieren, vogelen, kronkel-dieren, slangen en draken* (1660) by Jan Jonston [John Johnston], known as the "Zoological album". In his treatise on Western painting, *Gahô kôryô* (1778), Shozan insisted on the practical utility of painting as visual information and severely criticised the conventions of Oriental painting, which could not discern either the colour or the convexity and concavity. He also attacked the absurdity of translating the relation of far/near in terms of upper/lower on the pictorial plane. He boasted that by his study of the principles of Western painting he could clearly discern far/near, high/low and clear/obscure etc. It is also in this treatise that the horizon is singled out and defined for the first time in Japan as the base line of the linear perspective.[8]

画法綱領

A striking implementation of this manifesto is the *Pine Tree with Parrot* [ILL. 8]. Exaggerated foreshortening, clear-cut shading, contrast between the foreground in vivid colours and the background in pale blue, extremely lowered horizon line, dichotomic contrast between clear and obscure in the modelling, which gives the impression that the trunk of the pine tree is rectangular rather than round ... However the most striking feature in this work is the superposition of the gigantically magnified foreground on the extremely miniaturised background. It is possible that some etching illustration of fauna and flora in the Western encyclopaedias at his disposal had inspired this disproportionate contrast. Still the fact remains that Shozan perceived such a contrast as the essential lesson of Western science. In my view, here is the vital point in the reception and cultural translation of the linear perspective in Tokugawa Japan.

The linear perspective had been conceived and elaborated as a device for constructing a *homogeneous* pictorial plane. By projecting on a screen in front of the eye the configurations of the objects to be represented, it aims at determining their relative distances and positions on the two-dimensional plane by a series of purely geometrical and mechanical operations. By contrast, Shozan interpreted the same device as a tool of *differentiation*, exaggerating the heterogeneity between what is near and what is far away. Instead of establishing a unified homogeneous pictorial plane, Shozan took advantage of the linear perspective to realise polyvalent space, brought about by the contrast of heterogeneous elements, which he could clearly "discern", as he proudly put it.

Symptomatic in this regard was the term adopted by Shozan to designate linear perspective: "degree of far/near" (*enkin no dosû*). Shiba 遠近の度数
Kôkan (1748-1818) also spoke of the "reason of far/near" (*enkin no* 司馬江漢
ri) in his *Seiyô gadan* ("Essay on Western Pictures," 1799). Of course 遠近の理
"far/near" (*enkin*) was the only Chinese term at their disposal to trans- 西洋画談、遠近
late the idea of "perspective". Still, it is undeniable that this definition in dichotomic terms, typical of Chinese thought, contributed to underlining the effect of differentiation and contrast between the near and far. In Shozan's case, the pine tree in the foreground derives from the
Kanô school screen painting, while the landscape in the background is 狩野派
an imitation of Dutch etching. These two contrasting elements could be put in the same pictorial plane thanks to the lack of an intermediate joining space in the middleground, which by its absence served as the transitional space.[9] In this process of cultural translation, what was supposed to be faithful understanding of the Western linear perspective by Shozan himself, actually resulted in a remarkable deviation. What then does this deviation imply?

IV

As early as 1978 I proposed to call this typical composition of the Akita school a *still life-landscape* combination, which marked the birth of a new genre in the history of Japanese painting. *The Pond of Shi-*
nobazu [ILL. 9] by Odano Naotake (1749-1780), one of Shozan's samurai 小田野直武
subjects, shows more clearly the implication of this new combination. Here the peony flowers in the foreground "borrow," as it were, as their convenient "background" the Pond of Shinobazu. This "borrowing of
the natural background" inevitably reminds us of *shakkei*, or "borrow- 借景

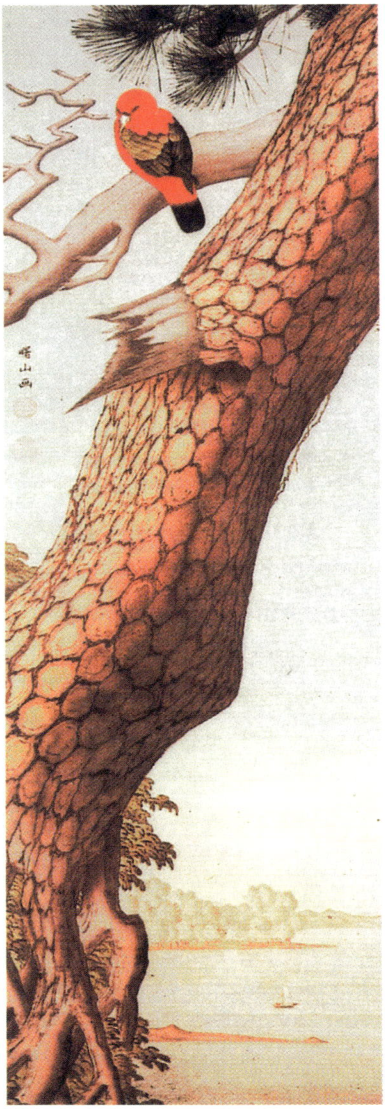

ILL 8

ILL 9

ILL 10

ing of the natural scenery," a technique practised in the garden design of East Asia [ill. 10]. Curiously enough, the adaptation of the linear perspective by the Akita school consequently amounted to the pictorial version of the *shakkei* garden, both of them sharing one and the same deep structure. Just as in the *shakkei* garden heterogeneous elements were set side-by-side in terms of far/near superimposition to constitute an entity for contemplation, the linear perspective, imported from the West, was transformed and reduced into the matrix that served the same effect of visual *montage*.[10]

Here is a case of circular interplay between the translation of culture and culture of translation. While the linear perspective helped the Akita school inventing a new Westernised pictorial plane, the new pictorial plane in question turned out to be the manifestation of the way Far Eastern culture on the archipelago had incorporated the exterior elements. Thus, the linear perspective as it was translated by the Akita school, paradoxically revealed one of the constants of Japanese culture as a culture of translation. Here we can probably detect a secret in the mechanism of "Japanization". And retrospectively, we now can see that not only the re-interpretation of the linear perspective by the Akita school, but also the spatial superimposition of Okumura Masanobu or the editing technique of wood block insertion practised by the Torii school were governed by the same principle, which consists of juxtaposing heterogeneous elements without respecting the governing principle of each of the incorporated elements.[11] 鳥居派

V

In connection with this "constant" in Japan's culture of translation, I want to advance my last hypothesis as to the combination of a human figure and a landscape in the ukiyo-e prints. Early examples of this combination can be traced back to Suzuki Harunobu's (-1770) series of *Eight Scenes of Edo* (*Fûryû Edo hakkei*). Just take one print, *Enjoying the Evening Cool at Ryôgoku-Bridge* (*Ryôgokubashi yûsuzumi*) [ill. 11], where the beauties in the foreground are directly superimposed upon the background of Ryôgoku Bridge in bird's-eye view. Still, it is impossible to find here any explicit trace of Western linear perspective. 鈴木春信 風流江戸八景 両国橋夕照

Harunobu is remembered as the inventor of the *nishiki-e*, or the polychrome prints, which he first produced around 1765, and Morishima Chûryô (1754-1808) has suggested Gennai's involvement in this inven- 錦絵 森島中良

春重

tion. Curiously enough there are several prints signed Harunobu, which show an awkward application of the linear perspective [ILL. 12]. Specialists unanimously agree that these pieces were made by Harushige, as a substitute, after the sudden death of Harunobu in 1770. In some cases, Harushige even intentionally and ostentatiously introduced the horizon line to a composition borrowed from Harunobu. Suzuki Harushige was, in reality, nobody else than Shiba Kôkan, a scholar of Dutch studies, who seems to have known Odano Naotake of the Akita school personally – a surprising connection.

見立て、品川帰帆
瀟湘八景
遠浦帰帆
矢橋帰帆
琵琶湖
弁天島

Characteristic in Harunobu's later works was the trope named *mitate*. In the case of *Returning Sail at Shinagawa* (*Shinagawa Kihan*), the title evokes one of the eight poetic views of Xiao-Xiang in China (Ch. *Xiao-Xiang bajing*, Jap. *Shô-shô hakkei*), i.e. *Returning Sail at Yuan Pu* (*Enpo Kihan*), which was transposed in a diminutive version into *Returning Sail at Yabase* (*Yabase Kihan*), a scenic spot on Lake Biwa, of which *Returning Sail at Shinagawa* is an Edo version pastiche. Here we see a kind of rebus. In the foreground, a girl is looking at the bowl floating on the basin, which evokes and is echoed by the returning fishermen's boat depicted in the background. In this way, there is a secret correspondence between the foreground and the background.[12] Although Harunobu himself does not seem to have been directly influenced by the Western linear perspective, this trope of *mitate*, or aesthetics of "seeing as," had something in common with the spatial combination invented by the Akita school of Western style painting. Curiously enough, a *vue d'optique* in a Westernised style, quite similar to the style of Odano Naotake, is known to exist, and seems to have served as a prototype of the Pond of Shinobazunoike in many *ukiyo-e* prints. An example by Toyoharu, with timid but typical small figures in procession in the foreground borrows for its background the same landscape around the islet Bentenjima, foreshortened in a strongly Western style.

磯田湖龍斎
風流江戸八景
忍び合い

But more striking is an example in a series by Isoda Koryûsai entitled *Eight Views of Edo* (*Fûryû Edo hakkei*) [ILL. 13]. The landscape of the Pond of Shinobazunoike, clearly reflecting the foreshortening in the style of Odano Naotake, is evidently chosen as a relevant background and associated with the human figures in the foreground to evoke the secret meeting of the couple (*shinobi-ai*), faithful to the tradition of wordplay (*shinobu / shinobazu* connoting "secrecy") that typifies the trope of *mitate*. If Odano Naotake has borrowed the scene of the Pond of Shinobazunoike for the background of his still life of peony flowers,

Koryûsai seems to have replaced the flowers by the human figures of beauties.

In this way the landscape depicted in Western linear perspective finally served as the background to the *ukiyo-e* prints of beauties as a "borrowed landscape". Thus, we may assume that the *still life-landscape* and the *figure-landscape* were as it were twins and both were born from the translation and re-adaptation of linear perspective into Japan's cultural climate.

It seems that one of the greatest influences that the Japanese ukiyo-e prints had on Western painters in the second half of the nineteenth century was exerted by the type of composition I have defined as *still life-landscape* and *figure-landscape* combinations. As I have tried to demonstrate in this paper, these combinations were mostly realised as the result of Japan's translation of the Western linear perspective into "Japanese" pictorial language. However, this translation did not mean a passive acceptance of things Western. On the contrary, in the process of its translation into Japanese, Western linear perspective lost the supremacy it had enjoyed for several centuries as the absolute grammatical canon in Western fine arts. However this does not mean, any more, that the Western linear perspective was reduced to a mere vocabulary at the disposal of the Japanese. On the contrary, the translation of linear perspective did change the syntax of spatial configuration in Japanese painting. And yet, it cannot be denied that this change followed the same "transformation rule" (N. Chomsky) as was observed in the *shakkei* aesthetics.

The way Japanese painters translated Western linear perspective seems to have imparted a far-reaching lesson to the avant-garde Western artists in search of modernist aesthetics. They were convinced they would find in Japanese art some of the keys to get rid of the yoke of the academic tradition. Let me close this paper by quoting a statement that Ernest Fenollosa made in the last year of his life:
"The Oriental influence was no accident, no ephemeral ripple on the world's art stream, but a second main current of human achievement sweeping around into the ancient European channel, and thus isolating the three-hundred years-long island of academic extravagance." [13]
"The Oriental influence" here must be understood not in its essentialist sense but in terms of the Oriental art of translating Western culture.

ILL 11

ILL 13

ILL 12

NOTES

1 This is an abbreviated and modified version of my "La transformation de la perspective linéaire au Japon (1740-1810) et son retour en France", *Actes de la recherche en sciences sociales* 49 (1983): 29-46, which was based on my M.A. dissertation presented to the University of Tokyo in 1979.
2 On the "politics" of the linear perspective, see a.o. Pierre Francastel, *Peinture et la société* (Paris: Denoël/Gonthier, 1951 [1977]); Hubert Damisch, *L'origine de la perspective* (Paris: Flammarion, 1987) and James Erkins, *The Poetics of Perspective* (Ithaca: Cornell University Press, 1994).
3 On Japonisme, see a.o. Yamada Chizaburoh 山田智三郎, ed., *Japonisme in Art* (Tokyo: Kôdansha International, 1981).

4 Julian J. Lee, "The Origin and Development of Japanese Landscape Prints", unpublished Ph.D., Washington State University, 1977.

5 Kishi Fumikazu 岸文和, *Edo no enkin-hô, ukie no shikaku* 「江戸の遠近法 浮絵の視覚」 (*The Perspective in Edo, the Perception of Uki-e*) (Tokyo: Keisôshobô, 1994).

6 Timon Screech, "The Meaning of Western Perspective in Edo Popular Culture", *Archives of Asian Art* 47 (1994): 58-69. Also by the same author, *The Western Scientific Gaze and Popular Imagery in Late Edo Japan* (Cambridge: Cambridge University Press, 1996).

7 Okano Keiichi 岡野圭一, "Eine Venezianische Vedute im japanischen Holzschnitt", *Pantheon* 27 (1969).

8 Reproduced in facsimile in Takehana Rintar 武塙林太郎, *Akita ranga* 「秋田蘭画」 (Akita: Akita Sakigake Shinpô-sha, 1992), 134-141.

9 Cf. Itô Teiji 伊藤ていじ, *Nihon dezain ron* 「日本デザイン論」 (Studies in the Design in Japan) (Tokyo: Kashima shuppankai, 1969) ch. 2.

10 For a recent criticism on the aesthetics of *mitate*, see Karatani Kôjin 柄谷行人, "Shakkei ni tsuite no kôsatsu" 「借景についての考察」 (Reflection on the Borrowed Landscape), *Hihyô kûkan* 「批評空間」 (Critical Space) 17 (1998): 35-47; an English version of this article in: Karatani Kôjin 柄谷行人, "Uses of Aesthetics: After Orientalism", *Boundary* vol. 25, no. 2 (1998): 145-160.

11 For a more generalised hypothesis on this function of "mediation" in Japanese space experience, see Augustin Berque, *Vivre l'espace au Japon* (Paris: Presses universitaires de France, 1981), last chapter.

12 As for the miniaturised *mitate* aesthetics in Japan, see Haga Tôru 芳賀徹, "Fûkei no hikaku bunkashi, Shôshô Hakkei to Ômi Hakkei" 「風景の比較文化史 –「瀟湘八景」と「近江八景」, *Hikaku bungaku kenkyû* 「比較文学研究」 50 (1986): 1-27.

13 Ernest Fenollosa, "The Place in History of Mr. Whistler's Art", *Lotus* 1 (Dec. 1903): 16.

ILLUSTRATIONS

ILL 1 *Enjoying the Cool at Ryôgoku*, by Okumura Masanobu, ca. 1745. Private collection.

ILL 2 *View of the Nagasaki Port*, attributed to Maruyama Ôkyo. Data from the Nagasaki Prefectural Museum; Hosono Masanobu, *Yôfû hanga* (*Western Style Prints*) (Tokyo: Shibundô, 1969).

ILL 3 Interior view of a Kabuki theatre in *trompe l'oeil*, by Torii Kiyotada. Data from the Riccar Art Museum, *Exhibition Catalogue Uki-e* (Tokyo, 1975) no. 4.

ILL 4 Interior view of a Kabuki Theatre in *trompe l'oeil*, by Okumura Masanobu, Kobe City Museum of Nanban Art. Data from the Riccar Art Museum, *Exhibition Catalogue, Uki-e* (Tokyo, 1975) no. 1. According to Kishi Fumikazu's hypothesis, Okumura committed intentionally a further deviation from the principle of linear perspective in his re-adaptation of Kiyotada's theatre interior (FIG. 3).

ILL 5 *The Vendetta of the 47 Samurai*, by Maruyama Ôkyo. Reproduced from Hosono Masanobu, *Yôfû hanga* (Western Style Prints) (Tokyo: Shibundô, 1969).

ILL 6 *View of a Western Port*, by Maruyama Ôkyo. Private collection.

ILL 7 *Enjoying the Evening Cool at the Shijôgawara in Kyoto*, by Maruyama Ôkyo. Data from the Riccar Art Museum, *Exhibition Catalogue Uki-e* (Tokyo, 1975) no. 18.

ILL 8 *Pine Tree with Parrot*, by Satake Shozan. Reprinted from Naruse Fujio, *Edo no Yôfûga* (*Western Style Painting in Edo Period*) (Tokyo: Shôgakkan, 1977).

ILL 9 *The Pond of Shinobazu*, by Odano Naotake. Reprinted from Naruse Fujio, *Edo no Yôfûga* (*Western Style Painting in Edo Period*) (Tokyo: Shôgakkan, 1977).

ILL 10 A photograph of Okôchi Sansô Garden in Arashiyama, Kyoto. The garden, located near the Tenryûji Temple, borrows as its background the natural sceneries of the Arashiyama Mountain, which extends across the Hozugawa River.

ILL 11 *Enjoying the Evening Cool at Ryôgoku*, by Suzuki Harunobu. Data from *Ukiyoe taikan* (Tokyo: Kôdansha, 1990).

ILL 12 *Calligraphy*, by Suzuki Harushige (signed Harunobu). Data from Hosono Masanobu, *Yôfû hanga* (*Western Style Prints*) (Tokyo: Shibundô, 1969).

ILL 13 *Evening Bell at the Hill of Shinobugaoka*, by Isoda Koryûsai. Data from *Ukiyoe taikan* (Tokyo: Kôdansha, 1990).

VIII

THE TOKUGAWA BAKUFU'S POLICIES FOR THE NATIONAL PRODUCTION OF MEDICINES AND DODONÆUS' CRUIJDEBOECK

Kazuhiko Kasaya
International Research Center for Japanese Studies, Kyoto

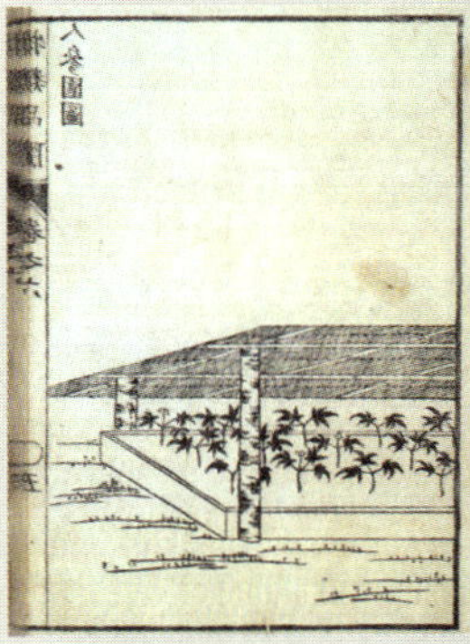

Introduction

Today the name of Dodoens has been all but forgotten in the West, but before Linnaeus founded the modern science of botany, he was considered a leading authority in the field of botany in Europe. The book he compiled under the title *Cruijdeboeck* was one of the best-known repositories of botanical information of his era. This important text was introduced in Japan by the Dutch traders in Nagasaki around the middle of the seventeenth century, and it was offered in tribute to the Shogun at that time. However, none of his subjects took interest in this book, written as it was in strange Western characters nobody could read or understand. Consequently, the book was put away in a Bakufu storeroom, where it gathered dust for many years. It was around 1740 that the book was rediscovered by the eighth Shogun, Tokugawa Yoshimune (1684-1751, reigned 1716-1745). I would like to start by explaining the background to this "rediscovery."

1 Society in the Genroku and Kyôhô Eras

The last decades of the seventeenth century, commonly referred to as the *Genroku* 元禄 period (1688-1703), witnessed an enormous growth in production and trade. This led to a blooming mercantile economy, which pervaded throughout the whole of society, both in urban as well as rural areas. This had an enormous impact on society in many ways and brought about many changes. The cultivation of cash crops stimulated the growth of a market economy in which competition and profit became ever more compelling motives. At the level of the cultivators, the gap between rich and poor widened spectacularly. At one extreme were parasitic landlords and wealthy peasants who monopolised resources and farming land. At the other extreme were increasing numbers of people who pawned their lands and became indebted to local landlords, or sought employment with wealthy farmers and merchants as indentured servants, or left for other lands and to the cities in search of employment.

All of these circumstances brought about structural changes on Japanese society. Social strains and turbulence developed in both urban and rural areas. Administrative and judicial responses to these unfamiliar problems became focal points of concern. The city of Edo expanded spectacularly in the seventeenth century, and in the *Kyôhô* period the population grew from 700.000 to one million. Due to the system of alternate attendance (*sankin kôtai*), *daimyô* and the retainers in their service came from throughout the country to live in Edo. To cater to the varied and voracious demands of the samurai, merchants, craftsmen, and day labourers, carpenters and workers in the building trade, grew steadily in number, and made their home in one of the wards of Edo. Further exacerbating the population problem were the impoverished farmers, who under increasing pressure from wealthy landowners, left their lands behind and flocked to Edo.

江戸
享保
参勤交代

II Currency Reforms and the Nagasaki Trade

In the *Genroku* period, the Bakufu devalued the gold and silver currencies at the order of the fifth Shogun, Tokugawa Tsunayoshi (1646-1709, reigned 1680-1709). Decreasing the amount of gold and silver in the coinage was originally intended to reap profits from reminting the currency and thus make up for the deficits in the Bakufu treasury. Re-minting was also meant to boost the supply of currency in order to offset the increasing demands for currency in connection with the upsurge in trade at that time. However, the simple act of devaluing the currency proved to be a strong temptation in the light of the short-term windfalls it brought, and it was difficult to stop the practice. Enacted as these measures were without consideration for the economic needs of society, the result of the continued devaluation of the currency was outright inflation. The trading system lapsed into a state of dangerous confusion.

徳川綱吉

The resolution of the currency problem was one of the most important achievements of Yoshimune's reign. Even if the Bakufu succeeded in stabilising the amount of precious metal in the coinage at a fixed level and thereby staunch inflation, the demand for currency had increased due to the development of trade and the commercialisation of society. At the beginning of the early modern period, Japan had been a leading country in terms of output of silver and gold. By the *Genroku* era, however, the output of its mines had dropped, and the enormous amounts of gold and silver that had already been mined were being drained out of the country through the port of Nagasaki into the foreign trade circuit.

In brief, the countrywide currency and economic crisis was linked to the issue of foreign trade. The initial measures that were taken to stop the outflow of currency from the country were reactive restrictions on trade. Later on, the focus of the efforts shifted to producing locally in Japan those goods that used to be imported, and proactive measures were enacted to systematically encourage domestic production. These measures were comparable to those that were enacted in the West with a view to promoting domestic production, a process that marked the shift from bullionism to mercantilism. In the West political and economic developments went hand in hand in a process that promoted and sped up modernisation. In eighteenth century Japan, the government of the eighth Shogun Yoshimune carried out similar policies, which proved to play a key role in Japan's modernisation in the political, economic and cultural fields. One of the most important of these policies was the long-term effort to produce medicinal plants in Japan.

III Arai Hakuseki's economic policy line

新井白石
徳川家宣
荻原重秀
正徳新例

Actually Yoshimune inherited these policies from Arai Hakuseki (1657-1725), the brain and *de facto* policy maker during the reign of the preceding Shogun, Ienobu (1663-1712, reigned 1709-1712). Hakuseki fiercely opposed the policy of coinage debasement pursued by the senior councillor Ogihara Shigehide (1658-1713) in the late seventeenth century.[1] In order to stabilise the economic order, he advocated the restoration of the currency standard. Keenly feeling the need to maintain adequate stocks of bullion, he enforced new rules for trade known as the *Shôtoku shinrei* ('New regulations of the *Shôtoku* era') in an effort to regulate and control the trade volume in the port of Nagasaki. In a further step Hakuseki laid out plans for import substitution and stimulated the development of domestic production, replacing for instance imported raw silk by domestic produce. In addition he conceived the plan to substitute imported medicinal drugs, including ginseng, sugar and other imports by domestic production.[2]

Particularly noteworthy in Hakuseki's policies is the fact that they were aimed at promoting the interest and wealth of the whole country. This was a totally new concept, a radical departure from the customary Bakufu-centred approach. In Hakuseki's view the Bakufu policies were not just for the benefit of the Bakufu itself, but they had to serve a public purpose, i.e., the stability and welfare of the whole society. This concern is manifest in his monetary policy, which did not begrudge the expenditure of 130.000 *kan-me* (ca. 500.000 kg) of silver from 貫目 the Bakufu's treasury in order to restore the currency debased by Ogihara.[3]

At the outset Hakuseki's economic policy views were inspired by the idea that the wealth of a nation lies in its gold and silver holdings, in other words by bullionism.[4] However, as he deepened his reflection he gradually reached the insight that the wealth of a nation lies in its economic products, such as raw silk, medicinal drugs, sugar and the like and he shifted his objectives towards their development. Hakuseki's political views were promising, but his power was short-lived, and he was forced to retire from the political stage before he could fully implement them.

iv The Kyôhô Reforms: national wealth and the encouragement of the domestic production of medicinal drugs

1 An outline of the policy for the encouragement of domestic production of medical drugs.

When Yoshimune took over the reins of government, many senior Bakufu councillors and hereditary vassals who had previously endorsed Hakuseki's policies, now began to reject them as too progressive. Yoshimune himself was also critical of the policies the advisor of the former Shogun had advocated. Hakuseki had devoted a lot of time and energy to the reception and entertainment of the Korean envoys as well as to the regulation of ceremonial functions and etiquette inside the Bakufu administration, but Yoshimune considered these matters to be inconsequential and merely cosmetic. However, as for Hakuseki's practical proposals regarding monetary matters, foreign trade and economic policy, Yoshimune found much in them to commend and carried them on. Especially the continuation of the *Shôtoku shinrei* became a major issue after Hakuseki's resignation. The majority of the Bakufu officials leaned towards abrogation, but when Yoshimune read the

大岡清相 memorial of the Nagasaki commissioner Ôoka Kiyosuke (1679-1717), he investigated the rules and regulations stipulated by the *Shôtoku shinrei* and concluded that they were good laws worth supporting and maintaining.[5]

In continuing Hakuseki's economic policies, Yoshimune became the inheritor of Hakuseki's economic policy line and concept of reform, and this would lead the Shogun to the actual implementation of Hakuseki's abortive policies for the promotion of domestic production, in particular the domestic cultivation of medicinal plants. At the heart of Yoshimune's policy concerning medicinal materials lay a concern akin to that of Hakuseki, i.e., the interest and the wealth of the country.

The countrywide venture aimed at producing medicinal plants inside Japan was a major undertaking covering almost the entire thirty years of Yoshimune's reign, and enacted under his personal guidance. The project focused on transplanting and growing Korean ginseng in Japan, as well as on stepping up the collection of medicinal plants growing in mountainous regions throughout the country.[6] Yoshimune valued medicinal plants highly. In the first place, medicinal plants are obviously indispensable to society and he sought to supply the people with valuable medicines at low prices. For that reason Yoshimune wanted to increase their cultivation and production. This policy was undoubtedly based on an idea of general welfare.

By the beginning of the eighteenth century, medicines had long been a valuable import item, and demand for them remained high in Yoshimune's time. One reason for this high demand was the economic development that Japanese society had witnessed during a century of lasting peace. There was also a widespread concern for better health and improved standards of living. Population in that period peaked at one million in Edo, and cities grew throughout Japan. The number of people living in urban areas rose dramatically, but sanitary conditions in these crowded cities were poor, and contagious diseases spread easily. Moreover, in contrast to rural villages where family and relatives all lived close at hand, life in the cities produced a situation where many more of the infirm and aged lived alone. All of these factors reveal the underlying causes of the high demand for effective medicines.[7] A second reason for indigenous production of medicinal plants was of a financial nature. Yoshimune wanted to stop the importation of foreign medicines. Ginseng was imported from Korea and most of the other medicines came from China or the Netherlands.[8] This resulted in a huge outflow of Japanese currency overseas and scarcity at home. Consequently, the policies aiming at domestic production were also inspired by mercantilism.

Yoshimune's policy of promoting the domestic cultivation of medical drugs started around the fourth year of *Kyôhô* (1719), about four years after taking the helm of state affairs. It was implemented in two projects, which although interrelated, had distinct objectives. One was the domestic cultivation of ginseng, the other was the dispatch across the country of "herborizers" (*saiyaku-shi*), whose mission it was to detect and collect the medicinal materials that remained untapped everywhere in the country, with a view to developing domestic medicinal drugs of a quality equal to that of the imported products. The two most important officials involved in these projects were Hayashi Ryôki (1695-1722), physician in ordinary to the Bakufu, confidant and pharmacological advisor to Yoshimune, and Uemura Saheiji (1695-1777), inspector of the Bakufu and informer of the Shogun, who toured the country as "herborizer". Other people involved were amateur pharmacologists and private surgeons such as Niwa Seihaku [Shôhaku] (1691-1756), Noro Genjô (1694-1761), Abe Shôô (?-1753) and Tamura Ransui (1718-1776). In response to the appeal launched by the Edo city magistrate (*Edo-machi bugyô*) they had become Bakufu advisor for medicinal herbs. Other people involved included apothecaries of the three major cities, Edo, Kyoto and Osaka.[9]

採薬使
林良喜
植村左平次
丹羽正伯
野呂元丈、阿部将翁
田村藍水
江戸町奉行

The year following his nomination to the position of Shogun, Yoshimune already manifested his interest in the problem of medical drugs, when he ordered the enlargement of the Shogunal herb garden at Koishikawa (in the sixth year of *Kyôhô*, this herb garden was enlarged to ten times its former size, covering now an acreage of 49.600 *tsubo*). In 1719 the Bakufu sent a letter to Lord Sô, daimyô of Tsushima, requesting drawings of the ginseng root and information about its properties. This proves that the matter of domestic cultivation of ginseng was being taken in hand. Formally the Bakufu was not seeking to smuggle live ginseng roots out of Korea, since the Korean authorities maintained a ban on the export of the product, but in private Hayashi Ryôki was nevertheless relaying the above-mentioned request to Lord Sô.[10]

小石川
坪、宗、対馬

During the 1720's Niwa Seihaku was busy collecting herbs in Hakone. In addition he invited Noro Genjô, who hailed from the same

箱根

伊勢 駿河、甲斐、信濃 日光山

province of Ise to come over and assist him on a herborizing tour of the provinces of Suruga, Kai and Shinano. Uemura Saheiji was sent on a herborizing mission on behalf of the Shogun to Mt. Nikkô.[11] All these facts are testimony that the policy for the search and collection of indigenous medical drugs was coming into full swing.

本草綱目 松岡恕庵 上方

At the same time medicinal materials from sources at home and abroad were being tested and identified with the herbs and drugs listed in Chinese herbals such as the *Bencao gangmu* (Jap.: ***Honzô kômoku***, 'Elements of Pharmacopoeia'). In connection with these activities pharmacologists and well-known physicians such as Matsuoka Joan (1668-1746) from Kyoto, as well as apothecaries from the Kamigata area were one after the other being summoned to Edo, to give instruction and guidance in the vetting and selection of medical drugs. Lectures on medical books were organised in the residence of Hayashi Ryôki, not only attended by physicians in ordinary of the Bakufu but also by private surgeons. However, these lectures grew so popular that Hayashi Ryôki's study could no longer accommodate the audience, necessitating the lease of one of the buildings of the Bakufu for use as auditorium.[12] It is clear that Yoshimune's policy for the promotion of the domestic production of medical drugs was not just a government policy, but involved a great number of private persons as well, and could count on a strong interest and enthusiasm in society at large.

倭館 釜山

As for the identification of the medical drugs, we learn from a recent study by Tashiro Kazui that in 1721 Hayashi Ryôki submitted to Lord Sô, daimyô of Tsushima, a list of drugs containing 104 Chinese and 74 Japanese names of medicinal materials, inquiring whether these existed in Korea or not, and asking to check the Japanese and Korean designations for these substances. At the same time a large-scale inquiry was reportedly being conducted from the Wakan (the residence of Japanese emissaries to Korea) in Pusan.[13] We may safely assume that similar inquiries for identification were being conducted through the Chinese merchants who came to Nagasaki.

和薬改会所

Identification and verification of indigenous drugs, as well as the assessment and control of their quality were handled by the *Wayaku aratame kaisho* (Japanese medicines vetting committee), which was set up in Ise-machi in Edo in 1722. This vetting committee was made up of 25 wholesale dealers of medicinal plants. Niwa Seihaku and other pharmacologists sat on this committee to vet medicinal herbs and minerals that were brought in by collectors from every region in the country.[14] This committee was to act as the driving force on the private sector side for the implementation of the domestic cultivation policy. The following year a similar committee was set up in Osaka. It did not limit its scrutiny to native medicines, but from 1724 onwards it also included medicines imported from China. The committee thus assumed the role of general inspectorate for pharmaceutical affairs. Its interest also included Western *materia medica*, which came to Japan via the Netherlands. We know that in 1725 Niwa Seihaku planted palm tree saplings and a number of Western medicinal herbs, which had been brought to Japan aboard a vessel of the Dutch East India Company.[15] These herbs were nursed in the herb garden of the Bakufu.

We know from previous research that as an extension of the policy for the development of medical drugs, sugar production and the cultivation of sweet potatoes were also taken in hand. Thus in various sectors of society all kinds of experiments and trials were undertaken towards the development of homegrown medical drugs. Some of these endeavours made a real contribution towards the practical use of the drugs, and succeeded in producing a quality that was not inferior to that of imported produce. In this context no issue was of greater importance than the domestic cultivation of ginseng.

II The domestic production of ginseng: the development of otane ninjin.

As I have already stated, the plan to promote the domestic cultivation of ginseng constituted one of Yoshimune's major projects. He initiated it right from the beginning of his reign and it occupied him for most of the goodly thirty years that he held power. We know that implementation started as early as 1719. The Bakufu issued a secret order to Lord Sô, the daimyô of Tsushima, instructing him over a number of times to procure live ginseng roots from Korea, ignoring the official Korean ban on exporting them. This marked the start of a decade of test planting of the ginseng roots in selected locations such as the herb garden of the Bakufu at Koishikawa and Nikkô (present day Tochigi prefecture), whose climate and environment were believed to resemble those of the Korean Peninsula. However, the attempts at transplanting the roots ended in failure. For almost ten years Yoshimune tried on numerous occasions to procure and transplant ginseng roots, but the experiment was not crowned with success, and realisation of the plan looked more remote than ever.

日光、栃木県

On the Korean Peninsula, the homeland of the ginseng, one found either potting on a very small scale or collecting the seeds of wild ginseng and sowing them in the same spot (so-called 山養 *san'yô* or 'hillside growing'). Although the growing process could be aided by forcing techniques, generally speaking, cultivation outside of a natural setting was considered to be next to impossible and there was no precedent of growing them outside their wild habitat. Had it been possible, ginseng would not have been such an extraordinarily expensive product. It is only under strict conditions that ginseng seeds germinate, and even then the rate of germination is very low.[16] In addition, the conditioning of the soil, the extent of sunshine and the degree of drainage are all very delicate. If any of these factors is mishandled, the roots tend to wither at once. Moreover, ginseng is a perennial plant taking between four and five years to develop a root of sufficient size. As if all this were not hard enough, the root is an easy prey for rats, weasels and pests, and one had also to be careful about the use of fertilisers. In addition, because the root is grown in the shade to ward off the direct rays of the sun, or on the northern slopes of hills and mountains, bacteria and mould could easily develop in the soil, with the possible risk of bringing the labour of many years to naught. Thus the greatest care was needed not only to grow the ginseng itself, but also to maintain the complex surrounding conditions, such as the selection of the right soil and the control of its composition.

Yoshimune's policy for fostering domestic production of ginseng was a long and winding road of trial and error. By identifying the various difficulties that prevented the cultivation of ginseng, creating the conditions to overcome them, he finally succeeded in devising an adequate nursing technique. The question now is at what point in time and by what means success was finally achieved. Yoshimune attempted to have ginseng cultivated from seeds, even though this approach was not thought to be very likely to succeed. After several attempts over a period of almost eight years some of the ginseng took root and produced seeds. This occurred in 1738, and this small success was attributed largely to the efforts and perseverance of botanical specialists such as Tamura Ransui and Abe Shôô. The Bakufu then distributed seeds in many areas to people who wanted to cultivate ginseng, and furthermore issued manuals with instructions for their cultivation, which were based on the Bakufu's long experience. Thus the project finally succeeded under a nationwide movement around 1745 – nearly thirty years after Yoshimune had initiated it.

In a decree issued in the eighth month of the third year of *Enkyô* (1746), the Bakufu allowed the free sale of Japanese grown ginseng, cultivated from the seeds of ginseng (afterwards known as *otane ninjin*). Hence we know that the successful cultivation must be prior to that date, but that is about all we can say with any degree of certainty. Several hypotheses have been formulated, but none has remained uncontested. Imamura Tomo's book entitled *Ninjin-shi* ('The History of Ginseng'), the classic study on this topic, proposes as date the year 1733. That we find no mention of procuring live roots or seeds from Korea via Tsushima after the year 1728, is supposed to imply that the live roots and seeds had struck root and started to grow. That Uemura Saheiji is mentioned travelling to Nikkô in late 1733 for the outdoor cultivation of ginseng is assumed to prove that the experiment was proceeding according to wish.[17] Yasue Masakazu on the other hand draws our attention to the ginseng cultivation within the jurisdiction of the Bakufu commissioner of Sado Island (*Sadogashima*), and claims that the first yields were recorded in the year 1725, that in 1728 seeds were sown in cases and henceforth the harvests were steady and secure.[18]

延享
御種人参
今村鞆
安江政一
佐渡島

Kumata Hajime claims that Uemura Saheiji and Tamura Ransui were sent as Bakufu experts to Nikkô in 1729 to start a round of growing experiments.[19] In old records preserved in the city of Utsunomiya, Kawashima Yûji has found evidence that in the year 1729 the Bakufu gave three ginseng roots to Ôide Denzaemon of Imaichi-machi in the foothills of Nikkô, instructing him to grow them. He surmises that these three roots were among the eight specimens Lord Sô of Tsushima had offered to the Bakufu the previous year. He contends that the three specimens planted by Ôide Denzaemon struck root and eventually led to the success of the ginseng-growing project.[20]

熊田一
宇都宮
川島裕次
大出伝左衛門

At any rate we find evidence in the records that around 1730 the Bakufu was distributing seedlings of ginseng to the daimyô that were related to the Tokugawa clan. This proves that by that time the experiments had at least reached the stage of nursing seedlings for the home production of ginseng. However, it is likely that in most cases the success was due to a stroke of good luck, and if we take into account that only a trifling quantity of seedlings and seeds was distributed, it is probable that the number of successful trials was limited and that a stable nursing technique was still a long way ahead.

According to the historical sources I will subsequently introduce, it would appear that the effort for breeding ginseng did not garner lasting success until the 1740's. We have a report dated 25th of the fifth month of the year 1737,[21] which proves that in 1736 the Bakufu had the Edo city magistrate's office distribute fifty ginseng seeds each to three horticulturists, with the instruction to try and breed them. One of 牡丹屋彦右衛門 the horticulturists, Botan'ya Hikoemon seems to have reported fairly good results. He had filled a box with a mixture of red earth and gravel and planted 25 seeds in it. The nursing went rather well and yielded forty roots. They measured 9 centimetres and grew trifoliate leaves. Even the remaining 25 seeds, which he had planted in the soil had yielded more than 20 roots. However, while the first trial had yielded a total of more than 90 roots, the subsequent yields had gradually dwindled, following a pattern that the other two horticulturists had 伊兵衛、小右衛門 reported. They were the gardeners Ihee of Somei and Shôemon of Akô. 赤穂、伊藤伊兵衛 Itô Ihee of Somei was known as the best gardener of Edo, and it is interesting to note that he was selected for the nursing experiment, although he failed sorely. According to his report, he had planted the seeds in jars filled with earth that he had had brought from various regions in Japan. About thirty had struck root, but they had gradually started to wither and in the end only thirteen remained. Shôemon of Akô had prepared a special parterre filled with a mixture of red and 彼岸 black earth. Judging that the equinoctial week (*higan*) would be the right season for planting, he had planted his seeds in the middle of the eighth month, but so far not one had struck root. Shôemon had probably erred on the safe side and planted the seeds too late.

When the above results were reported to the city magistrate, the latter entrusted Botan'ya Hikoemon, the one of the three horticulturists who had returned a report of success, with another 200 seeds on the fifth day of the seventh month of 1737, for further trials. In that year 1736 the above-mentioned horticulturists were not the only ones to whom seeds were distributed; they were also supplied to the physicians in ordinary with the instruction to plant them. We also know that in 1737 seeds were given to the pharmacologist Tamura Ransui and around the same time the Bakufu distributed seeds to a number of daimyô with the instruction to plant and nurse them.[22] This series of facts proves that the technique for growing ginseng had not been firmly established yet and that it was still in the stage of trial and error. It is also noteworthy that, when handing over the 200 seeds to Botan'ya Hikoemon in 1737, the city magistrate found it necessary to add that they were being "entrusted" to him, whereas the previous year they had simply been granted him. This too suggests that the ginseng seeds were certainly not a commodity in abundance. It is clear that the experiment of growing ginseng was not completed yet, and that the lessons learned from the breeding trials by Botan'ya Hikoemon, Tamura Ransui and many others indeed served as major stepping stones towards the accomplishment of the ultimate goal.

Although the harvesting of the seeds went well, the roots fell easily prey to vermin damage and blight. Maintenance and control of the soil mixture were difficult, and in order to guarantee a stable harvest of life roots, many technical problems still had to be overcome. In his book *Ninjin kôsaku-ki* ('On the Breeding of Ginseng') Tamura Ransui gives detailed proposals on methods to solve the problem of vermin damage, based on his experience with trial breeding around this period.

人参耕作記

Even after that date the Bakufu continued the distribution of seeds and the breeding trials. In 1743 it granted 100 seeds to Tamura Ransui, ordering him to grow them, and the following year it gave 515 seeds to Abe Shôô, with the instruction to submit a monthly report on the progress of the nursing.[23]

In view of what precedes, it would appear that the establishment of the breeding technique and the stable production of ginseng have to be dated after 1740, and that in effect we have to wait until around 1746, the year when *otane ninjin* were put on sale. If we take into account that it takes between three to five years for a ginseng root to fully grow, we may assume that the know-how gathered in the course of the many experiments undertaken by the public played a pivotal role in the realisation of the domestic production of ginseng root.

The cultivation of ginseng spread widely to various areas throughout Japan, which was ironic given the chronic shortage of ginseng in its native place Korea. This policy not only succeeded in providing ample income to the farmers who grew the ginseng, but it also met the medical demands for the plant, providing as it did an inexpensive and reliable product. The type of cultivated ginseng that Yoshimune developed was distinguished from the natural variety and called *otane ninjin*, and this is still its proper name today. The word *otane* means "seed" cultivated by the Shogun. That this variety of ginseng is named thus is largely in homage to Shogun Yoshimune who had it developed.

Yoshimune's policy for the encouragement of the domestic production of medicinal drugs was implemented on a nationwide scale, including both the Bakufu domains and the fiefs, and pursued with concerted efforts for over three decades. It constituted one of the most important enterprises of his reign. Moreover, this policy was not just implemented in pursuit of financial profit for the Bakufu. It was devised to provide society with medical drugs in abundance at low price; to nurture the know-how to develop good quality drugs on a level with the hitherto imported products, and thus to stop the outflow of bullion into the channels of international trade. Seen from these angles Yoshimune's policy was intended to promote the interests and well being of the whole of Japanese society. Because it was targeted at such broad base it was also possible to enlist the cooperation of not just the Bakufu officials, but also of private surgeons and amateur pharmacologists, apothecaries and herb merchants, farmers throughout the country, as well as daimyô in the various regions and their retainers and surgeons. Their voluntary collaboration constituted a major driving force from below that propelled the project forward.

That the Bakufu forwent its own profit is clearly manifested by the way in which the *otane ninjin* were put up for sale. When in 1738 the Bakufu sold ginseng seeds throughout the country to all those who wished to grow them, it published the results of its yearlong experience with cultivating ginseng and distributed leaflets with instructions among the buyers of the seeds. This shows that the Bakufu had no intention of monopolising the cultivation of ginseng, but that on the contrary it wanted to take on this challenge on a nationwide level, using the experience and know-how of tillers throughout the land. In point of fact the Bakufu made no profit on the sale of the seeds. Thus in 1746 the seeds that had been harvested in Nikkô and other places in Japan were put on sale. On that occasion the Bakufu issued an order to the Edo city magistrate instructing him that trade had to be conducted at prices set freely by the apothecaries, but that, apart from the commission withheld by the apothecaries, the entire proceeds of the sales had to go to the farmers.[24]

Yoshimune skilfully employed these specialists in a variety of fields, allowing them to freely demonstrate their specialised skills. He thus facilitated the success of the countrywide project for the domestic production of medicines. If Yoshimune's policies had originally been enacted to produce medicinal plants in Japan, he subsequently expanded them to include the production of more general and useful products. He stimulated interest in the natural products and goods to be found in the various regions of Japan. This was given concrete expression in the large-scale and comprehensive survey of domestic produce, known as the *Shokoku sanbutsu torishirabe*.[25] This survey reveals that at that time there was a keen interest in Japan in potential sources of national wealth.

諸国産物取調

The members involved in these projects studied Chinese traditional pharmacology (*honzô*). After they had completed this study, Yoshimune wanted to acquire new knowledge from the Western world, and this brought Dodoens's *Cruijdeboeck* to the intellectual forefront of the time.[26] Before that time Yoshimune had of course had no idea about who Dodoens might be. Yet, Dodoens's accurate reproduction of medicinal plants, carefully depicting in detail the petals, leaves, roots and fruit accorded with what Yoshimune knew from many years of actual observation of the real world. The illustrations in the *Cruijdeboeck* were incomparable to the botanical illustrations in Chinese and Japanese books, and Yoshimune found this book to be of great value. Trusting his intuition, he decided to attempt to have the contents of this Western book translated for the benefit of the scholars involved in the project of domestic cultivation of medicinal plants. This endeavour was to have a lasting effect, laying the groundwork for Western learning in Japan.

本草

Consequently, a project to translate Western books was planned, but up to that time nobody had experience translating Western texts into Japanese. Obviously there was no dictionary and the task had to start from the absolute beginning. There were translators at Nagasaki, which was the station for trade with the Netherlands, but their proficiency was limited to conducting a conversation. Reading books in Dutch was beyond their grasp. The herbalists Noro Genjô and Aoki Kon'yô (1695-1769) both worked on this project. Noro Genjô called at the Nagasakiya, the inn in Edo where the head of the Dutch delegation to the Shogun was staying, with his attendants and his Japanese interpreter. Noro brought Dodoens's book along and questioned the Dutch about its contents. He started his study of Dutch in 1741 and worked on it for many years. After ten years he was able to present Yoshimune with his "translation", which is known as *Oranda honzô wage* ('Japanese Translation of the Dutch Herbal'), a book in eight volumes. In this work he determined the appropriate Chinese and Japanese equivalents for the Latin and Dutch names of a variety of plants.

青木昆陽

長崎屋

阿蘭陀本草和解

It was a very simple compilation and apart from the listing of Chinese and Japanese equivalents for Western terms, it only included notes on the medicinal uses of plants. This book may be called a rather primitive attempt at translation but it nevertheless marked the first if humble step in the enterprise of translating Western books into Japanese, which was to have a deep impact on the future course of Japan. The successors of Noro and Aoki, including Hiraga Gennai (1728-1779), Sugita Genpaku (1733-1817) and Maeno Ryôtaku (1723-1803) would all make important contributions to the field of Western studies. Thus *Rangaku* would come into full bloom in Japanese society during the latter half of the eighteenth century. This happened at the same time that the conditions for experiential and objective analysis that had appeared earlier were evolving into powerful catalysts for the development of modern learning.

平賀源内、杉田玄白
前野良沢
蘭学

Yoshimune's policy with regard to medical drugs achieved the desired end towards the end of his reign. By that time however it had transcended the narrow bounds of the drugs issue *per se* and come to encompass natural objects in general, in short all matters that pertained to human economy and natural history. There was a growing interest in all kinds of products and natural objects found in Japan, which stimulated research aimed at discovering new potentially useful products, and increasing their yield. Non-official exhibitions were organised to display the specialties and rare products of various regions. They made an important contribution towards practical learning and the development of industry. This paved the way for the activities of pioneering figures of the next generation such as Hiraga Gennai. Incidentally, the teacher of Hiraga Gennai was none other than Tamura Ransui, who was so instrumental in the development of a nursing technique for ginseng. Ransui and Gennai also initiated new ventures such as produce exhibitions. In this respect we are warranted to consider Gennai as the rightful heir of Yoshimune's policies.

Apart from the interest in the possibilities of industrialisation, the comprehensive investigation of products and natural objects throughout Japan also stimulated interest in natural history, which eventually evolved into a purely academic interest. People were fascinated by the multifarious attraction natural objects exerted on them and classified the different species.

Yoshimune's enlightened policies were not limited to medical drugs.

He also encouraged practical learning, which could be helpful in the development of industry in general, and in an effort to acquire new scientific and technological know-how, he loosened the ban on the importation of Chinese translations of Western books. He also set great store by the study of astronomy and geography. He employed Takebe Katahiro (1664-1739), a leading disciple of Seki Takakazu (?-1708), in order to have him draw up a highly accurate general map of Japan (*Nihon sô-ezu*), and promoted the reform of the calendar. Yoshimune himself did not tried his hand at astronomical and meteorological observation and even designed an improved version of the simplified armillary sphere (*kantengi*). The series of policies pursued by Yoshimune were implemented in the ways described above. They were practical, rational and very successful. The contribution they have made towards Japan's modernisation cannot be overrated.

建部賢弘
関孝和
日本総絵図
簡天儀

NOTES

1 Arai Hakuseki 新井白石, "Kaika-gi"「改貨議」, in *Hakuseki kengi, Arai Hakuseki zenshû dai 6 kan*「新井白石全集』第六巻「白石建議」(Kokusho kankôkai, 1907).

2 Kurita Mototsugu 栗田元次, *Arai Hakuseki no bunchi seiji*「新井白石の文治政治」(Ishizaki shoten, 1952); Kasaya Kazuhiko 笠谷和比古, "Arai Hakuseki to Tokugawa Yoshimune. Tokugawa jidai no seiji to honzô"「新井白石と徳川吉宗 – 徳川時代の政治と本草」, in *Mono no imêji* (*image*): *Honzô to hakubutsugaku e no shôtai*「物のイメージ／本草と博物学への招待」, ed. Yamada Keiji 山田慶兒編 (Asahi shinbunsha, 1994); Kasaya Kazuhiko笠谷和比古, *Tokugawa Yoshimune*「徳川吉宗」(Chikuma shobô, 1995).

3 Miyazaki Michio 宮崎道生, *Arai Hakuseki no kenkyû, zôteiban*「新井白石の研究　増訂版」(Yoshikawa Kôbunkan, 1969), 141.

4 Arai Hakuseki 新井白石, "Goji-ryaku"「五事略」, in *Arai Hakuseki zenshû* vol. 3. 新井白石全集」第三巻

5 Muro Kyûsô 室鳩巣, "Kenzan hisaku, dai 3 satsu"「兼山秘策」第三冊, in *Nihon keizai taiten*, vol. 6「日本経済大典」第六巻 (Shishi shuppansha, 1928).

6 Shirai Kôtarô 白井光太郎, *Kaitei zôho Nihon hakubutsugaku nenpyô*「改訂増補日本博物学年表」(Ôokayama shoten, 1934); Imamura Tomo 今村鞆, *Ninjin-shi*「人参史」(Chôsen sôtokufu, 1936; reprint Shibunkaku shuppan, 1971); Grant Kohn Goodman, *The Dutch Impact on Japan* (1640-1853) (Leiden: Brill, 1967); Ueno Masuzô 上野益三, *Nihon hakubutsugakushi*「日本博物学史」(Heibonsha, 1973); Ueno Masuzô 上野益三, *Nenpyô Nihon hakubutsugakushi*「年表日本博物学史」(Yasaka shobô, 1989); Ueda Sanpei 上田三平, *Kaitei zôho Nihon yakuenshi no kenkyû*「改訂増補日本薬園史の研究」, ed. Miura Saburô 三浦三郎 (Watanabe shoten, 1972); Kimura Yôjirô 木村陽二郎, *Edo-ki no nachurarisuto* (*naturalists*)「江戸期のナチュラリスト」(Asahi shinbunsha, 1985); Sôda Hajime 宗田一, "Nihon no baiyaku, kansei saibai Chôsenshu ninjin (otane ninjin) no hanbai"「日本の売薬・官製栽培朝鮮種人参（オタネニンジン）の販売」, *Iyaku jânaru* (*journal*)「医薬ジャーナル」27 (1991), nos. 1-4; Ôishi Manabu 大石学, *Kyôhô kaikaku to jikata seisaku*「享保改革と地方政」(Yoshikawa Kôbunkan, 1996); Kawashima Yûji 川島裕次, *Chôsen ninjin hishi*「朝鮮人参秘史」(Yasaka shobô, 1993).

7 Tsukamoto Manabu 塚本学, *Tokai to inaka*「都会と田舎」(Heibonsha, 1991).

8 Sôda Hajime 宗田一, *Toraiyaku no bunkashi. Oranda-sen ga hakonda yôyaku*「渡来薬の文化史 – オランダ船が運んだ洋薬 –」(Yasaka shobô, 1993) 44ff.

9 For biographical details see the following works: Shirai Kôtarô 白井光太郎, "Shina oyobi Nihon honzôgaku no enkaku oyobi honzôka no denki"「支那及日本本草学の沿革及本草家の伝記」, in *Honzôgaku ronkô*「本草学論攷」vol. 1 (Shun'yôdô, 1933); Matsushima Hiroshi 松島博, *Kinsei Ise ni okeru honzôgakusha no kenkyû*「近世伊勢における本草学者の研究」(Kôdansha, 1974); Yasuda Takashi 安田健, *Edo shokoku sanbutsuchô. Niwa Seihaku no hito to shigoto*「江戸諸国産物帳 – 丹羽正伯の人と仕事-」(Shôbunsha, 1987).

10 Imamura, *Ninjin-shi*, 4:231ff.
11 For personal details, year and region of dispatch of these "herborizers", see Ôishi, *Kyôhô kaikaku to jikata seisaku.*
12 (*Kyôhô rokunen*) *rokugatsu muika-zuke, Noro Genji shojô.* [*Takahashi Jûzô hoka nimei ate*] (享保六年)六月六日付、野呂源次書状「高橋十蔵他二名宛」. Ônishi Gen'ichi 大西源一, "Noro Genji den"「野呂源次伝」, in *Mie-ken shidankai kaishi, shiryôhen*「三重県史談会会誌」資料編」, vol. 5 no. 12, 3.
13 Tashiro Kazui 田代和生, *Edo jidai Chôsen yakuzai chôsa no kenkyû*「江戸時代朝鮮薬材調査の研究」(Keiô gijuku daigaku shuppankai, 1999).
14 *Oyakusô goyô aitsutome sôrô omomuki môshiage sôrô kakitsuke utsushi*「御薬草御用相勤候趣申上候書付写」(Kokuritsu kôbunshokan naikaku bunkozô).
15 *Niwa Seihaku bussan nikki*「丹羽正伯物産日記」(Kokuritsu kokkai toshokan Shirai bunko zô).
16 Kawashima, *Chôsen ninjin hishi*, vol. 19, "Ninjin saibai no seikô"「人参栽培の成功」.
17 "Nihon ni okeru saibai"「日本における栽培」, in Imamura, *Ninjinshi*, vol. 4, chap. 4, 254.
18 Yasue Masakazu 安江政一, "Edo jidai no Sado ni okeru Chôsen ninjin no saibai kenkyû ni tsuite"「江戸時代の佐渡における朝鮮人参の栽培研究について」, in *Yakushigaku zasshi*「薬史学雑誌」17 (1982), no. 1.
19 Kumata Hajime 熊田一, "Nikkô goryô chiiki ni tenkai sareta Chôsenshu ninjin no saibai ni tsuite"「日光御料地域に展開された朝鮮種人参の栽培について」, in *Kanuma shirin*「鹿沼史林」3 (1956).
20 "Ninjin saibai no seikô" in Kawashima, *Chôsen ninjin hishi*, vol. 19.
21 "Yakushu no bu"「薬種之部」36, in *Kyôho sen'yô ruishû*「享保撰要類集」(Kokkai toshokan zô, kyû Bakufu hikitsugi bunsho).
22 Sôda, *Nihon no baiyaku, kansei saibai Chôsenshu ninjin* (*Otane ninjin*) *no hanbai*, 1; Tamura Ransui 田村藍水, *Ninjin Kôsakuki*「人参耕作記」(Kokkai toshokan Shirai bunko zô).
23 *Oyakusô goyô aitsutome sôrô omomuki môshiage sôrô kakitsuke utsushi.*
24 *Ofuregaki Hôreki shûsei*「御触書宝暦集成」no. 1384.
25 Yasuda, *Edo shokoku sanbutsuchô. Niwa Seihaku no hito to shigoto.*
26 Shirai, *Kaitei zôho Nihon hakubutsugaku nenpyô*; Goodman, *The Dutch Impact on Japan* (1640-1853); Ueno, *Nihon hakubutsugakushi;* Kasaya, *Tokugawa Yoshimune.*

ILLUSTRATIONS

ILL 1 Sprout of Korean ginseng. Hiraga Gennai, *Butsurui hinshitsu* (1763), maki 5 (zue:sanbutsu zue). Collection Université catholique de Louvain, Bibliothèque générale et de sciences humaines, Donation japonaise.

ILL 2 Picture of ginseng root and plant, illustrated by Kusumoto Sekkei (So Shiseki). Hiraga Gennai, *Butsurui hinshitsu* (1763), maki 5 (zue:sanbutsu zue). Collection Université catholique de Louvain, Bibliothèque générale et de sciences humaines, Donation japonaise.

ILL 3 Ginseng garden. Hiraga Gennai, *Butsurui hinshitsu* (1763), maki 5 (zue:sanbutsu zue). Collection Université catholique de Louvain, Bibliothèque générale et de sciences humaines, Donation japonaise.

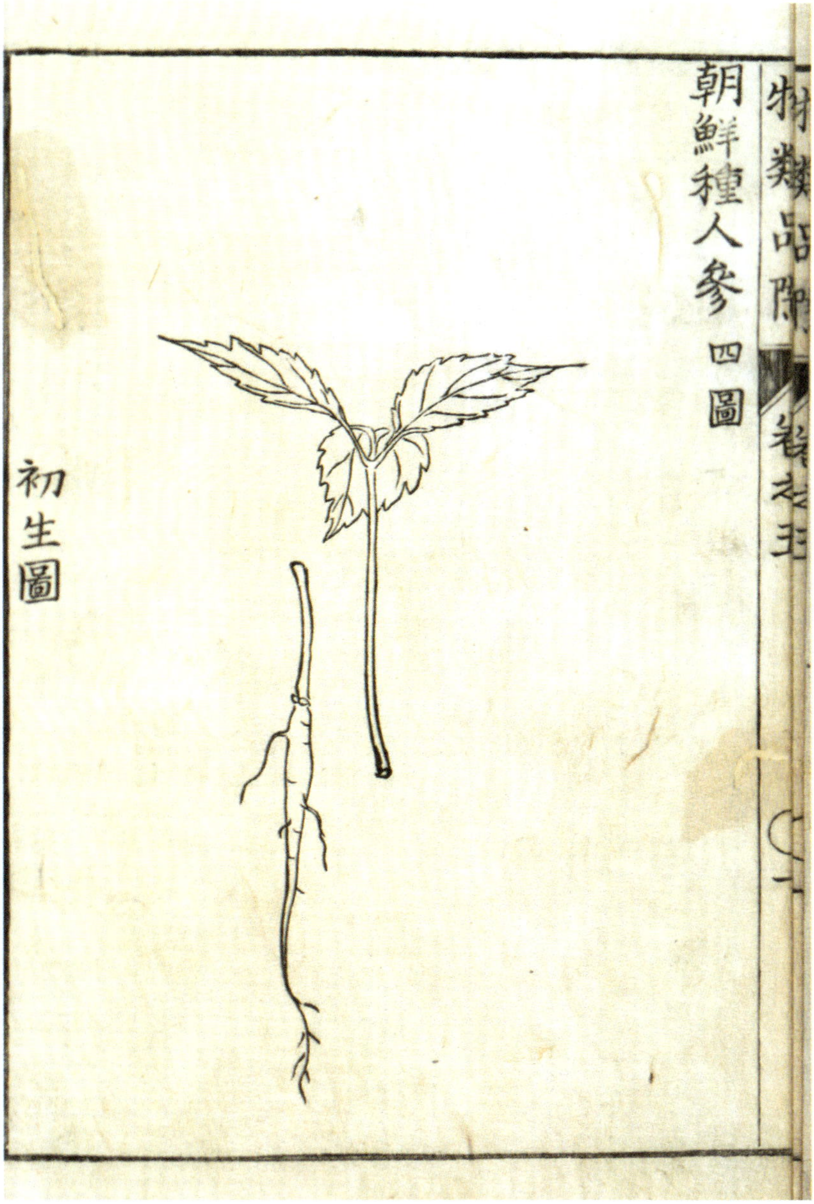

ILL 1

ILL 2

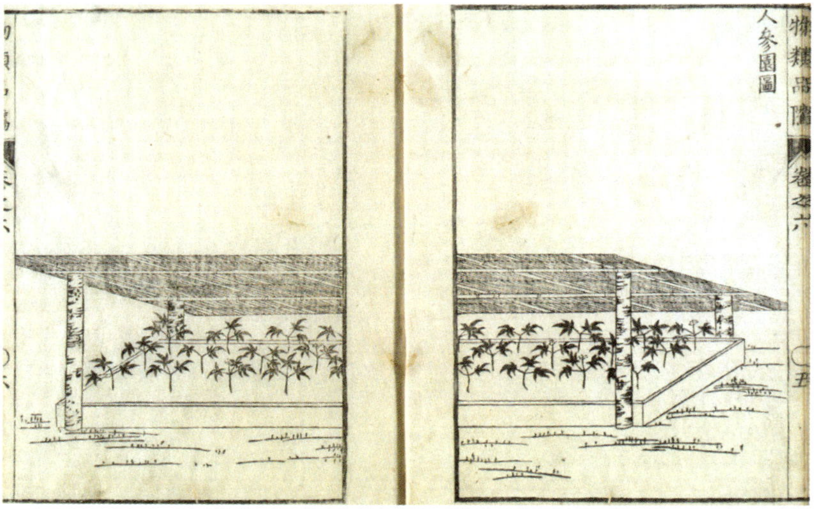

ILL 3

PART III

JAPANESE RENDITIONS OF DODONÆUS

IX

THE RECEPTION AND SPREAD OF DODONÆUS' CRUYDT-BOECK IN JAPAN.

Kiyoshi Matsuda
Faculty of Integrated Human Studies, Kyoto University

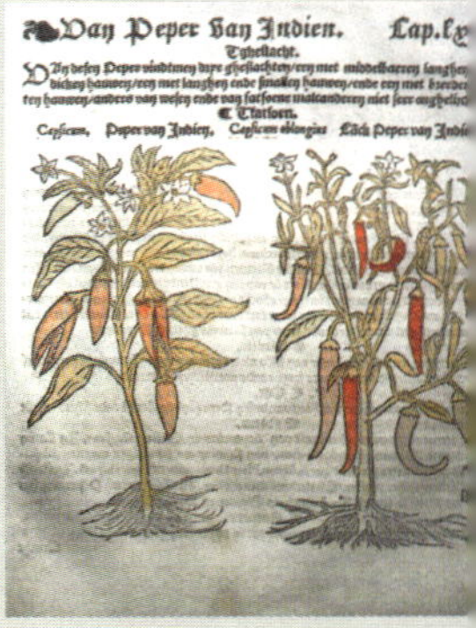

This contribution endeavours to give a brief outline of the history of the translation of Dodonæus' *Cruydt-Boeck* in early modern Japan. I would like to start with a survey of the existing copies of Dodonæus' herbal that were imported in the Edo period (*Edo jidai*, 1600-1868). The Mechelen-born Dodonæus is alive in present-day Japan in the form of books. I had the opportunity to examine three copies of Dodonæus' herbal that were imported in the Edo period. The 1644 edition of the *Cruydt-Boeck* in the possession of the National Museum of Tokyo,
蕃書調所 bears the ownership seal of Bansho shirabe-sho, the Research Institute for Western Studies, established by the Bakufu in 1856, but the period of importation is unknown. On the lower part of the title page is a note: "M. Dominicus Vlammen; A.o 1646 10/13." This probably means that this person purchased the book on October 13, 1646. Who this man was, we do not know. The only existing copy of the *Cruydt-Boeck* of which we know the period of importation is the one in the possession of the library of the Medical Faculty of Kanazawa University [ILL. 1, 2 & 3]. I will introduce the third copy of the *Cruydt-Boeck* at the end of this paper. The Kanazawa copy of the *Cruydt-Boeck* unfortunately lacks the title page, but there is no doubt that it is the third edition, printed in 1644 in the *Plantijnsche Druckerije*, the print shop of the Antwerp
壮猶館 printer Balthasar Moretus. The ownership seal of the Sôyûkan, the
加賀藩 research institute for Western Studies of the Kaga fief, established in 1854, can be discerned. It is a luxurious edition bound in gilded leather, gilt-edged at top and bottom and decorated with arabesque engravings. The woodblock illustrations are all coloured by hand [ILL. 4 & 5]. Regarding this *Cruydt-Boeck* in the possession of the Kaga fief, we know that the
前田綱紀 *daimyô*, Maeda Tsunanori, a scholar and book collector, who devoted
阿蘭陀国本草之書 himself to the preservation of medieval documents, acquired the book
甲比丹 *Oranda-koku honzô-no-sho* ('Dutch Herbal') from the head (Jap.: *kap-*
出島 *itan*, Dutch: *Opperhoofd*) of the Dutch Factory in Deshima, Nagasaki

in the winter of 1682. The head at that time, Andreas Cleyer [Andries Cleyer], was an enthusiastic student of Oriental medicine and botany. One of his subordinates at Deshima was Georg Meister, who later wrote *Der orientalisch-indianische Kunst- und Lust-Gärtner* (Dresden, 1692). In the Dagh-register of Cleyer the entry of November 14, 1682 reads: "Matzendairo Cange-zamma [Matsudaira Kaga Sama] has sent through the chief interpreter Kitzizeymon [i.e., Kafuku Kichizaemon], in return for the illuminated *Cruydt-Boeck* of Dodonæus (het geschilderde kruydeboek van Dodonæus), ten gold coins, two big barrels of sake and two very big stone-breams (10 goude coubanghs, 2 groote baelijs met sackij en 2 seer groote steenbraessems)." It is clear from this text that the above-mentioned 'Dutch Herbal' was a present from Cleyer.

松平加賀様
加福吉左衛門

The two most representative Japanese herbalists of the end of the seventeenth century, Kaibara Ekiken (1630-1714) and Inô Jakusui (1655-1715) were both active in Kyoto. Jakusui entered the service of Maeda Tsunanori in 1692 and started the edition of *Shobutsu-ruisan* ('Nomenclature of All Things Natural'). There are no traces indicating that Jakusui consulted Dodonæus' *Cruydt-Boeck*. Ekiken, who served the Kuroda family of the Fukuoka fief, on the other hand refers to plants from "barbarian countries" (*bansan*) in his *Yamato honzô* ('Japanese Herbal') while an inventory of books in his library lists an *Oranda honzô* ('Dutch Herbal'). The possibility that he had recourse to Dodonæus' original or a translation (not identified) cannot be discarded. It is interesting to note that Jakusui, who completed a thoroughly revised edition of Li Shizhen's, *Bencao gangmu* (Jap.: *Honzô kômoku*, 'Elements of Pharmacopoeia,' 1590; the so-called "Jakusui edition") and wanted to lay the foundation for an international scholarship in East Asia, showed no interest in Dodonæus, while Ekiken, who strove for the Japanisation of traditional Chinese botany, turned his attention towards the Western herbal. We may say that this is a manifestation of the practical character of Ekiken's scholarship.

貝原益軒、稲生若水
庶物類纂
黒田氏、福岡藩
蛮産、大和本草
李時珍
本草綱目

As I mentioned above, the title page is missing from the *Cruydt-Boeck* that Maeda received from Cleyer. There is perhaps a connection with the fact that the *Rangakusha* Udagawa Genshin, who was invited to the Kaga fief in 1808, made a translation of the title page only. The photograph of Genshin's autograph manuscript, which remained in the possession of the Maeda family until the Second World War, is shown in appendix no. 1, together with my translation. It is clear that Genshin's proficiency in Dutch was limited.

蘭学者、宇田川玄真

ILL 1

ILL 2

ILL 3

ILL 4

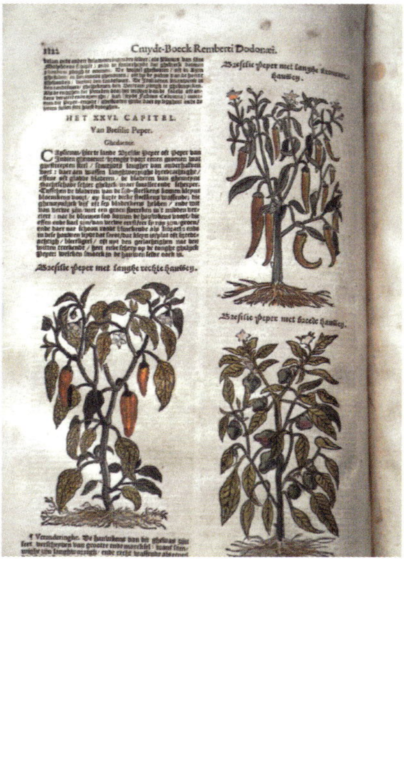

Cruydt-Boeck Remberti Dodonæi.

HET XXVI. CAPITEL.

Van Breſilie Peper.

ILL 5

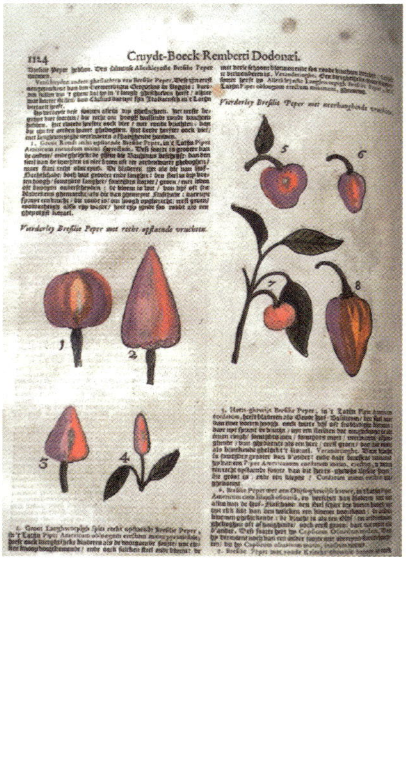

1124 Cruydt-Boeck Remberti Dodonæi.

It is well known that the first "translation" of Dodonæus' *Cruydt-*
野呂元丈 *Boeck* is Noro Genjô's, *Oranda honzô wage* ('Translation of the Dutch
阿蘭陀本草和解 Herbal,' completed between 1741 and 1750), made by order of Toku-
gawa Yoshimune. This is however not a translation in the strict sense
of the word, as it was not based on the original Dutch text, but a
通詞 summary of oral reports by Dutch language interpreters (*tsûji*), who
オランダ商館 accompanied the head of the Dutch Factory (*Oranda shôkan*) on his
江戸参府 yearly court journey to Edo (*Hofreis*, Jap.: *Edo sanpu*). This kind of
本木良意 translation may be compared to Motoki Ryôi's *Oranda keiraku kin-*
阿蘭陀経絡筋脈臓 *myaku zôfu zukai* ('Dutch Illustrations of Nerves, Muscles and Intes-
図解 tines'), said to be the first translation of a Western anatomical work,
based on the Dutch version of J. Remmelin's *Pinax Microcosmo-*
graphicus (1667). This translation was presented to the Bakufu in
和蘭全躯内外 1682 and was published in 1772 under the title *Oranda zenku naigai*
分合図 *bungô-zu* ('Dutch Tables of the Exterior and Interior Parts of the
Human Body').

The Japanese translation based on the original Dutch text of
Dodonæus' *Cruydt-Boeck* was started in Nagasaki in the wake of the
蘭学 rise of Dutch Studies (*Rangaku*) during the last quarter of the eighteenth century. The translators were exclusively Dutch language interpreters. One salient feature of this early translation is the fact that, besides a number of varieties of medicinal herbs imported at that time, the translators selected a number of curious animals and creatures such as elephants, unicorns, alligators, mermaids, dolphins, sloths, mummies, in other words material for curious and fantastic stories. Therefore, as a translation of Dodonæus' *Cruydt-Boeck*, these writings were still very selective and fragmentary. One also finds an introduction
志筑忠雄、万国管闚 about curious animals in *Shizuki Tadao's* reading notes *Bankoku kanki*
('Peeking at the World,' completed in 1782). It seems that this was
a popular topic of conversation among the interpreters of Dutch in
吉雄耕牛 Nagasaki at that time. Chief interpreter (*oppertolk*) Yoshio Kôgyû's
松村元綱、松浦静山 close friend, the interpreter Matsumura Genkô, translated for Mat-
平戸藩 suura Seizan, lord of the Hirado fief, the Dutch abridged version of
Pliny's [C. Plinius Secundus] *Natural History* (C. Plini Secundi, *Des wijdt-vermaerden natuurkondigers vijf Boecken*, Amsterdam, 1662). This original deals exclusively with animals. Unfortunately, Matsumu-
プリニウス略解 ra's draft translation *Puriniusu ryakukai* ('Summary of Pliny') is no
蛮産諸品訳稿 longer extant. Matsumura's autograph manuscript *Bansan shohin*
yakukô ('Translations on Various Products from Barbarian Countries,' in the possession of Nagasaki Prefectural Library, completed in 1787),

lists seven varieties of imported medicinal herbs, besides the animals mentioned above.

1 カシヤヒストル（羅天語）トロンムルストツコ（紅毛語）
阿勃勒（本草綱目）Cassie oft Cassia fistula (xxx capitel; p. 1233) Trommelstok
[APPENDIX NO. 2, ILLUST. F]

2 サアレツプ（亜剌皮亜語）スタンデルコロイト（紅毛語）サテイリヨン（紅毛語、一名）
Orchis Serapias (p. 374) Standel-cruydt Satyrion (xxix capitel; p. 372)

3 レーリヨム（羅天語）レーリイ（紅毛語）百合（漢名）Lelium Lelie

4 アマンドル（紅毛語）アメンドウス（蛮語）巴旦杏（本草綱目）Amandel-boom (p. 1250)
[APPENDIX NO. 2, ILLUST. E]

5 胡椒ペイプル（紅毛語）一種ランガペイプル（長胡椒）Peper (p. 1440) Langh Peper
(pp. 1140-1441) [APPENDIX NO. 2, ILLUST. A2]

6 サフラーン（紅毛語）コロークス（羅天語）雑腹蘭（本草綱目又職方外記）
Saffraen (p. 329)

7 安産草　ローズハンヱリゴ（阿蘭陀語）Roose van Hierico (p.1447)
[APPENDIX NO. 2, ILLUST. C2]

Matsumura gives descriptions of these herbs based on his own observations and on the related articles in Johann Jacob Woyt's medical dictionary *Gazophylacium medico-physicum, of schat-kamer der genees- en natuur-kundige zaaken* (Amsterdam, 1741). For the four varieties "fistula," "orchis," "lily," and "pepper" however, he refers to Dodonæus, although he does not translate the description.

Incidentally, a manuscript entitled *Oranda honzô tekiyô-kai* ('Summary Explanation of Dutch Botany,' bearing the outer title (*gedai*) *Dodoniusu honzô-shô*, in the possession of Kyoto Prefectural Botanical Garden, Ômori Collection), which belongs to the same lineage as *Bansan shohin yakukô*, contains very similar articles on "saffran," "roos van Hierico," "peper," "amandel," "serapias," "lelium," and "fistula," as well as two new articles, namely: "Japanese translation of the olive-tree in the Dutch Herbal: Olijf-boom, i.e., a Dutch word, page 1287" and "Dodonæus, page 842, Ghehoeckte Ciceren." [APPENDIX NO. 2, ILL. G] Moreover, for the eight varieties except the lily, illustrations are added, fairly faithfully copied from the original *Cruydt-Boeck*. I have indicated in the foregoing list the page numbers of the illustrations in the Dutch original (edition of 1644). In appendix no. 2 I have added some photographs of the copied illustrations [APPENDIX NO. 2, ILL. A1, B1, C1, D1] as well as of the illustrations in the 1618 and 1644 editions. [APPENDIX NO. 2, ILL. A2, B2, C2, D2] The two added articles appear to be literal translations of

和蘭本草摘要解
外題
筭々仁宇須本草抄、
京都府立植物園・
大森記念文庫

ポルトガル樹の和解
加福喜蔵
加福安次郎

the original text. In the article *Porutogaru-ju no wage* ('Japanese Translation of the Olive Tree'), the name of the translator is mentioned as "Kafuku Kizô". Among the interpreters who collaborated with Noro Genjô on his *Oranda honzô wage* ('Dutch Botany Explained in Japanese'), figures the chief interpreter, Kafuku Kizô, who participated twice in the court journey to Edo in 1743 and 1747, but here his son is meant, Kafuku Yasujirô (1735-?), who became assistant interpreter (*ondertolk*) in 1788 and chief interpreter (*oppertolk*) in the following year. *Oranda honzô tekiyô-kai* is probably Matsumura's revision of his *Bansan shohin yakukô*, supplemented with Kafuku's draft translation. "Ghehoeckte Ciceren" was most likely copied from Kafuku's translation draught. It appears that the credit for the first faithful translation of the original *Cruydt-Boeck*, albeit fragmentary, goes to Kafuku Yasujirô. The beginning and the original text of *Porutogaru-ju no wage* are shown in appendix no. 3.

ドドネウス本草
草木名彙
アベセ類聚
伊藤圭介
泰西本草名疏
茵蔯
牛尾蒿
九牛草

Kafuku Yasujirô's illustrious predecessor Yoshio Kôgyû (1724-1800), who accompanied Yasujirô's father Kizô as assistant interpreter on the court journey of 1747 and collaborated with Noro Genjô, was promoted to chief interpreter in that year. He is the author of *Dodoneusu honzô sômoku mei-i* ('Lexicon of Plant Names Excerpted from Dodonæus' Herbal,' also called: *Abc Ruiju*, 'Alphabetical Enumeration'). The manuscript I examined in the Central Library of Nagoya University (Nagoya Daigaku) is Kôgyû's autograph. This manuscript was formerly in the possession of Itô Keisuke, who was a pupil of von Siebold's and authored *Taisei honzô meiso* ('The Nomenclature of the Western Herbal,' 1829), a commentary on C.P. Thunberg's *Flora Japonica* (Lipsiae [Leipzig], 1784). Kôgyû's manuscript simply lists the Dutch names of sixty-three varieties with their page numbers, together with the corresponding Japanese and Chinese names. Concerning the three mugwort varieties "groote averoone. 23. Chinese: Inchin [i.e. in *pinyin* 'Yinchen'], Japanese: Kawara-yomogi"; "kleine oft roomsche alssen. 27. Chinese: Gyûbikô [i.e. in *pinyin* 'Niuweihao'], Japanese: Hime-yomogi"; and "alssen met breede bladeren. 25. Japanese: Oka-yomogi, Chinese: Kyûgyûsô [i.e. in *pinyin* 'Jiuniucao']," a fragment of the original is copied and a translation of the terms attached (see appendix no. 4). Only in the case of "groote averoone" an annotated translation in *kanbun* (classical Chinese) is attached (compare appendix no. 2). It is evident that in order to match Japanese and Chinese equivalents with the Dutch terms, not only language ability but also sufficient botanical knowledge are required. Yoshio Kôgyû's grandson,

Yoshio Jôzan was, as will be stated below, the central figure in the enterprise of translating Dodonæus' herbal, known as *Ensei Dodoneusu sômoku-fu* ('The Western Herbal of Dodonæus Illustrated') ordered by Matsudaira Sadanobu. Speaking of the hardships his grandfather encountered in translating Dodonæus' *Cruydt-Boeck*, he writes:

吉雄常三 遠西ドドネウス 草木譜 松平定信

"Translating terms means translating pertinent Western terms in the relevant article and finding the matching Japanese and Chinese terms. As a rule, this is the most difficult task in the study of translation. This is because there are no fixed equivalents for the medicinal herbs used by earlier and contemporary authors and it is difficult to follow any of these. My grandfather, master Kôgyû deplored this, and late in his life he set about translating the great herbal written by the Western physician Dodonæus. He worked day and night, but because of the magnitude of this book, he could not finish the enterprise. Halfway through the work, in the fall of the twelfth year of the *Kansei* period (1800), he fell ill at his desk and passed away" (*Oranda sômoku-fu*, preface, 1833).

寛政

As he mentions in this quotation, when the aged Kôgyû set about translating Dodonæus, probably in the 1790's, he focussed his efforts on the identification of matching Japanese and Chinese names, rather than on the translation of the text.

It was Ono Ranzan (1729-1810), the greatest herbalist of the second half of the Edo period, who at the age of seventy-one was summoned from Kyoto to the *Igakukan* ('Institute for Medicine') of the Bakufu and embarked upon the translation in earnest. Ranzan wrote *Dodoniusu buppin-kô* ('Essay on the Identification of Plants in Dodonæus' Herbal') based on the 1618 edition of Dodonæus' *Cruydt-Boeck*. For 668 varieties he listed the Chinese and Japanese names and the page numbers of the original. I have examined a manuscript, which was written by Gunpô-en Chinjin (not identified) in the thirteenth year of *Bunka* (1816), formerly in the possession of Makino Tomitarô. I believe that this was a work in preparation of the translation ordered by Matsudaira Sadanobu. Ranzan consulted not only Dodonæus' *Cruydt-Boeck*, but also J.W. Weinmann's *Taalryk register der plaatofte figuur-beschryvingen der bloemdragende gewassen* (Amsterdam, 1736-1748). Although only partially, Ranzan's autograph manuscript with copies of some of Weinmann's colour illustrations is extant.

小野蘭山 医学館 檪度涅烏私物品考 群芳園陳人 文化、牧野富太郎

曽昌啓（占春） 薩摩藩 島津重豪 宗田一 海椰子 椰子 職方外記 職方外紀 艾儒略

Matsudaira Sadanobu not only ordered the cooperation of Ono Ranzan, but also of Sô Shôkei (Senshun), a herbalist of the Satsuma fief. It seems that the Weinmann consulted by Ranzan was in the possession of Shimazu Shigehide, lord of the Satsuma fief. Recently, the former collection of the great historian of medicine, the late Sôda Hajime, was donated to the International Research Center for Japanese Studies (Kyoto). In this collection, there is a fragment of a letter in Ranzan's autograph. It is a note, written in a fluent and beautiful hand, about the fruit of a tree washed ashore: "The fruit of a tree washed ashore. It is a kind of sea coconut (*umi-yashio*), one that is also listed in the article on "coconut" (*yashi*) in the Dutch Herbal. Ranzan." The *Dodoniusu buppinkô* mentions: "Sea coconut". In *Shokuhô gaiki* it is said that there is a tree that grows in the sea and that its fruit, which is tiny, cures all kinds of diseases. p. 1409." From this passage we can infer that he consulted Giulio Aleni's *Zhifang waiji* (Jap.: *Shokuhô gaiki*, 'Chronicle of Foreign Countries') and Dodonæus' *Cruydt-Boeck*. Giulio Aleni (Chinese name Ai Rulue) was a missionary active in China in the beginning of the seventeenth century. *Zhifang waiji* (*Shokuhô gaiki*) was widely read from the second half of the eighteenth century

小野蘭山自筆

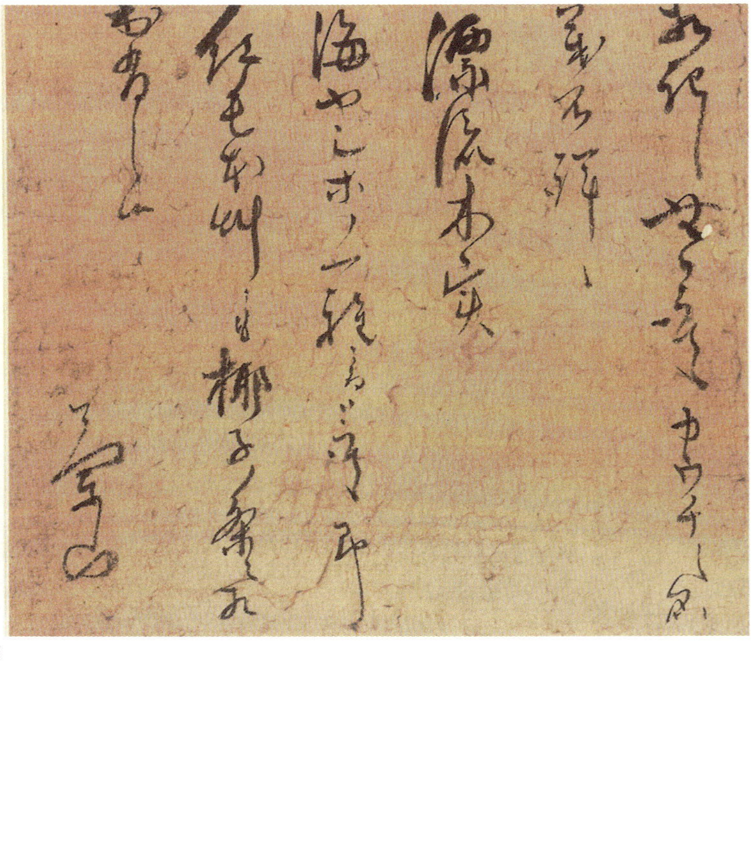

ILL 6

until the end of the Bakufu period as a source of information on foreign countries. Appendix no. 5 reproduces the illustration on page 1409 of the 1644 edition of Dodonæus' *Cruydt-Boeck* (*Rechte afbeeldinghe van het Tweede Indiaensch Notken*).

After Matsudaira Sadanobu had resigned as *Rôjû shuza* (chief councillor) in the fifth year of *Kansei* (1793), he ordered the complete translation of Dodonæus' *Cruydt-Boeck*. It was completed around the year 1823 under the title 'Illustrated Herbal of the Westerner Dodonæus' and in the twelfth year of *Bunsei* (1829) plans were made to publish it, but due to Sadanobu's death in that year and the loss of the woodblocks in a fire, the publication was never realised. The original *Cruydt-Boeck* (edition of 1618), the translation draft and a part of the printing proofs are preserved in the Library of Waseda University. The publishing department of Waseda University has recently released a facsimile edition, which facilitates the use of the manuscript.

老中首座

文政

According to the notice at the beginning of each chapter in *Ensei Dodoneusu sômoku-fu*, the "original translation" by the erstwhile interpreter Ishii Shôsuke (Tôkô) was furnished with a "supplementary translation" by Haguri Hi from Nagasaki, "revised" by Arai Kôjun of Aizu and "corrected and finalised" by the Edo herbalist, Yoshida Seikyô. Ishii Shôsuke completed the *Edo Haruma* ('Edo Halma') in the eighth year of Kansei (1796), but judging from the level of this first Dutch-Japanese dictionary, it seems that his proficiency in Dutch was not outstanding. Tanoi Genshin's, *Ijôsho hôtei ka'an* ('Draught for the Presentation of the Bequeathed Book,' fourteenth year of *Tenpô*, 1843), a text attached to the manuscript in Waseda University, relates the details of the translation and planned publication, but strangely enough, it fails to mention Haguri Hi, i.e., Yoshio Kôgyû's grandson, the *Rangakusha* Yoshio Jôzan (1787-1843). According to Tanoi, the translation of Ishii was confined to the main text, but Yoshida mustering his knowledge of botany and his proficiency in Dutch translated "the supplement of Clusius" and the preface of the original text. However, the heading of the first chapter clearly mentions: "Translated by Haguri Hi." Yoshida explains that the *Kurusiusu no furoku* ('Supplement of Clusius') contains new information added to the old articles of Dodonæus, but this so-called "supplement" is probably a mistake for the supplement (*anhangsel*) that was appended by the translator of the Dutch edition, Joost van Ravelingen. In my hypothesis, it was precisely Yoshio Jôzan who not only taught Dutch to Yoshida but

石井庄助（当光）
羽栗費、荒井行順
会津、吉田正恭
江戸ハルマ

田井元陳、
遺上書奉呈下案
天保

クルシウスの附録

also corrected and supplemented Ishii's "original translation" and basically was the driving force behind completion of *Ensei Dodoneusu sômoku-fu*. Indeed, the draught translation which can be viewed as the "original translation" by Ishii is still extant.

杏雨書屋
和蘭本草翻訳
田藩文庫
田安府芸堂
品類、形状、出処
時候、名義
性功主治、附言
ヱノキクサ一種
鏨菜キセワタイ

The *Kyôu sho'oku* in Osaka owns a draft translation in three fascicles, entitled *Oranda honzô honyaku* ('Translation of the Dutch Herbal'), which corresponds to pp. 47-191 and 271-360 of Dodonæus'original. It contains precise copies of the illustrations of the original for about two thirds of the articles. Every article is a translation of the main text only. Ravelingen's supplement is not translated. Moreover, the manuscript bears the seals of *Denhan bunko* and *Tayasu fugeidô*, which indicates that it was in the possession of the Tayasu family, Matsudaira Sadanobu's parental family. Let us compare one article in this manuscript with the corresponding article in the manuscript of Waseda University (see appendix no. 6). This is the translation of "Het VII. Capitel. Van de twee soorten van Molucca oft Vremde Melissen," taken from pages 125-126 in the original. In the manuscript of Waseda, the supplement is also translated and the translation as a whole is more accurate. This translation gives the following subtitles: "Hinrui" (category), "Keijô" (shape), "Shutsujo" (place of origin), "Jikô" (flowering season), "Meigi" (name), "Seikô-shuchi" (potency and workings), "Fugen" (supplement). The proofs of these articles are not extant, but they are closer to the format of other existing proofs. The notes "enokikusa isshu" (a kind of hackberry) and "sansai kisewatai" in the margin are identifications added by Ono Ranzan. From the above characteristics, we have to conclude that there is an extremely high probability that the manuscript in Waseda University is to be attributed to Yoshio Jôzan.

尾張

The above-mentioned *Oranda sômoku-fu* by Yoshio Jôzan is a complete and precise translation of P. Nylandt's *De Nederlandse herbarius of Kruydt-boeck* (Amsterdam, 1670), which may be called an abridged and revised edition of Dodonæus' *Cruydt-Boeck*, but Jôzan also added the Linnaean classification for each plant, which he culled from his friend Itô Keisuke's *Taisei honzô meiso*, and for the corresponding Japanese and Chinese names, he consulted his grandfather Kôgyû's posthumous manuscripts and sought confirmation from Itô. Yoshio Jôzan, after laying the groundwork for the translation of Dodonæus' *Cruydt-Boeck* in Edo, engaged in research in the field of Dutch studies together with Itô Keisuke in Owari. He certainly deserves a reappraisal in the future.

Finally, we have to touch upon another translation of Dodonæus'

Cruydt-Boeck, which was carried out in approximately the same period as the translation in Edo. This translation has gone practically unnoticed so far. As is well known, the Rangakusha from Osaka, Hashimoto Sôkichi is the author of *Seiyô iji shûsei hôkan* ("Treasure Book of Western Medical Matters," author's preface dated 1821, introductory notice dated 1819). This is a complete translation of Woyt's *Gazophylacium*, a book which the interpreters in Nagasaki had translated partially to gather information on the imported medicines. Woyt's book is a medical dictionary in alphabetical order, listing the Latin names of the medicine as entry, followed by the Dutch name, description of the recipe and the effects of the medicine in Dutch. Sôkichi divides his integral translation in two sections "Honzô-hen" (botany, 20 chapters) and "Chibyô-hen" (treatment of diseases, 15 chapters). The section "Honzô-hen" follows the traditional classification of Li Shizhen's *Bencao gangmu* (*Honzô kômoku*). Shibaoka Gizen, Sôkichi's disciple, was in charge of the compilation of the second half of this botanic section. According to his testimony, entitled *Honzô-bu kôhen fugen* ('Preface to the Second Part of Botany'), his master Sôkichi translated Dodonæus' *Cruydt-Boeck* in order to know the shape of the medicinal herbs listed in Woyt's original. He also writes that Sôkichi only translated the sorts (Sôron, Geslachte) and shapes (Keijô, Ghedaente) in each chapter, and omitted the other items, i.e., place of origin (Plaetse), flowering season (Tijdt), name (Naemen), dispositions (Aerd) and potency and workings (Kracht ende Werckinghe), as well as the supplement (Furoku shohin, Biivoeghsel), added at the end of each chapter by the editor of the revised edition, J. van Ravelingen. In total, this translation covered more than thirty chapters. He also translated Clusius' supplement "Beschriivinghe van de Indiaensche oft VVtlandische Boomen, Heesteren ende Cruyden" as "Shohan no ishu-hen" ('Lexicon of the Various Sorts in Barbarian Countries') in "20 chapters." This translation draught, counting more than 50 chapters in total, is unfortunately not extant. It is noteworthy that this translation was made with the intellectual framework of Li Shizhen's *Bencao gangmu*.

橋本宗吉
西洋医事集成宝函
本草篇
治病篇
柴岡宜全
本草部後編附言
総論、形状
附録諸品
諸藩ノ異種篇

In 1823, the year of the completion of the translation of Dodonæus' *Cruydt-Boeck* in Edo, von Siebold arrived in Nagasaki. The books he brought with him, notably Thunberg's *Flora Japonica*, M. Houttuyn's *Natuurlyke historie volgens het samenstel van den Heer Linnaeus* (Amsterdam, 1761-1785) and C[arl] L[udwig] Willdenow's *Handleiding tot de kennis der planten* (Amsterdam, 1819), had a great influence on his disciples in Nagasaki, particularly Itô Keisuke from

賀来佐之、豊後 桂川甫賢 山本亡羊

Owari and Kaku Sukeyuki from Bungo. Itô Keisuke wrote *Taisei honzô meiso* and Kaku Sukeyuki translated Willdenow's modern textbook on botany. For them, Dodonæus' *Cruydt-Boeck* already belonged to the past. Also the Bakufu court physician, Katsuragawa Hoken, who put at the head of *Ensei Dodoneusu sômoku-fu* his translation of the biography of Dodonæus into classical Chinese from a chapter of Willdenow's textbook on the history of botany, was actually familiar with the botany of contemporary Europe. But, on the other hand, Dodonæus continued to hold the authority among the traditional herbalists, in the same way as Li Shizhen's *Bencao gangmu* did. Yamamoto Bôyô, a herbalist living at the end of the Bakufu period, under whom Kaku Sukeyuki studied in his youth, treasured the first Dutch edition (1554) of Dodonæus' herbal. This book is still in the possession of his descendants today. I add some reproductions from that copy in appendix no. 7.

ILLUSTRATIONS

ILL 1 Gilt-edged copy of Dodonæus' *Cruydt-Boeck* in possession of the Medical Faculty of Kanazawa University.

ILL 2 Half title page of Dodoens' *Cruydt-Boeck* bearing the ownership seal of the erstwhile Kanazawa Medical College. Medical Faculty of Kanazawa University.

ILL 3 Gilded leather cover, decorated with arabesque engravings of Dodonæus' *Cruydt-Boeck* in the possession of the Medical Faculty of Kanazawa University.

ILL 4 Page 1122 of the *Cruydt-Boeck* (1644), reproduced from the copy owned by the Medical Faculty of Kanazawa University.

ILL 5 Page 1124 of the *Cruydt-Boeck* (1644), reproduced from the copy owned by the Medical Faculty of Kanazawa University.

ILL 6 Autograph of Ono Ranzan, in the collection of the International Research Center for Japanese Studies, Kyoto.

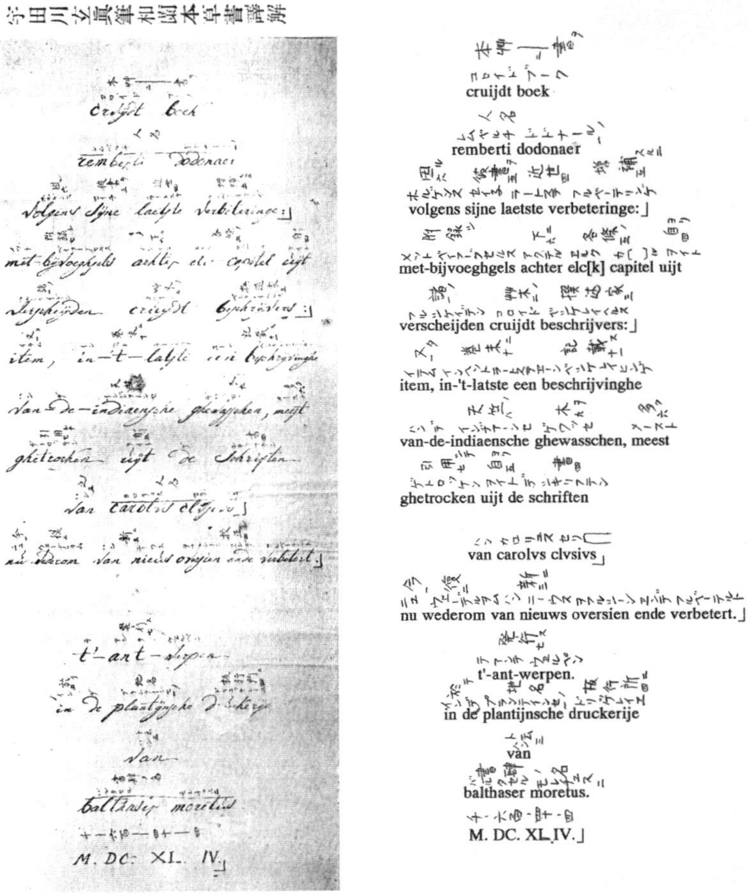

Translation of the title-page of Dodonæus' *Cruydt-Boeck* by Udagawa Genshin.

ILL A1
Illustration from *Oranda honzô tekiyô-kai*. Kyoto Prefectural Botanical Garden, Ômori Collection.

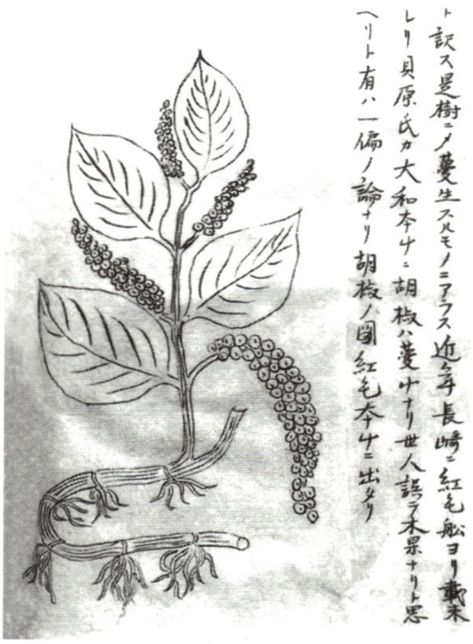

ILL A2
Page 1440 of the *Cruydt-Boeck* (1644), reproduced from the copy owned by the Medical Faculty of Kanazawa University.

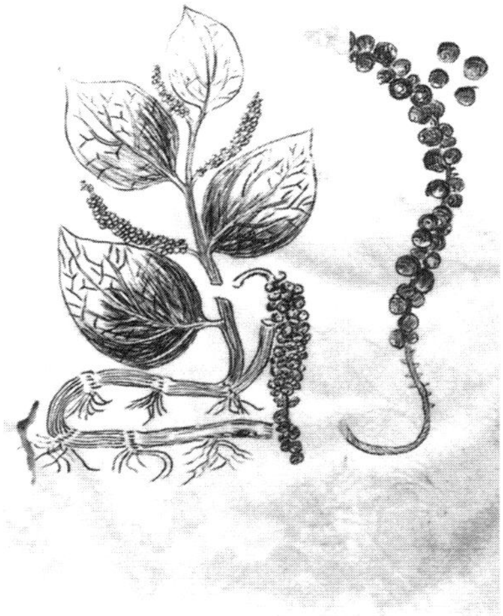

ILL B1
Illustration from *Oranda honzô tekiyô-kai*. Kyoto Prefectural Botanical Garden, Ômori Collection.

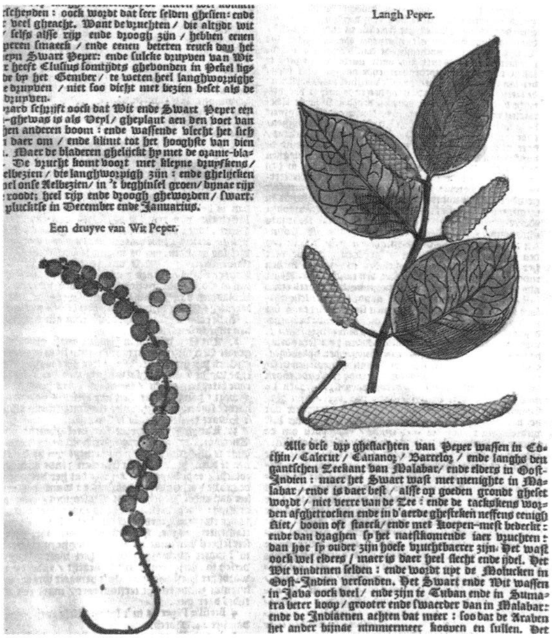

...schenden : oock wordt dat seer selden ghesien: ende
... veel gheacht. Want de vruchten / die altijdt wit
... selfs alsse rijp ende droogh zijn / hebben eenen
...peren smaeck / ende eenen beteren reuck dan het
...eyn Swart Peper: ende sulcke druyven van Wit
... heeft Clusius somtijdts ghevonden in Pekel ligs
...de by het Gember / te weten heel langhworpighe
...e druyven / niet soo dicht met bezien beset als de
... druyven.
...ard schrijft oock dat Wit ende Swart Peper een
...-ghewas is als Veyl / gheplant aen den voet van
...en anderen boom : ende wassende vlecht het sich
... daer om / ende klimt tot het hooghste van dien
... Maer de bladeren ghelijckt hy met de oranie-bla=
... De vrucht komt voort met kleyne druyfkens /
...elbezien / die langhworpigh zijn : ende ghelijcken
...el onse Aelbezien / in 't beghinsel groen / bynae rijp
... roodt; heel rijp ende droogh gheworden / swart:
... pluckt se in December ende Januarius.

Een druyve van Wit Peper.

Langh Peper.

Alle dese drij gheslachten van Peper wassen in Co=
chin / Calecut / Cananor / Barcelor / ende langhs den
gantschen Zeekant van Malabar / ende elders in Oost-
Indien : maer het Swart wast met menighte in Ma=
labar / ende is daer best / alsse op goeden grondt gheset
wordt / niet verre van de Zee : ende de tacxkens wor=
den afghetrocken ende in d'aerde ghesteken neffens eenigh
kiet / boom oft staeck / ende met Koeyen-mest bedeckt :
ende dan draghen sy het naestkomende iaer vruchten :
dan hoe sy ouder zijn hoe se vruchtbaerder zijn. Het wast
oock wel elders / maer is daer heel slecht ende ydel. Het
Wit vindtmen selden : ende wordt uyt de Molucken in
Oost-Indien versonden. Het Swart ende Wit wassen
in Java oock veel / ende zijn te Tuban ende in Suma=
tra beter koop / grooter ende swaerder dan in Malabar:
ende de Indiaenen achten dat meer : soo dat de Araben
het ander bijnae nimmermeer koopen en sullen. Het

ILL B2
Page 1441 of the *Cruydt-Boeck* (1644), reproduced from the copy owned by the Medical Faculty of Kanazawa University.

ILL C1
Illustration from *Oranda honzô tekiyô-kai*. Kyoto Prefectural Botanical Garden, Ômori Collection.

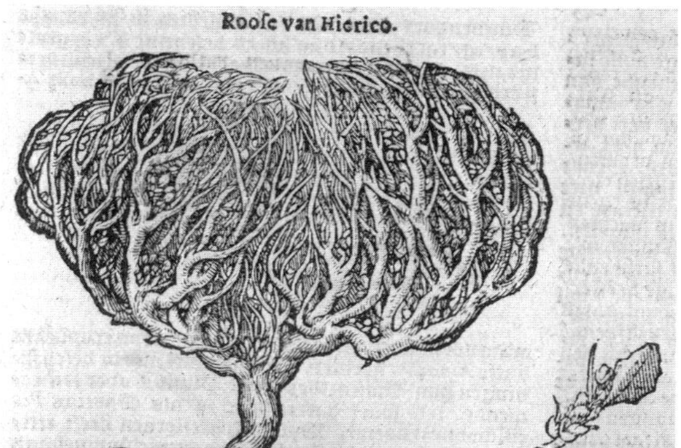

ILL C2
Roose van Hierico, *Cryudt-Boeck*, 1618 edition, p. 1447. Collection Katholieke Universiteit Leuven, Central Library, Tabularium, B 3998.

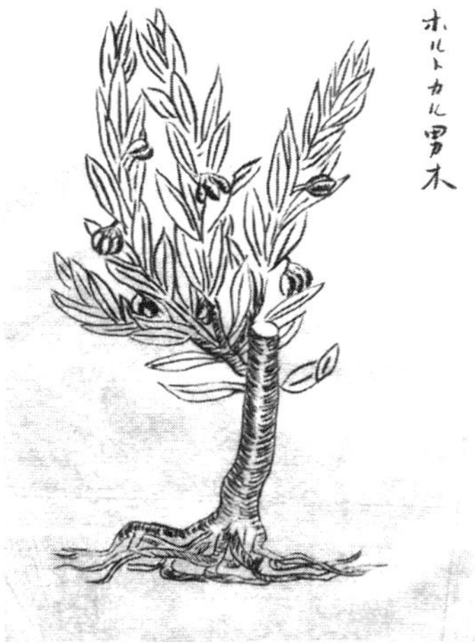

ILL D1
Illustration from *Oranda honzô tekiyô-kai*. Kyoto Prefectural Botanical Garden, Ômori Collection.

ILL D2
Olijfboom (olive-tree), *Cruydt-Boeck*, 1644 edition, p. 1287.
Collection Katholieke Universiteit Leuven, Central Library, Tabularium C1827

ILL D3
Olijftak met de bloemen ende vruchten (olive-twig with blossoms and fruits), *Cruydt-Boeck*, 1644 edition, p. 127. Collection Katholieke Universiteit Leuven, Central Library, Tabularium C1827

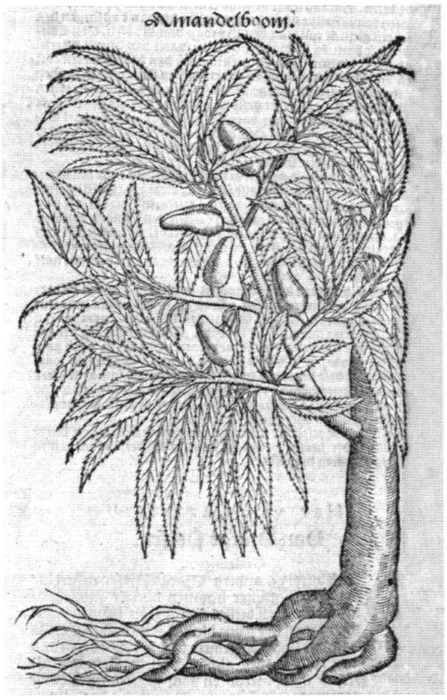

ILL E
Amandelboom (almond-tree), *Cryudt-Boeck*, 1618 edition, p. 1250. Collection Katholieke Universiteit Leuven, Central Library, Tabularium, B 3998.

ILL F
Cassie ofte Cassia fistula, *Cryudt-Boeck*, 1618 edition, p. 1233. Collection Katholieke Universiteit Leuven, Central Library, Tabularium, B 3998.

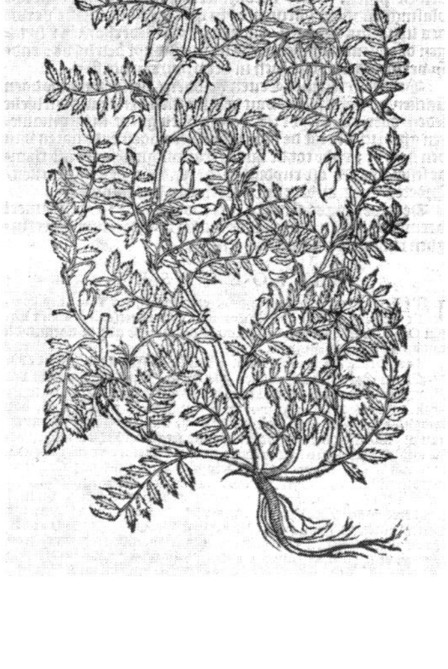

ILL G
Ghehoeckte of kantighe Ciceren, *Cryudt-Boeck*, 1644 edition, p. 842. Collection Katholieke Universiteit Leuven, Central Library, Tabularium c 1827.

阿蘭陀本ナ之ホルトカル樹之和解
ヲレイヒボーム　但阿蘭陀詞
千二百八十七枚目之紙数
拾八冊目之內　二十九番目
一ボルトガル樹女男有之男樹ハ実ヲ結ヒ大木ニ成ル枝モ大ニ繁
リ葉ノ形長ク堅ク白ミ有花モ白ミアリ花ハ葡萄之花ノ形ニメ
実ハ丸ク長ク初メハ青色熟ノ黒ミニナル実内ニ油ヲ貯ヘ核ハ堅
ク形ハ実ノ形リ應テアリ
一ボルトカルノ女樹ハ木ノ形チ男樹ヨリハ小ク枝モ細長ニ生立千葉
モ男樹ヨリ細ク実モ男樹ヨリ細ク繁ク結ヒ然モ肉薄ク油モ少ク有
之イタリヤ国フランス国スパンヤ国其外之嶋々ニモ有之多クハ海濱
ニ出生ス　或説ニ海濱遠有ル処ニハ生立スト実ヲ結ヒ時節阿蘭陀国
之九月比ヨリ十一月十二月ニ熟シ其節ニ採少日乾シ実ニ皺立節
水ヲ少シ打テ油船ニ入レシメ取ル其后水ノ上ニ浮ミ油ヲスクヒ採最
右之実ヲ食物ニスル節ハ未熟之節採リ醋漬等ニス
一ポルトカル樹之名
アレイヤ　〇ヲレヤサトア　〇ヲレヤユルバ子
ヲレイビホーム ホルトカルノ樹 阿蘭陀詞　〇シルバアユム

Description of the olive-tree (olijfboom): *Porutogaru-ju no wage*.
Illustration from *Oranda honzô tekiyô-kai.*
Kyoto Prefectural Botanical Garden, Ômori Collection.

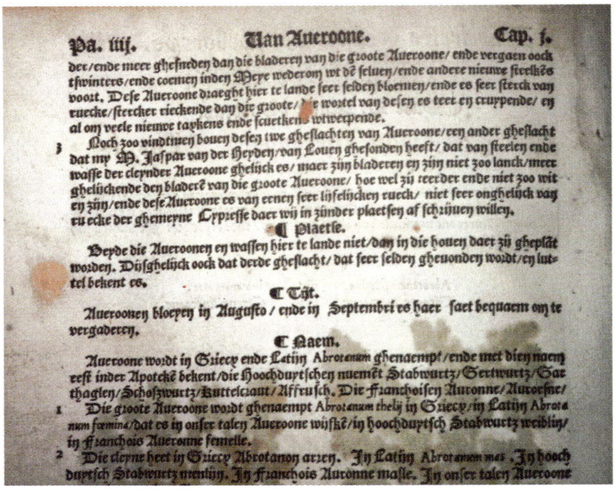

Pa. iiij. Van Averoone. Cap. j.
der/ende meer ghesneden dan die bladeren van die groote Averoone/ ende vergaen oock
tswinters/ende coemen inden Meye wederom wt dē seluen/ende andere nieuwe steelkēs
voort. Dese Averoone draeght hier te lande seer selden bloemen/ende es seer sterck van
rueche/stercker rieckende dan die groote/ die wortel van desen es teer en cruypende/ en
al om veele nieuwe taykens ende scuetkens wtwerpende.
3 Noch zoo vindtmen bouen desen twe gheslachten van Averoone/een ander gheslacht
dat my M. Jaspar van der Heyden/van Louen ghesonden heeft/ dat van steelen ende
wasse der cleynder Averoone ghelijck es/ maer zijn bladeren en zijn niet zoo lanck/meer
ghelijckende den bladerē van die groote Averoone/ hoe wel zij teerder ende niet zoo wit
en zijn/ende dese Averoone es van eenen seer lijfelijcken rueck/ niet seer onghelijck van
ru ecke der ghemeyne Cypresse daer wij in zijnder plaetsen af schrijuen willen.

Plaetse.

Beyde die Averoonen en wassen hier te lande niet/dan in die houen daer zij gheplāt
worden. Dijsgheliijck oock dat derde gheslacht/ dat seer selden gheuonden wordt/en lut-
tel bekent es.

Tijt.

Averoonen bloeyen in Augusto / ende in Septembri es haer saet bequaem om te
vergaderen.

Naem.

Averoone wordt in Grieck ende Latijn Abrotanum ghenaempt/ende met dien naem
eest inder Apotekē bekent/die Hoochduytschen nuemēt Stabwurtz/Gertwurtz/Gar
thaglen/Schoszwurtz/Kuttelcraut/Affrusch. Die Franchoisen Auronne/Auroesne/
1 Die groote Averoone wordt ghenaempt Abrotanum thelij in Grieck/in Latijn Abrota
num fœmina/dat es in onser talen Averoone wijfkē/in hoochduytsch Stabwurtz weiblin/
in Franchois Averonne femelle.
2 Die cleyne heet in Grieck Abrotanon arren. In Latijn Abrotanum mas. In hooch
duytsch Stabwurtz menlijn. In Franchois Auronne masle. In onser talen Averoone

Page with description of Averoone in the first Dutch edition of Dodoens's *Cruijdeboeck* (1554), reproduced from the personal copy of Yamamoto Bôyô, presently in the possession of Yamamoto dokushoshitsu, Kyoto.

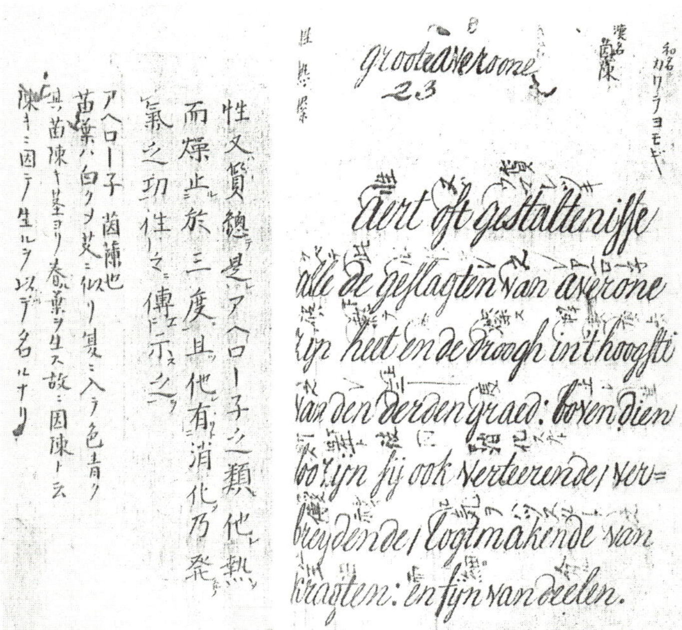

漢名 茵陳

和名 カワラヨモギ

Groote Averoone
23

Aert oft gestaltenisse
alle de geslagten van averone
zijn heet en de droogh in thoogsti
van den derden graed: boven dien
zoo zijn sij ook verteerende / ver=
breydende / logtmakende van
kragten: en fijn van deelen.

性又質總是アヘロー子之類他熱
而燥止於三度且他有消化乃發
氣之功性傳示之
アヘロー子 茵蔯也
茵蔯ハ白クシテ艾ニ似リ夏ニ入テ色青ク
其茵蔯ハ茎ヨリ春ニ葉ヲ生ス故ニ因陳トス
陳キニ因テ生ルヲ以テ名ルナリ

Fragment dealing with "Groote Averoone" from Yoshio Kôgyû's manuscript *Dodoneusu honzô sômoku mei-i*. Central library of Nagoya University.

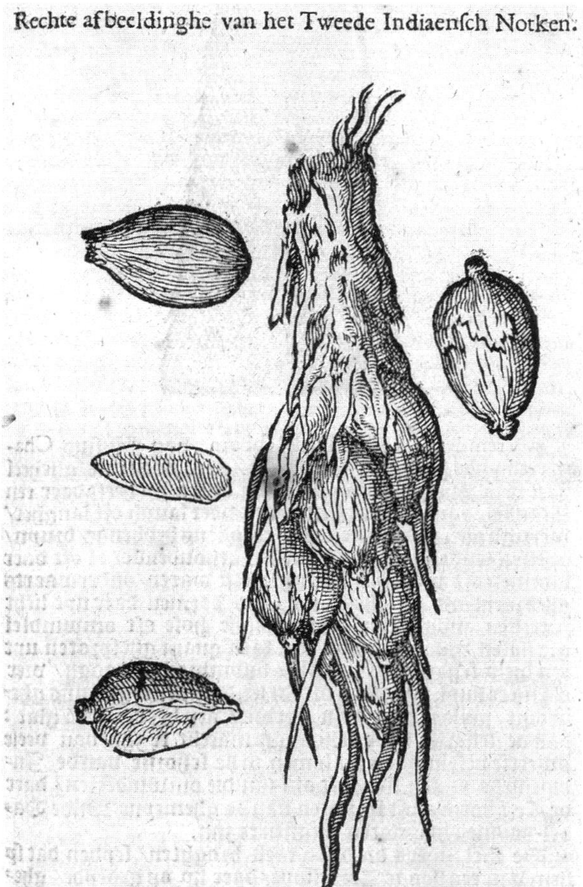

Tweede Indiaansch Notken, *Cruydt-Boeck*, 1644 edition, p. 1409. Collection Katholieke Universiteit Leuven, Central Library, Tabularium C1827

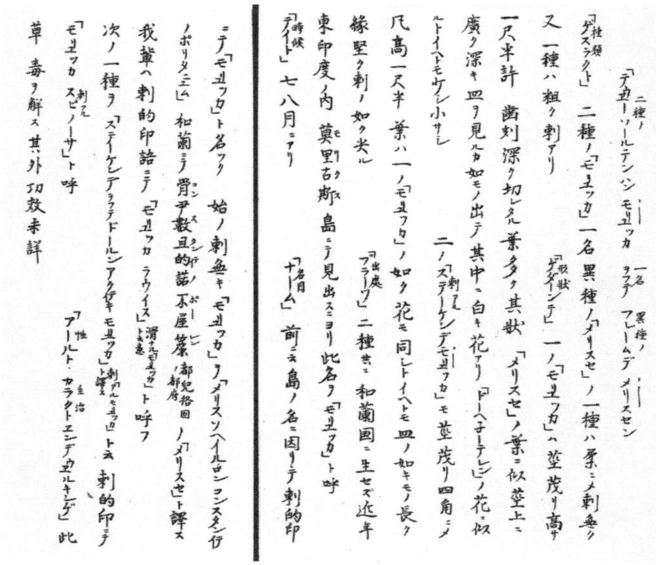

Fragment from *Oranda honzô honyaku*. Kyôu sho'oku, Osaka.

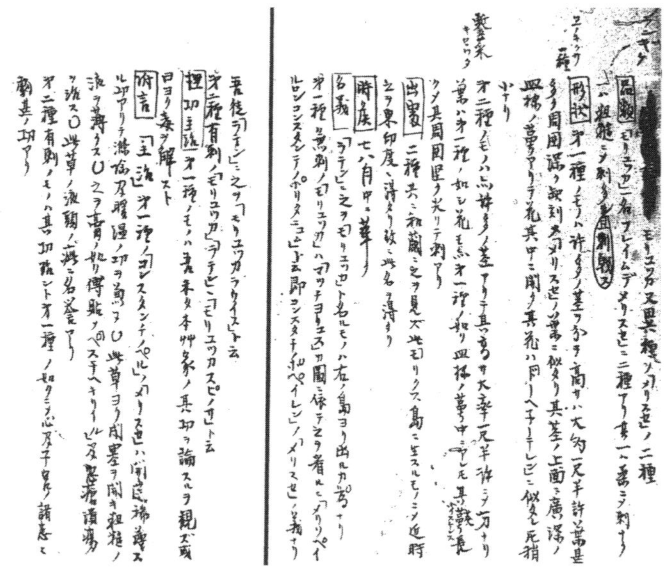

Fragment from *Ensei dodoneusu sômoku-fu*. Library of Waseda University.

Cruijdeboeck.

In den welcken die gheheele historie/dat es Tgheslacht/tfatsoen/naem/natuere/cracht ende werckinghe/ van den Cruyden/ niet alleen hier te lande wassende / maer oock van den anderen vremden in der Medecijnen oorboorlijck/met grooter neersticheyt begrepen ende verclaert es/met der seluer Cruyden natuerlick naer dat leuen conterfeytsel daer by ghestelt.

DEr hoochgheborene ende alder doorluchtichste Coninghinne ende Vrouwe/Vrouw Marien Coninghinne Douaigiere van Hungheren/ende Bohemen 2c. Regente ende Gouuernante van des K. M. Neerlanden/toegheschreuen.

Duer D. Rembert Dodoens/Medecijn van der stadt van Mechelen.

Title page of the first edition of Dodoens's Cruijdeboeck (1554), reproduced from the copy owned by Yamamoto Bôyô, presently in the possession of Yamamoto dokushoshitsu, Kyoto.

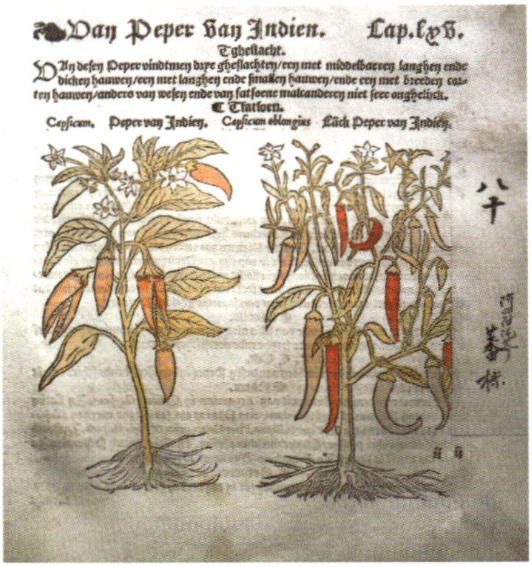

Page from the first Dutch edition of Dodoens's *Cruijdeboeck* (1554), reproduced from the personal copy of Yamamoto Bôyô, presently in the possession of Yamamoto dokushoshitsu, Kyoto.

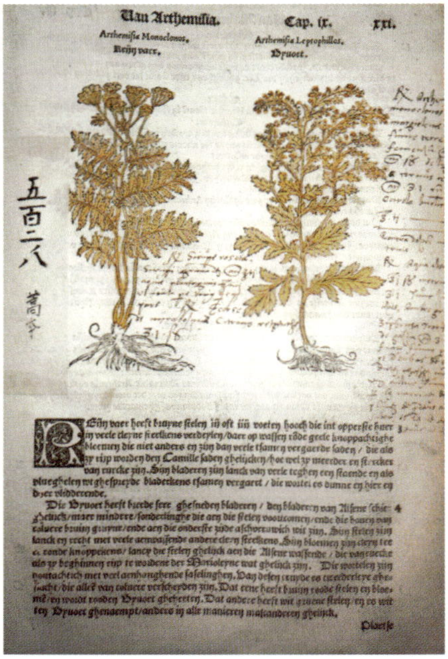

Page from the first Dutch edition of Dodoens's *Cruijdeboeck* (1554), reproduced from the personal copy of Yamamoto Bôyô, presently in the possession of Yamamoto dokushoshitsu, Kyoto.

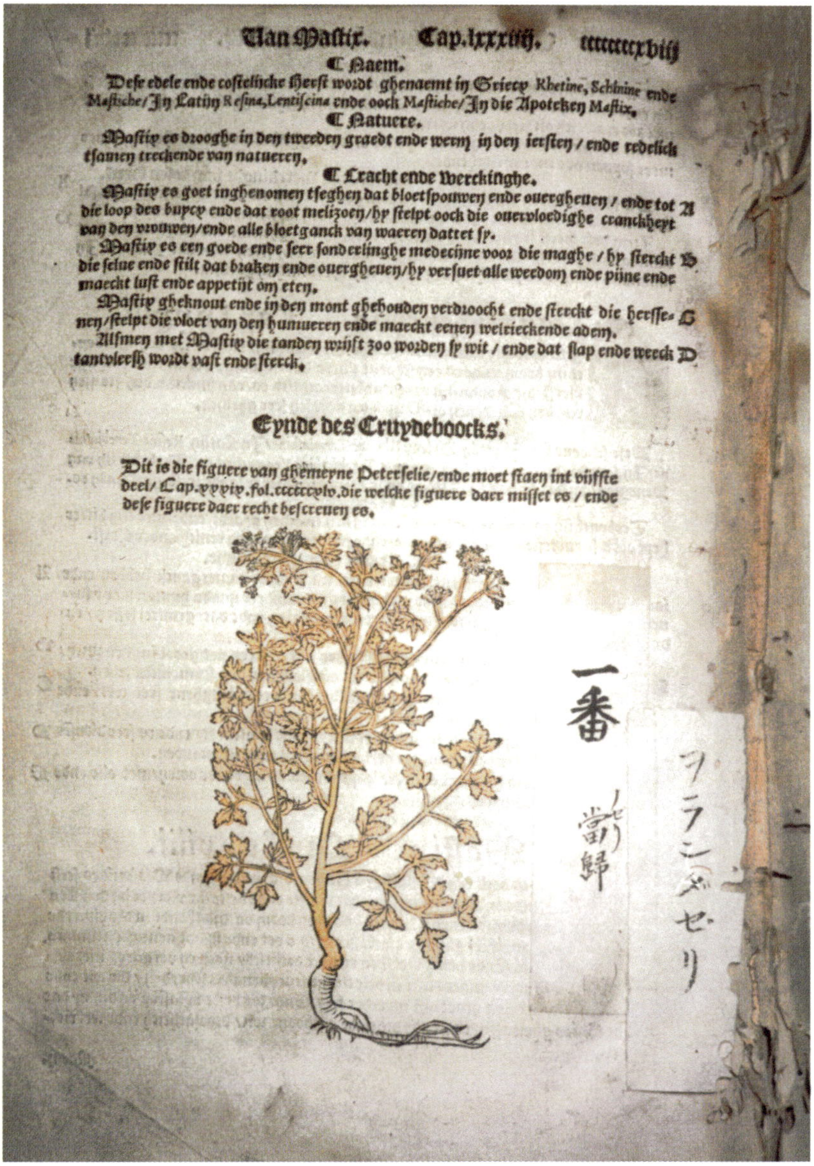

Van Mastix. Cap.lxxxiiij. ccccccxviij

Naem.

Dese edele ende costelicke herst wordt ghenaemt in Griecx Rhetine, Schinine ende Mastiche/In Latijn Resina, Lentiscina ende oock Mastiche/In die Apotecken Mastix.

Natuere.

Mastix es drooghe in den tweeden graedt ende werm in den iersten / ende redelick tsamen treckende van natueren.

Cracht ende Werckinghe.

Mastix es goet inghenomen tseghen dat bloetspouwen ende overgheuen / ende tot die loop des buycx ende dat root melizoen/hy stelpt oock die overvloedighe cranckheyt van den vrouwen/ende alle bloetganck van waeren dattet sy. A

Mastix es een goede ende seer sonderlinghe medecijne voor die maghe / hy sterckt die selue ende stilt dat braken ende overgheuen/hy versuet alle weedom ende pijne ende maeckt lust ende appetijt om eten. B

Mastix gheknout ende in den mont ghehouden verdroocht ende sterckt die herssenen/stelpt die vloet van den humueren ende maeckt eenen welrieckende adem. C

Alsmen met Mastix die tanden wrijft zoo worden sy wit / ende dat slap ende weeck tantvleesch wordt vast ende sterck. D

Eynde des Cruydeboecks.

Dit is die figuere van ghemeyne Peterselie/ende moet staen int vijffte deel/ Cap.xxxix.fol.cccccxlv.die welcke figuere daer misset es / ende dese figuere daer recht bescreuen es.

Page from the first Dutch edition of Dodoens's *Cruijdeboeck* (1554), reproduced from the personal copy of Yamamoto Bôyô, presently in the possession of Yamamoto dokushoshitsu, Kyoto.

x

THE VISUAL LEGACY OF DODONÆUS IN BOTANICAL AND HUMAN CATEGORISATION

Timon Screech
SOAS & Sainsbury Institute, London

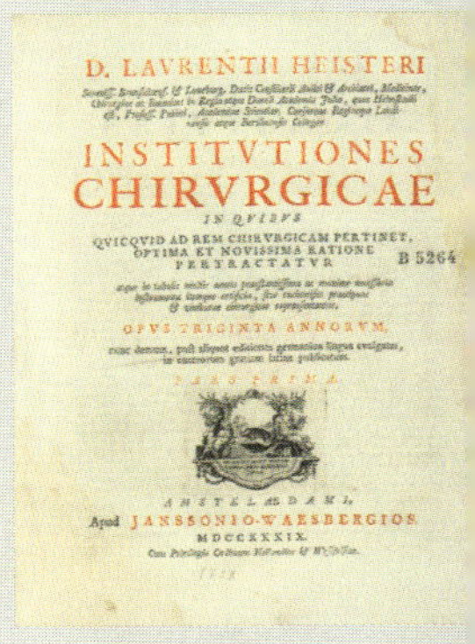
D. LAVRENTII HEISTERI

INSTITVTIONES
CHIRVRGICAE
IN QVIBVS
QVICQVID AD REM CHIRVRGICAM PERTINET,
OPTIMA ET NOVISSIMA RATIONE
PERTRACTATVR

OPVS TRIGINTA ANNORVM,

AMSTELAEDAMI,
Apud JANSSONIO-WAESBERGIOS
MDCCXXXIX.

This collection of essays establishes, I think, that the name of Dodoens was famous in eighteenth-century Japan. Other essays deal succinctly with the impact of his work in the fields of experimental science, and with how his important book altered Edo period (1600-1868) perceptions of the flora of the world, of its make-up, and generally with scientific categorisations. It is the smaller task here to pursue a different, twin line of inquiry. Firstly, I shall look at the reactions to Dodoens among those interested in his pictures, as pictures. He offered advances in plant scholarship, but his data was largely put across visually, and he demonstrated how perceived truths about the world might be put down pictorially, offering concomitantly a style in which to do so. The first issue, then, has to do with his and his illustrators' achievements within a certain order and technology of pictorial production – within *representation*. The second, somewhat different line, grows out of this: for all the cogency of Dodoens's pictures, his illustrations looked, to Edo period viewers, *foreign*; this was inescapable. The extent to which these pictures did, or could, match up to local expectations, and even local empirically-based studies, was moot. Here we will look at how foreignness entwined with plant categorisation. It was foreigners who brought the new knowledge, and so, imbricated in botanical questions were matters relating to the categorisation of humanity too.

To begin at the beginning, we may note that, following the continental Northeast Asian lead, Japan has a tradition of plant study that went back many centuries. European thinking began to infiltrate this from the end of the sixteenth century, although it was not to be until the eighteenth that the most serious effects of international influence were felt. As for Dodoens, the earliest reports of import of his book can be found in the seventeenth century, but it was during the eighteenth that its (and hence his) status rose highest. As late as 1795, for example,

we learn that the book had attained a name-recognition value whereby it was being used as a basis for Dutch language study, and being read quite apart from any interest or otherwise in its subject-matter. The matter of Dodoens's name as a mantra is something to which we will return. We read in the *Dagh-register* (*Deshima diary*) of the VOC captain (opperhoofd), Gijsbert Hemmij, however, that in the spring of 1795 some of the Nagasaki interpreters were given three months to familiarise themselves with Dodoens's book, and were then made to answer questions on it, as part of their proficiency examinations. According to Hemmij, this was a public event, unlike the normal in-house paedagogies, for 'the city magistrate (*machi-bugyô*) himself 町奉行
examined some interpreters and apprentices about some parts of Dodonæus.' This particular governor was an amateur student of Dutch, and if whether he used Dodoens or not is unsure, he seemed, to Hemmij 'to be making some progress'.[1]

This was fully 241 years after the first publication of the *Cruijdeboeck* in Antwerp, which occurred in 1554. An enlarged and revised edition was published in Antwerp in 1563. This was subsequently reissued in Leiden in 1608, in which and subsequent editions it came to Japan. It is important not to over-stress the significance of this time-lag, however, for although Dodoens may have lost most of its currency in cutting-edge European academic circles over time, it continued to be used by village-based pharmacists in the Low Countries right into the twentieth century, for the admirable clarity of his explanations. In Japan, Dodoens continued to figure in his role as something of a cypher throughout the Edo period. Dodoens and his book were not just to learn from, but were icons of what it meant to study in the European way, that is, to engage in the discipline that came to be called *Rangaku* 蘭学
('Dutch Studies'); this was only peripherally linked to botany per se. Those who considered themselves '*Rangaku* specialists' (*Rangakusha*) 蘭学者
would try by all means to own a copy, or at least to thumb through one. Similarly, those who wished to lampoon *Rangaku*, might sometimes do it via a jibe at Dodoens. Few would have read the *Cruijdeboeck*, but many would have seen it, known it by vague report, or had access to a torn-out page or copy of a picture. It is worth recalling how decimated imported books might become as they went from hand to hand. Leaves became detached, lost, or were surreptitiously torn out and never returned, or were removed for giving as gifts. This may have hindered scholarly study, but since few were engaged in *Rangaku* anyway, this was not too critical. The destruction of books in

fact enhanced their reputations and permitted greater circulation of the parts. When in 1724, David Drinkman was shown a Dutch edition of Ambroise Paré's famous *de Chirurgie, ende Opera* ... of 1649, it was thoroughly stained and of the 900 pages of the original, only 128 remained.[2] And this was the Shogun's own copy, which one might have expected to be kept most carefully – on the other hand, the Shogunal house was under some of the heaviest obligations of gift-giving. Pages of text made little sense, so it was the pictures that became the chief medium of exchange. When the Japanese-speaking leader of the Dutch オランダ商館 Factory (*Oranda shôkan*), Isaac Titsingh found to his fury that a fine illustrated book had been stolen from his rooms in Nagasaki early in 1781, he was nothing mollified when it was returned, as 'all the pictures [had] been torn out';[3] the thief removed the illustrations, but had no use for the text.

The book, as object, and generally as an object containing pictures, was the public hallmark of the scholar, and especially, the Western book – so differently bound and so other in appearance – was the 司馬江漢 symbol of the *Rangakusha*. When the Westernist artist Shiba Kôkan 吉雄幸作 painted a rough image of the Nagasaki translator Yoshio Kôsaku (or 耕牛 Kôgyû), in whose house he was then staying, he supplied him with all the paraphernalia of a European apotheosis. But since Kôsaku was known as a physician, and indeed had become extremely rich through dispensing cures for venereal disease made from saffron (a technique he had learned from the Carl Peter Thunberg we will encounter again below), Kôkan gives him a book entitled simply *Heelkonst* ('medicine') [Fig. 1]. The Nagasaki translators did obeisance to Western books as part of their new year rituals, for it was the practice for all officials to make offerings "before whatever bears the strongest analogy to their profession," wrote Titsingh; all fighting samurai did homage to the Shogun's personal suit of armour, while the interpreters "place a Dutch book on a table, and set their offerings before the book."[4] Such books filtered out via the interpreters. European sailors often sold them under the counter, "not at the sale," noted Thunberg, but secretly, "and that to considerable advantage," although the interpreters' collections need 杉田玄白 not have been the final resting places.[5] In 1766, the physician Sugita Genpaku obtained by barter a copy of Laurens Heister's *Heelkundige onderwyzingen*[6] ('Instructions in Surgery' 1741) [Fig. 2 and 3] at a cost of twenty kegs of good sake.[7] The Europeans might obtain local items, or even Japanese books, for European books, and, for example the Dutch leader Jan Crans swapped copies of Dodoens, another important work, Jonston's encyclopaedia of animals, and the shell compendium *D'Amboinsche rariteitkamer* ... (Amsterdam, 1705)[8] for the Japanese herbal *Honzô kômoku.*[9]

Pictures had independent value, but even out of context, they pointed to the postulates of the book from which they came, and were the means by which these postulates were more or less clearly expounded. What could not be understood by an independent picture, in all probability, was not understood at all. Pictures, as it were, moved signification off the Dodonaean page, and planted it within Edo's intellectual space. The kinds of questions asked of plant-life in this new setting were not the same as those asked in Europe, and as such, botanical debate evolved quite differently. But Dodoens's book remained as a peg on which to hang a whole perception of the episteme of Europe.

The *Cruydt-Boeck* shared this symbolic role with one other imported work, which must also be cited, and with which Dodoens's work can usefully be compared. This was the natural history of quadrupeds by the Polish physician Jan Jonston (also called Johannes Jonstonus or John Johnston, 1603-1675) that was known in Edo in the Dutch edition with the title *Naeukeurige beschryving van de natuur der viervoetige dieren, vissen en bloedloze water-dieren, vogelen, kronkel-dieren, slangen en draken* (Amsterdam 1660); this was reissued over the generations, as Dodoens was, and currency was pretty high. Pan-European attention was assured by the Amsterdam Latin version of 1755, *Theatrum universale omnium animalium quadrupedum* [Fig. 4]. Among the many works of Western scholarship important to the development of Edo period thought, Dodoens and Jonston alone, I would maintain, acquired cultic status.

The mythic first encounter of the Japanese authorities with European books is a well-known story – though surely not an accurate one. According to Sugita Genpaku, the first historian of Western studies, it was this encounter that sparked the entire *Rangaku* movement. Genpaku invented a lofty initiation for *Rangaku* as a way of enhancing the marginalised field socially. Nevertheless, his book, *Rangaku kotohajime* (蘭学事始) ('The Beginnings of *Rangaku*,' 1815) became so widely read that, to this day, his myth enjoys the presumptions of fact among many. It is written that the then-Shogun, Tokugawa Yoshimune, at an undated moment in the mid-eighteenth century, asked to be shown 'for his august viewing pleasure' a copy of a 'book containing illus-

ILL 1

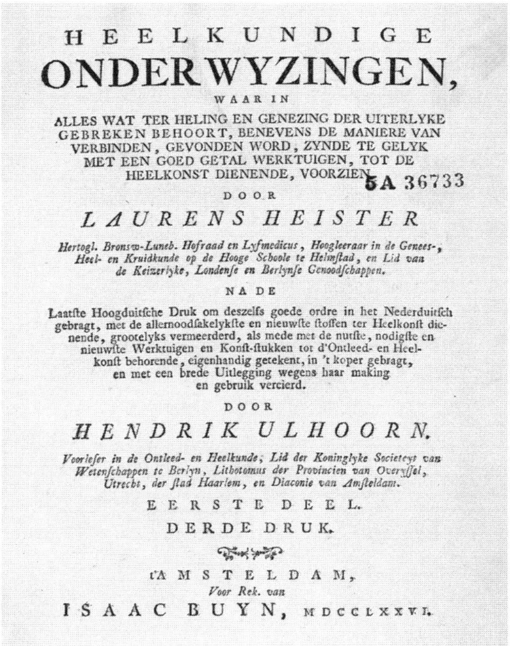

HEELKUNDIGE

ONDERWYZINGEN,

WAAR IN

ALLES WAT TER HELING EN GENEZING DER UITERLYKE GEBREKEN BEHOORT, BENEVENS DE MANIERE VAN VERBINDEN, GEVONDEN WORD, ZYNDE TE GELYK MET EEN GOED GETAL WERKTUIGEN, TOT DE HEELKONST DIENENDE, VOORZIEN.

DOOR

LAURENS HEISTER

Hertogl. Bronsw-Luneb. Hofraad en Lyfmedicus, Hoogleeraar in de Genees-, Heel- en Kruidkunde op de Hooge Schoole te Helmstad, en Lid van de Keizerlyke, Londense en Berlynse Genoodschappen.

NA DE

Laatste Hoogduitsche Druk om deszelfs goede ordre in het Nederduitsch gebragt, met de allernoodsakelykste en nieuwste stoffen ter Heelkonst dienende, grootelyks vermeerderd, als mede met de nutste, nodigste en nieuwste Werktuigen en Konst-stukken tot d'Ontleed- en Heelkonst behorende, eigenhandig getekent, in 't koper gebragt, en met een brede Uitlegging wegens haar making en gebruik vercierd.

DOOR

HENDRIK ULHOORN.

Voorleser in de Ontleed- en Heelkunde, Lid der Koninglyke Societeyt van Wetenschappen te Berlyn, Lithotomus der Provincien van Overyssel, Utrecht, der stad Haarlem, en Diaconie van Amsteldam.

EERSTE DEEL.

DERDE DRUK.

t'AMSTELDAM,

Voor Rek. van

ISAAC BUYN, MDCCLXXVI.

ILL 2

ILL 3

ILL 4

trations;[10] although the record does not state so, it had always been assumed what he was furnished with was a copy of Jonston's natural history.[11] Yoshimune realised the significance of European learning. Genpaku's telling goes like this:

> "The [illustrations] are extremely precise, and if only someone were able to decipher the theories contained in [the book] they would assuredly find them to be detailed too. Here in Edo, someone should study the matter."

野呂元丈 青木文蔵 精妙

Noro Genjô and Aoki Bunzô stepped forward. Note how the still unknowing Yoshimune is alerted to the whole field of animal science by seeing the *pictures*. It is significant that he as yet has no inkling of the content of the text. He therefore commands study of the Roman alphabet. Genpaku suggests he went through the same kind of enlightenment, for, after obtaining his copy of Heister, he noted, 'although unable to read the book's content letter by letter or line by line, I could see from the pictures that it was totally different from continental or Japanese interpretations; when I saw the precision (*seimyô*) of the illustrations, I felt as if my mind had been suddenly opened'[12] [Figs. 5 & 6a-e].

徳川家康 名君 東照大権現 本地 垂迹, 薬師

Genpaku's history was published in 1816, which was a special year, as it marked the 200th anniversary of the death of Tokugawa Ieyasu, founder of the Shogunal line, which was the subject of extended celebrations. Genpaku is saying, in effect, that Rangaku is good for the state, was supported by good rulers, and that its validity is demonstrable, *ab initio*, via its pictures. The story could have been pinnable on any Shogun, but by Genpaku's time, Yoshimune had come to be seen as one of the (few) good Tokugawa Shoguns, and was popularly known by the accolade of a 'notable prince' (*meikun*). One further twist may be added. Ieyasu was deified posthumously as the Tôshô Daigongen, or Great Shining Avatar of the East; the 'root-ground' Buddha (*honji*) of this manifested Shinto god (*suijaku*) was Yakushi, the Buddha of Medicine. Anything to do with medical learning could be, and was, subsumed into the mythology of the physic of the state. The mythology of a 'body politic' is well known in Northeast Asia.[13] Genpaku saw all *Rangaku*, from Dodoens to Jonston and beyond, as primarily medical. He was himself a doctor. In his narrative of the origin of *Rangaku*, the reader is taken from Yoshimune's wise setting of the field in motion, to a group of physicians all in senior government employ (including Genpaku himself) to witness their first autopsy, having been stimulated by another imported book, Johann Adam Kulm's (or Johannes Kulmus') *Ontleedkundige tafelen* of 1725.[14] The Shoguns, and especially Ieyasu, were sorts of 'medicine kings' (*iô*) and could be venerated via icons of the Buddha of Medicine [Fig. 7]. Many holy images of Yakushi were produced in 1816 for the bicentennial jamborees. Genpaku terminates his history of *Rangaku* with his own pilgrimage to the shrine to Ieyasu/Tôshô Daigongen/Yakushi, in time for the anniversary.[15]

医王

We are getting a little off course, but I think it is necessary to stress the link between the immediacy and appeal of pictures and the generalised, feel-good aura cultivated politically for *Rangaku* by its adherents, in which the solace of medical cure played a major part. Jonston's book had a somewhat modern agenda, and was founded on the sense that knowledge in itself was good. But Dodoens's book was a practical pharmacopoeia, that is, not a botany as such, but a medicinal botany. The title of the English translation of 1578 made this crystal clear: the book was called *A New Herball*.

Yoshimune, supposedly, thought Jonston's pictures 'precise' (*seimitsu*), and it was this that he applauded. The same homage was offered in different, more extreme guise when Genpaku claimed that the greatest *Rangaku* scholar of the next generation, Hiraga Gennai, had sold the entire contents of his house, right down to his bedding, in order to raise funds to secure a copy. The pictures are not specifically mentioned in the Gennai anecdote, but it is stated that Gennai also owned a copy of Dodoens, and (using similar terminology to Yoshimune) said he found its pictures 'amazingly precise' (*hanahada kuwashiki*).[16] [Fig. 8]

精密

平賀源内

甚だ詳しき

Jonston's work was some century later than Dodoens, and necessarily its pictures were superior. The quality of book illustration improved, and to the critical, Dodoens began to compare unfavourably with Jonston. The Western-style artist and associate of both Genpaku and Gennai, Shiba Kôkan, remarked on the development of European image printing. Kôkan viewed a copy of Jonston, and compared it with Dodoens. After seeing the newer work, Kôkan decried the older images as 'remarkably coarse' (*zugyô hanahada sôsotsu*) and 'failing to resemble the real objects' (*shinbutsu ni nizu*).[17] Some century lay between the two books. The principal intervening innovation was the invention of etching. This radically altered the fineness of drawing, the density of markings on the page, and the number of impressions that could be taken before lines wore out. The search for the secret of etching pre-

図形甚粗率
真物に似ず

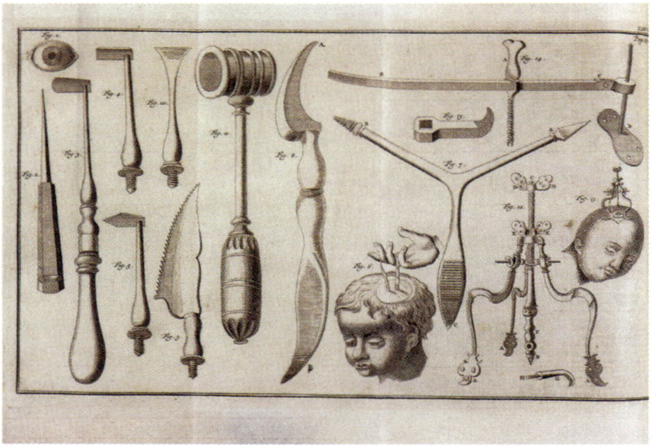

ILL 5

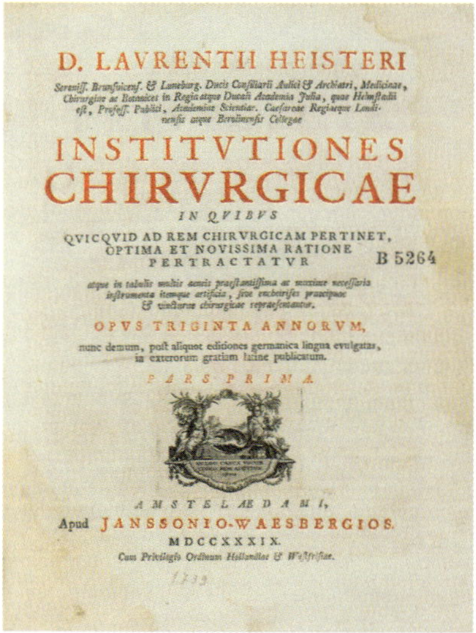

D. LAVRENTII HEISTERI

Sereniss. Brunsuicens. & Luneburg. Ducis Consiliarii Aulici & Archiatri, Medicinae, Chirurgiae ac Botanices in Regia atque Ducali Academia Julia, quae Helmstadii est, Profess. Publici, Academiae Scientiar. Caesareae Regiaeque Londinensis atque Berolinensis Collegae

INSTITVTIONES CHIRVRGICAE

IN QVIBVS

QVICQVID AD REM CHIRVRGICAM PERTINET, OPTIMA ET NOVISSIMA RATIONE PERTRACTATVR

atque in tabulis multis aeneis praestantissima ac maxime necessaria instrumenta itemque artificia, sive encheirises praecipuae & vincturae chirurgicae repraesentantur.

OPVS TRIGINTA ANNORVM,

nunc demum, post aliquot editiones germanica lingua evulgatas, in exterorum gratiam latine publicatum.

PARS PRIMA

AMSTELAEDAMI,

Apud JANSSONIO-WAESBERGIOS.

MDCCXXXIX.

Cum Privilegiis Ordinum Hollandiae & Westfrisiae.

ILL 6A

ILL 6B

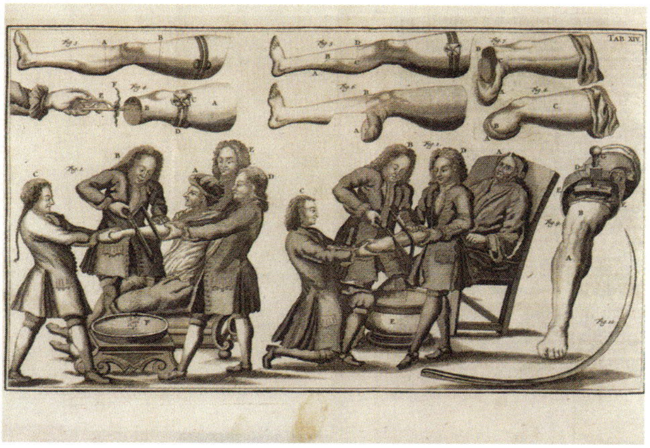

ILL 6C

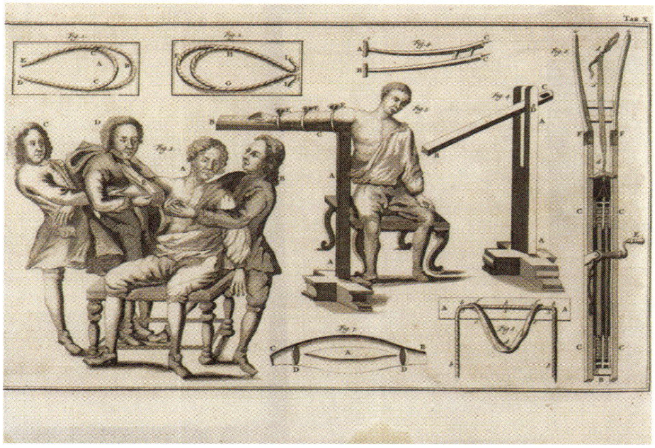

ILL 6D

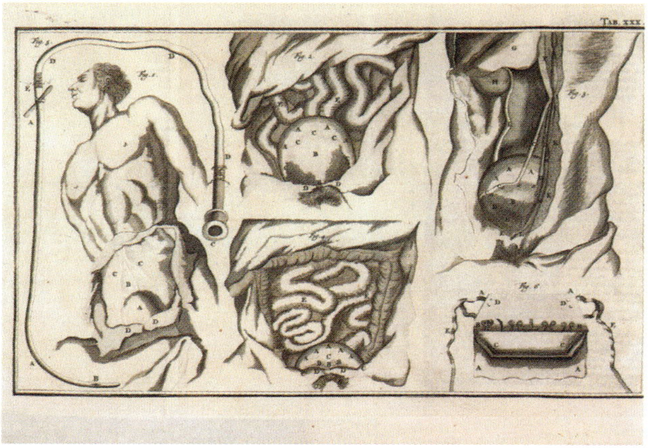

ILL 6E

ILL 7

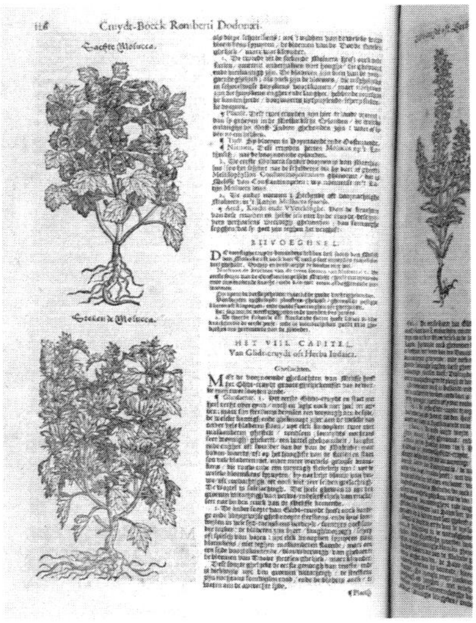

ILL 8

47　英泉画　「江戸日本橋ヨリ富士ヲ見ル図」

ILL 9

ILL 10

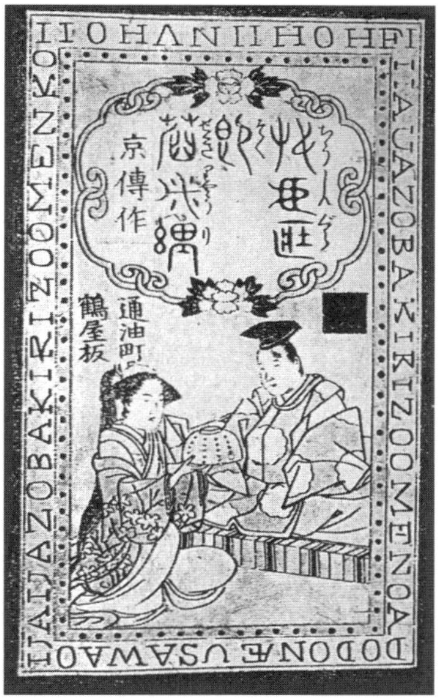

ILL 11

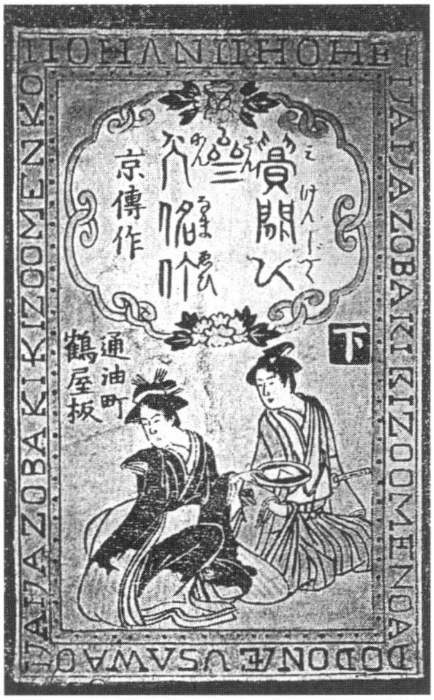

ILL 12

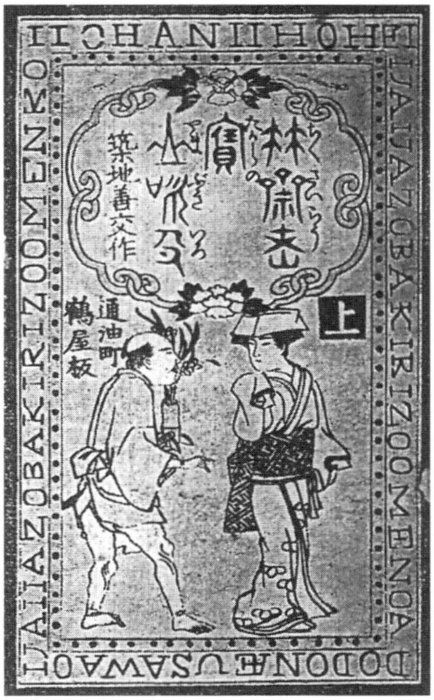

ILL 13

occupied many *Rangakusha*, until in 1784 Kôkan cracked the secret, and thereafter proudly referred to himself as First Maker in Japan 日本創製 (*Nihon sôsei*);[18] steps were being taken by others, with one of the phy- 桂川甫周 sicians present at the first autopsy mentioned by Genpaku, Katsura- 森島中良 gawa Hoshû, and his younger brother Morishima Chûryô (Nakay-oshi), offering one of the first public demonstrations of the use of *aqua forte* in 1786.[19]

If Jonston's copperplates stole the laurels from Paré's and Dodoens's earlier woodblocks, he did not oust them in popularity or fame. In 松平定信 1787, Matsudaira Sadanobu commanded the arduous task of a translation of Dodoens, and entrusted the project to the Nagasaki interpret- 石井庄助 ers Umada Seikichi and Ishii Shôsuke.[20] This was to be entitled 遠西本草攬要 *Ensei honzô ran'yô* ('The Essentials of European Pharmacology') – observe how the medical function is stressed. The order to translate was issued almost immediately upon Sadanobu's coming to power as head of the Shogunal council with a stated mission to enhance the prestige of the Tokugawa family. By commanding production of a medical work, 家斉 Sadanobu was aligning himself and the newly inducted Shogun Ienari with the old but largely exhausted Tokugawa mythology of the 'medicine king'; Sadanobu was attaching himself to the figure of Ieyasu and to the 'notable' Yoshimune, who happened to be his own grandfather.

It seems unlikely that Sadanobu's commission of the translation was known outside a small group of Shogunal advisors and translators, but the star of Dodoens continued to rise, and appears in an interesting context in 1794, while translation work was still underway. That 鶴屋、通油町 year, the popular publishing house, the Tsuru-ya, of Tôriabura-chô in central Edo, issued several of their comic books in the 'yellow-cover' 黄表紙 (*kibyôshi*) genre with a running title page ornamented with Roman letters. It was the custom for the garish yellow cover to have pasted onto it a square cartouche with some image that would be shared for all that year's productions, over which was set an additional element particular 絵題簽 to the specific book and also the title; this was called the *edaisen*. The use of Roman letters as motifs was not novel, for the little-understood graphs enjoyed a degree of prominence, and were often drawn on shop signs to attract trade, or deployed as purely ornamental motifs, in a 英泉 way that was to continue well into the nineteenth century (Eisen's series *Famous Views of Edo* of c. 1830 is well known) [Fig. 9]. *Kibyôshi* specialised in lampooning contemporary fads and foibles, and as such foreign-derived affairs were part of their pool of potential themes. In the case of the Tsuru-ya series, though, the decoration included not only random letters, nor even the one or two terms that would probably have been recognised by a few who passed by (like the 'Holland' or 'VOC' [Dutch East India Company] mark that Eisen used). The Tsuru-ya used the name of Dodoens. In the context of book covers, did the Tsuru-ya expect the letter to be actually read? At least four books shared this cover, and since *kibyôshi* all have three volumes, this gave a possible one dozen airings.[21]

It is not possible to assert the order of the books' actual publication, and in fact, they may well have been issued all together; most *kibyôshi* came out at New Year. First it is seen on Shiba Zenkô's 芝全交 *Zenkô-hôshi tsunezune-gusa* 全交法師常々艸; the title identifies the work as a parody of the mediaeval classic *Tsurezure-gusa* ('Essays in Idleness') of Kenkô-hôshi 徒然草、兼好法師; the name of Dodoens appears on the cover to all volumes [Fig. 10]. Second was *Chûshingura sokuseki ryôri* 忠臣蔵即席料理 ('The Treasury of Loyal Retainers: Dinner While You Wait'), by the great humorist Santô Kyôden 山東京伝; it took its thematic gist from the famous theatrical retelling of a vendetta of 1701 that transfixed Edo for months, *Kanadehon chûshingura* 仮名手本忠臣蔵 ('A Syllabary of the Treasury of Loyal Retainers') [Fig. 11]. In both cases the name of Dodoens prefigures, in a humorous way, retellings in modern urbane guises of important vernacular literary works, one from the distant, one from the recent past. Third was also by Kyôden, *Mikenjaku sannin namaei* 箕間尺参人酩酊 ('Mikenjaku and the Three Drunkards'), which weaves the story of O-shichi and her lover Kichiza into a travelogue of hell [Fig. 12]. This in itself represents a formidable line-up of popular writing talent, for Zenkô and Kyôden were lionised by the popular reading classes. All three books were moreover illustrated by the premier popular picture-maker, Kitao Shigemasa 北尾重政. (The designer of the edaisen is unknown, for such things were always anonymous). Kyôden, it may be added here, had already satirised Dodoens in 1777, so that the *edaisen* might have been specially appropriate for him: a joke book (*sharebon* 洒落本) of that year includes the stray line *Oranda honzô dodoneusu* ('Dutch pharmacopoeia Dodonæus'). This latter book would have been written just months after the visit to Edo of Carl Peter Thunberg, a Swedish physician of some prominence, who had a great impact on Edo scientific discourse, not just the treatment of venereal disease, and not least via the buying and selling of books; Thunberg met Katsuragawa Hoshû, became friendly with him, and noted the titles of some of the books he used: 'Johnston's [sic] *Historia Naturalis*, and Dodonæus' *Herball*, and in physic Woyt's *Treasury* (*Gazophylacium*), which books they had purchased from the Dutch. In surgery they had Heister, translated into Dutch, and … ' – Thunberg is unable to resist – ' … I sold to them at this time, amongst other books, a very fine edition of Muntingius' [Muntingh] *Phytographia.*'[22]

Shigemasa also illustrated the third book put out by the Tsuru-ya, and it is this that is the most interesting to us here. The work was entitled *Chikusai rôtakara yamabuki iro* (竹斎老宝山吹色) (*Chikusai's Ancient Treasure the Colour of Kerrias*), a somewhat opaque label that refers to the stock country sage character Chikusai (竹斎) [Fig. 13]. The author signed himself in this instance Tsukiji Zenkô. He was also known as Fûrai Sanjin Nisei (風来山人二世). Hiraga Gennai had taken the penname Fûrai Sanjin ('mountain man blown in the wind'), and this author was his literary successor. We have encountered him above as Morishima Chûryô (森島仲良), brother of Hoshû. Chûryô was known dynastically as Katsuragawa Hosan (桂川甫粲), and as son and then brother of the hereditary Shogunal court physician, quite a dynasty it was.[23] Chûryô had authored many *Rangaku* works, one of which had discussed his public demonstration of *aqua forte*. This was now some twenty years since Hoshû had befriended Thunberg, when the two were all young men, but the Swedish and Japanese doctors remained in touch, possibly right up to their deaths (Hoshû in 1809, Thunberg not until 1828).[24]

The subject matter of the stories need not detain us here, as the running covers could never be more than generalistically related to the specific stories, but the covers can be read to elicit a Japanese phrase:

Irohanihohei
Iyaiya soba kiri soumen oa
Dodoneusu awao
Iyaiya soba kiri soumen ko

It does not really make more than very partial and absurd sense, but it is just about translatable as something like, 'ABCDE, how terrible to cut up wheat noodles and thin noodles too, Dodonæus, oh! How terrible to cut up wheat noodles and thin noodles too.'

Arguments about which books might have the better or contain the more accurate pictures might be – and were – undertaken; it is interesting to reconstructs these debates. But my first intention here has been merely to note how the impact of the welter of foreign books was shared and subsumed under the name of Dodoens, served up here as a kind of mantra to the barely-knowing Edo public.

To return to the second point. The full systematics of animal and plant ordering were to reach fruition only after Dodoens, in the work of Linnaeus. His binomial system of nomenclature, contrived in the early-to-mid-eighteenth century, endures into the present and remains the norm today. Linnaeus was Swedish and had worked at Upsala University. So did Thunberg. In fact, the latter was a pupil of Linnaeus who had come under his wing in the 1760's, when still in his early twenties and shortly before leaving in 1771 on the trip that would eventually land him in Japan some five years later.[25] Thunberg went East principally with the aim of furthering Linnaean classifications, and a decade after his return he was to complete his great work on the subject, a *magnum opus* entitled *Flora Japonica*: this was published in Leipzig in 1784 [Fig. 14]. The following year Gustav III of Sweden ennobled him with the Order of the Wasa.

Thunberg came to Japan in the capacity of physician with the Dutch East India Company, and during his stay of about a year he spent nearly a month in Edo in the spring of 1776 – May 1-26 to be exact.[26] His friendship with Hoshû blossomed. They were about the same age, and Hoshû was famous as a rake and *bon vivant*, as well as being proficient in Dutch. Thunberg referred to Hoshû in his writings as 'my beloved pupil'.[27] Much has been written about Thunberg's impact in Japan, including by myself.[28] He spent a colossal amount of money on plant collecting, and by his own estimate parted with 1200 *rixdollars* before he returned to Europe.[29] But I would like to draw attention to an observation he made on binomial nomenclature. The remark pertains not to Japanese animals or plants, but to Japanese people. He recorded in the book of his travels published in 1788-91 (and in English in 1795):

> "There is likewise this singularity in the affair that the family name is not put after, but always before the adscititious name, in like manner as in botany, where the generic name of a plant precedes the specific."[30]

Thunberg's repeated sense of strangeness which he felt again and again in Japan, is here presented as definitely mitigated by how well the people fitted with the Linnaean categories; it is a source of assuas-

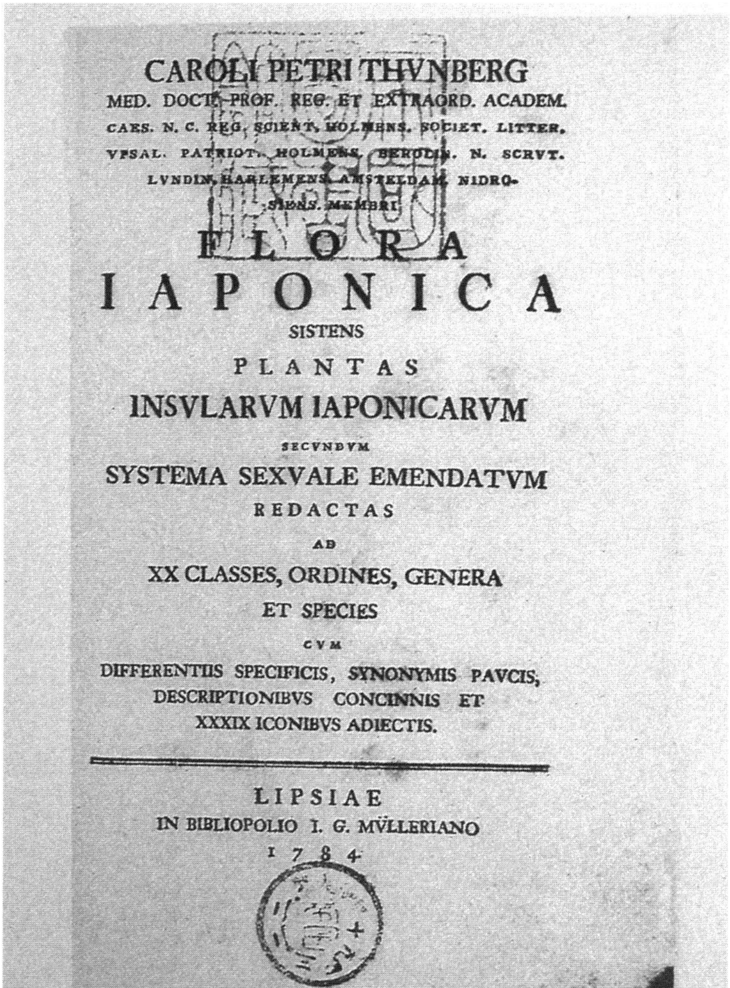

CAROLI PETRI THVNBERG
MED. DOCT. PROF. REG. ET EXTRAORD. ACADEM.
CAES. N. C. REG. SCIENT. HOLMENS. SOCIET. LITTER.
VPSAL. PATRIOT. HOLMENS. BEROLIN. N. SCRVT.
LVNDIN. HARLEMENS. AMSTELDAM. NIDRO-
SIENS. MEMBRI

FLORA
IAPONICA
SISTENS
PLANTAS
INSVLARVM IAPONICARVM
SECVNDVM
SYSTEMA SEXVALE EMENDATVM
REDACTAS
AD
XX CLASSES, ORDINES, GENERA
ET SPECIES
CVM
DIFFERENTIIS SPECIFICIS, SYNONYMIS PAVCIS,
DESCRIPTIONIBVS CONCINNIS ET
XXXIX ICONIBVS ADIECTIS.

LIPSIAE
IN BIBLIOPOLIO I. G. MVLLERIANO
1784

ILL 13

ion and delight that Japanese human life broke down into specific and generic parts, the latter being named first, in a way that did not work in Europe. Thunberg assumed that his plant and animal categories would serve in Japan, since his teacher's system was avowedly universal. But the way that a *cultural* system of nomenclature also seemed to match, appeared to vindicate the structure even further. People as social entities could be seen through a Linnaean prism – or rather, Japanese people could (he does not mention that this order of names is common across North- and Southeast Asia).

A parallel appears here between the imposition of an order upon nature by scholars, and the way in which a human society might order itself. Europe alone, to him, was 'enlightened by science', which had been possible through the distancing of internal and external observations, and through discovery of a difference between subjectivity and objectivity.[31] To Thunberg, Japan was not in this state yet. Nevertheless, he frequently repeated how strict, but also fair, Japanese laws were. Japan was not a despotism, but, as he put it,

> "[...] the right and liberties of the higher and lower class of people are equally protected by the laws; & the uncommon severity of these laws, joined to the inevitable execution of them, serves to keep every one with proper bounds."[32]

Only, this was limited by a lack of division between the person and the world. He went on that Japan progressed as far as it could, without "the light of science, by whose brighter rays it has not yet had the good fortune to be illuminated".

The bounds which scholars had imposed on discordant nature, were placed on the Japanese people by their rulers. This was evinced in the most basic formulae of naming. The sense that Linnaean structures might also work for cultural categorisation fitted with a wider sense that Japan, specifically, among all the new-found world regimes, was closest to Europe, while at the same time, being not quite there. During the 'age of discovery' Japan occupied a unique place as a country generally deemed quasi-equal to Europe, – a point made by Thunberg and many others. The myth of Japanese uniqueness was already emerging. But this quasi-equality took the form of reversal. Japan was the mirror that, paradoxically, authenticated European norms and that proved the sole ability of Europe to embrace the objectivity of science. Japan had never sought to apply its terminology outside its own borders, and so Japan's botanical levels of imposed order, whether imposed on the natural world or on its own people, contrasted on the one hand with the India, China or Africa, that simply *defied* order and logic, and on the other with the European privilege of objectivity to impose order *externally*.

NOTES

1 Cynthia Viallé & Leonard Blussé, eds., *The Deshima dagregisters: Their Original Tables of Contents*, vol. 10 (Leiden: Institute for the History of European Expansion, 1997), 84.
2 Grant K. Goodman, *Japan: The Dutch Experience* (London: Athlone Press, 1986), 59.
3 Viallé & Blussé, eds., *Deshima dagregisters*, vol. 9, 17.
4 [Isaac] Titsingh, *Illustrations of Japan*, trans. Frederic Shoberl (London, 1822), 121. The ritual took place on the 11th day of the New Year.
5 Carl Peter Thunberg, *Thunberg's Travels*, trans. F. & C. Rivington (London, 1795), 49.
6 Full title: *Heelkundige onderwyzingen, waarin alles wat ter heling en genezing der uiterlyke gebreken behoort, benevens de maniere van verbinden, gevonden word ... , na de Laatste Hoogduitsche Druk om deszelfs goede ordre in het Nederduitsch gebragt ... door Hendrik Ulhoorn.* Eerste deel. Derde druk (t'Amsteldam: Isaac Buyn, 1776).
7 Sugita Genpaku, *Rangaku kotohajime*, Nihon koten bungaku taikei, vol. 95 (Tokyo: Iwanami, 1963), 483.
8 Full title: *D'Amboinsche rariteitkamer, behelzende eene beschryvinge van allerhande zoo weeke als harde schaalvisschen, te weeten rare krabben, kreeften, en diergelijke zeedieren, als mede allerhande hoorntjes en schulpen, ... die men in de Amboinsche zee vindt: daar beneven zommige mineraalen, gesteenten, en soorten van aarde, die in d'Amboinsche, en zommige omleggende eilanden gevonden worden.*
9 Sugita Genpaku, *Rangaku kotohajime*, 486. It is not quite clear what book this was, as many had similar titles, see *Kokusho sômokuroku*, vol. 7 (Tokyo: Iwanami, 1970), 385-7. The work is not Ono Ranzan's famous *Honzô kômoku keimô* which was not published until 1803.
10 Ibid., 478.
11 Ibid., loc. cit., note 11.
12 Ibid., 483.
13 I have elaborated on this in my *Shogun's Painted Culture: Fear and Creativity in the Japanese States, 1760-1824* (London: Reaktion Books, 2000).
14 Sugita Genpaku, *Rangaku kotohajime*, 487-95. The book was known in Japan via the Dutch edition of 1734; the original German was entitled *Anatomische Tabellen.*
15 Ibid., 516.
16 Hiraga Gennai, *Butsurui hinshitsu*, Nihon koten zenshû 3, vol. 16 (Nihon koten zenshû kankôkai, 1928), 208.
17 Shiba Kôkan, "Seiyô gadan," in *Yôgaku* 2, Nihon shisô taikei, vol. 64 (Tokyo: Iwanami, 1976), 493.
18 This designation is inscribed on Kôkan's print of an artist in his study of 1794, now in the Kobe City Museum; for a convenient reproduction, see Calvin French, *Shiba Kôkan: Artist, Innovator and Pioneer in the Westernization of Japan* (London & Tokyo: Weatherhill), 133.
19 Morishima Chûryô, *Kômô zatsuwa*, Bunmei genryô sôsho, vol. 1 (Kokusho kankôkai, 1913), 480.
20 Fukui Hisazô, *Shodaimyô no gakujutsu to bungaku no kenkyû*, vol. 1 (Hara shobô, 1976, 2nd ed.), 324.
21 For discussions of the individual books and their contents, see Tanahashi Masahiro, *Kibyôshi sôran*, vol. 2, 391-93, 394-94, 404-6 & 414.
22 Thunberg, *Thunberg's Travels*, 184. See also Blomberg's contribution in the present volume.
23 See Imaizumi Genkichi, *Katsuragawa-ke no hitobito*, 3 vols. (Chikusaki shorin, 1969), and Tozawa Yukio, *Katsuragawa-ke no sekai* (Chikuma shokan, 1994).
24 Thunberg, *Thunberg's Travels*, 206. Thunberg says 'intercourse' was maintained for 'several years' after 1776.
25 See paper by Catharina Blomberg in present volume.

26 Paul van der Velde & Cynthia Viallé, eds., *Deshima dagregisters*, vol. 8, 151-152. Dates here are in the European calendar.
27 Thunberg, *Thunberg's Travels*, 198. Nakagawa Jun'an was included in the compliment.
28 See Timon Screech, *Edo no shintai o hiraku*, trans. Takayama Hiroshi (Sakuhinsha, 1997).
29 Thunberg, *Thunberg's Travels*, 30.
30 Ibid., 266.
31 Ibid., 251.
32 Ibid., 254.

ILLUSTRATIONS

ILL 1 Shiba Kôkan, Yoshio Kôsaku with paraphernalia. Hanging scroll, 1788. Private collection.
ILL 2 Title page of Laurens Heister, *Heelkundige onderwyzingen* (third impression of the Dutch edition, Amsterdam, 1776). Collection Katholieke Universiteit Leuven, Central Library, Tabularium 5 A 36733.
ILL 3 Frontispice of Laurens Heister, *Heelkundige onderwyzingen*, (third impression of the Dutch edition, Amsterdam, 1776), featuring portraits of Heister and Ulhoorn in medallions held by resp. the goddess of Medicine personified and Aesculapius, the God of Medicine. Collection Katholieke Universiteit Leuven, Central Library, Tabularium 5A 36733.
ILL 4 Elephant, unsigned illustration to Johannes Jonstonus, *Theatrum universale omnium animalium quadrupedum*, Amsterdam, 1755. Private collection.
ILL 5 Laurens Heister, *Heelkundige onderwyzingen*, eerste deel; derde druk (third impression of the Dutch edition, Amsterdam, 1776), tab. V (between pages 170 and 171). Collection Katholieke Universiteit Leuven, Central Library, Tabularium 5A 36733.
ILL 6 Title page of a Latin edition of Laurens Heister 's *Chirurgie* (1718): *Institutiones chirurgicae in quibus quicquid ad rem chirurgicam pertinet, optima et novissima ratione pertractatur* ... ,(Amstelaedami, Apud Janssonio-Waesbergios, 1739). Collection Katholieke Universiteit Leuven, Central Library, Tabularium B 5264.
ILL 6B Frontispice in the 1739 Latin edition of Laurens Heister's *Chirurgie*. Collection Katholieke Universiteit Leuven, Central Library, Tabularium B 5264.
ILL 6C Gatefold in the 1739 Latin edition of Laurens Heister's *Chirurgie* (pars prima, tab. XIV, between pages 514 and 515). Collection Katholieke Universiteit Leuven, Central Library, Tabularium B 5264.
ILL 6D Gatefold in the 1739 Latin edition of Laurens Heister's *Chirurgie* (pars prima, tab. X, between pages 266 and 267). Collection Katholieke Universiteit Leuven, Central Library, Tabularium B 5264.
ILL 6E Gatefold in the 1739 Latin edition of Laurens Heister's *Chirurgie* (pars secunda, tab. XXX, between pages 954-955). Collection Katholieke Universiteit Leuven, Central Library, Tabularium B 5264.
ILL 7 Kimura Ryôtaku, *the Buddha of Medicine*, colour on silk, Rinnôji temple, Nikkô.
ILL 8 *Cruydt-Boeck*, 1644 edition, p. 126. Collection Katholieke Universiteit Leuven, Central Library, Tabularium C1827
ILL 9 Keisei Eisen, *View of Mount Fuji from the bridge of Ryôgoku in Edo, from the series Famous Places of Edo* (c. 1830), private collection.
ILL 10 cover to vol. 3 of Shiba Zenkô, *Zenkô-hôshi tsunezune-gusa* (1794), private collection.
ILL 11 cover to vol. 2 of Santô Kyôden, *Chûshingura sokuseki ryôri* (1794), private collection.
ILL 12 cover to vol. 3 of Santô Kyôden, *Mikenjaku sannin namaei* (1794), private collection.
ILL 13 cover to vol. 1 of Tsukiji Zenkô, *Chikusai rôtakara yamabuki iro* (1794), private collection.
ILL 14 Carl Peter Thunberg, *Flora Japonica*, Leipzig, 1784.

XI

DODONÆUS AND TOKUGAWA CULTURE: HIRAGA GENNAI AND NATURAL HISTORY IN EIGHTEENTH-CENTURY JAPAN

Tôru Haga
International Research Center for Japanese Studies, Kyoto

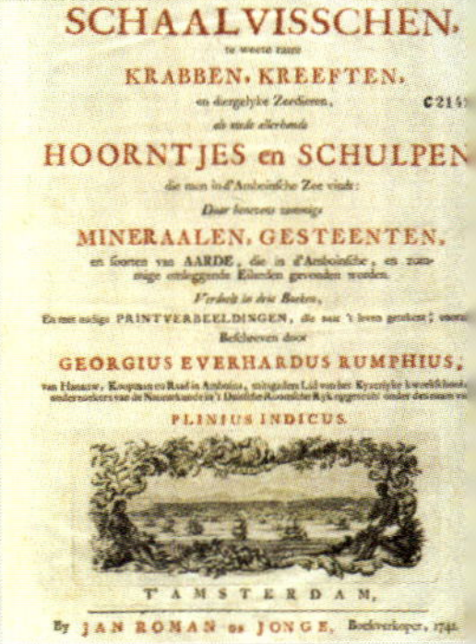
SCHAALVISSCHEN,
KRABBEN, KREEFTEN,
HOORNTJES en SCHULPEN
MINERAALEN, GESTEENTEN,
PRINTVERBEELDINGEN,
GEORGIUS EVERHARDUS RUMPHIUS,
PLINIUS INDICUS.
T'AMSTERDAM,
By JAN ROMAN de JONGE, Boekverkoper, 1741.

平賀源内
長崎屋

This paper examines the influence of Dodonæus on Tokugawa culture, in particular the importance of his work for Hiraga Gennai (1728-1779) and the study of natural history in eighteenth-century Japan. Dodonæus' herbal is impressive both by its physical voluminosity and its cultural importance. The sheer size and weight of the book must have had a deep cultural and psychological impact on the Japanese mind, when the eighth Shogun, Tokugawa Yoshimune, opened it for the first time in 1717, in a reading room of Edo castle. Yet it was not until fifty years later that Hiraga Gennai, an ex-samurai and scholar of natural history, purchased a copy of the same book for his own collection in the spring of 1765, in a drawing room of the Nagasakiya inn in the middle of the city of Edo.

The edition of Dodonæus' book I examined, was published in Leiden in 1618. The book measures 34 by 22 centimetres, is about 14 centimetres thick and weighs at least three kilograms. It contains 29 pages of preface and table of contents, 1.495 pages of description of plants with innumerable wood-block print illustrations and 56 pages of alphabetical index, totalling 1.580 pages. No Japanese nor other Asian scholar had ever laid eyes on such a bulky and heavy book before, nor imagined that any publication could be of such a dignified format. The sheer magnitude of the book must have aroused the interest and imagination of Tokugawa intellectuals, and surely reflected the prominence of the European tradition of scientific scholarship.

Once the Japanese turned over the leather cover, they must have been even more impressed by the intriguing beauty of the copperplate print title page. At the very bottom of the page we find in tiny characters the mention *W. Swan fecit* 1608. This refers to the Leiden engraver Willem Swanenburgh (1580-1612), who designed this lovely picture. The illustration shows a magnificent porch or gate with a heavy roof supported by four marble pillars through which a beautiful herb garden, laid out in perfect symmetry, is visible in linear perspective. On the top of the roof, there are garlands of exotic flowers and fruits, which hold the cartouche with the inscription *Cruydtboeck van Rembertus Dodonæus* from behind. Furthermore, at the entrance of the gate there are two cornucopias supporting a cartouche with the name of the publisher, Françoys van Rafelingen. Portraits of the author, Dodonæus, and another great Flemish humanist scholar, Carolus Clusius (1526-1609), whose descriptions of exotic plants were supplemented to the original edition, are shown at the base of the front pillars on either side of the gate. We can only speculate whether the Japanese scholars understood that the triangular shape in the middle of the garden was not a bird-cage, but a fountain.

We can, however, easily imagine how deeply impressed the Tokugawa scholars must have been at their first look at this title page when they opened the huge book. The view in perspective of the herb garden with its alignment of shadowy trees must have seemed to allure and entice them into the remote depths of European scholarship. Even at present one feels tempted to enter and stroll in this quiet garden of science. In fact, it is my intention to use the entire picture for the title page of my next book, replacing only the title and names of author and publisher. Hiraga Gennai did the same thing for his book on the thermometer, adopting the title page of an unknown European book. Odano Naotake, a disciple of Hiraga Gennai in western-style painting, followed his suit for his illustrations of the famous *Kaitai shinsho*, the *Ontleedkundige tafelen*, using the title page of Juan de Valverde's anatomy *Vivae imagines partium corporis humanis aereis formis expressae* (Antwerp, 1566).

Hiraga Gennai purchased Dodonæus' book in the spring of 1765, in the third month of the second year of the Meiwa period. This is obvious from a list of Western books on natural products in his collection entitled *Bussan shomoku*, which he drew up in the year 1769 and 物産書目 probably presented to someone of higher rank. All eight books on the list, except one, Emanuel Sweerts's (ca. 1552-1612) *Florilegium, tractatus de variis floribus et aliis Indicis plantis ad vivum delineatum* (Amsterdam, 1631), were purchased in the third month of different years during the 1760s. This implies that between the second and the sixth year of the Meiwa period (1765-1769), Gennai was regularly present at what was referred to as "conversations with the Dutch" (*Oranda taiwa*) at the Nagasakiya inn in Edo. It was in this inn that the members of the Dutch factory of Nagasaki and their Japanese interpreters stayed during their annual visit to the Shogun, which usually coincided with the season of cherry blossoms in the capital. Therefore the arrival of the Dutch later came to be used as a seasonal reference in *haikai* (haiku) poetry. From around the late 1750's on, the Hollanders and their Japanese interpreters made it a habit to bring a sampling of Dutch books with them to show at this annual meeting with a circle of Edo scholars.

Hiraga Gennai, a humble ex-samurai who had forsaken his lord in order to be able to devote himself to the study of natural history, was probably one of the best clients of the Dutch book trade. After purchasing Dodonæus' *Cruijdeboeck* in 1765, he consecutively acquired Georgius Rumphius' *D'Amboinsche rariteitkamer* ... (Amsterdam, 1705)[1] in 1766 (Meiwa 3), Jan Swammerdam's *Historia insectorum generalis ofte Algemeene verhandeling van de bloedeloose dierkens* (Utrecht, 1669) in 1767, and in 1768 (Meiwa 5), three books at one time, namely Francis Willughby's *De historia piscium* (Oxford, 1686)[2]; Isaac Bruckner's[3] *Zee-atlas* and *Nieuwe Atlas* (1759); and another huge and important work, Jan Jonston's fauna entitled *Naeukeurige beschryving van de Natuur der Viervoetige Dieren, Vissen en Bloedloze Water-Dieren, Vogelen, Kronkel-Dieren, Slangen en Draken* (Amsterdam, 1660). Finally in 1769, he managed to lay hands on Noel A. Pluche's *Schouwtoneel der natuur, of Samenspraaken over de bysonderheden der natuurlyke histori* ... (*Spectacle de la nature*, 14 vols.).[4]

About Sweerts's *Florilegium* of 1631, Gennai mentioned that he acquired the book in the summer of 1761 (Hôreki 11). It may have been a gift sent from Nagasaki by the Dutch surgeon Bauer, who had agreed to exchange it for a Japanese *slangensteen* ('snake stone'), which Gennai had shown and presented to him at a session of "conversations with the Dutch" in the spring of the same year.

Gennai's list of Western books is of great interest since it bears testimony to a history of encounter by a representative Edo intellectual with European scientific scholarship. How should one appraise this list of works? One could remark that an ignorant samurai in isolated Japan fell victim to the shrewdness of the Dutch traders, who sold him useless old books dating from the sixteenth and seventeenth century, which had become totally out-dated since the brilliant start of the taxonomy of nature by Carolus Linnaeus (1707-78) in the middle of the eighteenth century. But one could equally contend that thanks to the goodwill of the Dutch surgeon, Gennai, in the short period of four to five years, was able to constitute an excellent collection of the most basic European classics of natural history, a "canon" of European scientific tradition since the Renaissance.

We are inclined to opt for the latter claim. Gennai was well aware of the distance that these books had travelled in both time and space since their first publication. In the short comments he added to the list of European books on natural products, he noted down the number of years between the original publication date and his acquisition. For Jonston's fauna of 1660, for example, he commented that 108 years had passed before he bought it in 1768, and in the case of Sweerts's *Florilegium*, 135 years. This point of the time lag between the West and Japan is highly interesting for the study of the cultural and intellectual history of East Asia. Compared to the forty years of delay between the original Dutch edition of Johannes Adam Kulmus' book (1734) and its Japanese translation as *Kaitai shinsho* (1874), most of the European works in Gennai's collection lagged on average 100 years behind at the time of purchase; and 211 years had already passed since the publication of Dodonæus' herbal, if we count from its first original Flemish edition of 1554. Can the simple geographical distance to sail from Antwerp or Leiden to Edo via Batavia and Nagasaki explain this lag? Are the disturbing effects of political and economic isolation of Tokugawa Japan to be blamed, or rather the backwardness of Japanese scholarship in natural history?

Probably all of these factors played a considerable part, but Gennai's awareness of this time lag may be considered an important first step towards dealing with the problem. Kaibara Ekiken, the great encyclopaedic Confucian scholar of the late seventeenth and early eighteenth century, endeavoured to Japanise the up to that time China-centred herbal studies in Japan. Tokugawa Yoshimune, the eighth Shogun, wanted to promote a general survey of the nation's natural products by utilising Dodonæus and Jonston as a model. It was Hiraga Gennai's ambition, as a spiritual successor to these two individuals, to transform the whole system of research from pharmacopoeia to "natural history" by widely opening up the vista to European scholarship and even to foreign products. It can only be considered unfortunate for him that 貝原益軒

the European system of nature studies was further being transformed at a fast pace by the introduction of Linnaeus' system, just around the time that Gennai was trying to establish a new scholarship based on the works of Dodonæus, Jonston, Rumphius and others.

Compared to other books of natural history, the problem of time lag had decreased in importance when Gennai acquired Bruckner's atlases in 1768; he proudly and delightedly noted in his comment to the atlases: "This is a rare book of all times in which one can read all geographical details of the world, recently published in Holland nine years ago; last year, eight years after its publication, I could procure this precious thing from a place 13.000 *li* away."[5] The same goes for Noël Pluche's *Spectacle de la nature* of 1748, which Gennai purchased in 1769, with a delay of twenty years. Profiting from one chapter of Pluche's book he managed to build a miniature windmill to the surprise of many.

As for Dodonæus' herbal, Gennai made good use of the work even before acquiring the book in 1765. A note in the short comment on Dodonæus in Gennai's list of Western books mentions that five copies of the herbal were imported at the time of Yoshimune. One was still in the possession of the Shogun, one was in the hands of Tamura Gen'yu, a naturalist and specialist of Korean ginseng in Edo, two were held by Nagasaki interpreters and lastly one was owned by Gennai himself. Gennai may have had a look at the giant book and admired it at the house of Tamura Gen'yu, who had been his closest friend and most respected master and mentor in herbal studies since his arrival in Edo in 1756. When in 1762 Gennai organised a nation-wide exhibit of natural products of the Japanese Islands, he already quoted the name of Dodonæus and the title of the book in a long manifesto that stated his ideas and announced his enterprise, saying that with the nation-wide collaboration of specialists and amateurs alike it would be possible to discover most items contained in Dodonæus and other Dutch herbals. The following year, selecting about 360 items from the 1762 exhibit and the preceding ones, Gennai published a large work of his own on 物類品隲 natural history in Edo, entitled *Butsurui hinshitsu* ('An Examination into Natural Products').

Gennai's book was much less impressive than Dodonæus': it contained 186 folded sheets of paper, i.e. 372 pages according to the modern way of counting, including two prefaces, one by his teacher

Tamura Gen'yu, and one by a senior colleague of his. The book was divided into six unequal volumes, the last two being an appendix of wood block illustrations. The 360 kinds of natural products were classified into thirteen categories faithfully following the classical method of Li Shizhen (1518-93), the great Ming Dynasty herbalist whose important book *Bencao gangmu* (Jap.: *Honzô kômoku*, 'Elements of Pharmacopoeia,' 1590), was imported in Japan soon after its publication and exerted a long-lasting influence on Tokugawa herbals. Li's thirteen categories were

李時珍
本草綱目

1 water, 2 earth, 3 metal, 4 jade, 5 stones, 6 grasses, 7 grains, 8 vegetables, 9 fruits, 10 trees, 11 insects, 12 scaled animals and 13 animals.

However queer this classification may appear in our view, it had the merit of being synthetic rather than analytic, and practical. Staying on the whole within this framework of traditional East Asian herbal scholarship, Hiraga Gennai gave each product its Japanese and Chinese names, its Dutch name or dialectical alias, if necessary, and explained its shape, colour, properties, benefits, usage and places where to find it. Entries are irregular in length, are often critical about the author's Chinese and Japanese predecessors and, even more interesting, sometimes contain references to the feelings he had when acquiring a product. Because one of the principles that guided Gennai while compiling his book, was to omit banalities and to prize the rather rare products that had been presented at previous exhibits, the category "water", for instance, has only one entry, that of "rose water". In the same category, *Yamato honzô* ('Japanese Herbal,' 1708) by Kaibara Ekiken (1630-1714), the great founder of Japanese herbal studies and natural history, lists not only "hot water" and "hot spring" but also "hail", "chalybeate", "salt" etc., while *Honzô kômoku keimô* ('Commentaries on *Bencao gangmu*,' 1803) by Ono Ranzan (1729-1810), a great scholar and educator of Kyoto, discusses "puddle", "ice of summer", "water in the hollow of a tree or bamboo" and even "rain leaking from the roof" as independent entries in the same category. Hiraga Gennai excluded all these queer, old-fashioned, albeit interesting, entries and preferred to show off himself as a new type of naturalist by choosing the unique and exotic "rose water" for the first page of his book. We will quote the entry as an example of Gennai's style of description:

大和本草
本綱目啓蒙
小野蘭山

"Rose water (Sôbiro): mentioned under the entry of "dew" in [Li's] *Bencao gangmu. Bara no tsuyu* in Japanese, *rôzu wâtoru* (roze water) in Dutch. The *Red-Haired* [Hollanders] call "rose" all sorts of thorny plants; *wâtoru* means "water". This is a liquid distilled from

> rose flowers by an alembic. Among a variety of roses, the wild rose is said to be best suited [for distillation]. Li Dongbi says: "Barbarian countries produce rose water, which has a strong fragrance, and is said to be dewdrops gathered on rose petals, but we do not know whether this is correct or not." Again in the entry on fragrant hedge plants Li states that rose dew is a product of the Southern Barbarian countries, dew water of the rose, which has an unusual fragrance. Doctor Li wrote in this [vague] way since apparently he did not know anything about [the use of] alembic, which is an ingenious invention of the Westerners. The liquid has a remarkable efficacy in surgical treatment. The Hollanders always bring it with them to Nagasaki. In recent times [some] countrymen have learned how to make it and have tried to produce it. However, because of lack of expertise, the water stales easily and does not keep long. It can last for several decades if only a bit of sal ammoniac is added in the process of producing it. The same holds for distillations from Japanese plum and other flowers. In order to keep it, put it in a flask, then cork the flask and wrap its mouth in a sheet of paper. If cork is not available, seal the mouth with wax. Sal ammoniac and cork are explained under their respective headings.[6]"

Although Hiraga Gennai began his entry with a quotation from *Bencao gangmu*, as was customary with Tokugawa herbalists, his reference was in reality intended to demonstrate that even the great Chinese scholar could no longer be considered an authority in the field as far as European products were concerned. Therefore he wrote: "Doctor Li wrote in this way since he apparently did not know anything about [the use of] alembic, which is an ingenious invention of the Westerners", while he himself was very proud of knowing not only this distilling agent, but also the flask, the cork and even such exotic-sounding chemical substance as sal ammoniac (though he mistakenly rendered it as "sal almoniaca"). It is doubtful that he was well acquainted with these European recipients and chemicals: the entry on "sal ammoniac" in his *Butsurui hinshitsu*, for instance, is too brief and too superficial to allow us to draw any conclusion in that respect.

Yet, we still have to admit that these frequent references to things European were an apparent strategy used by the new-generation scholar Gennai in an attempt to unsettle traditional authorities of the China-oriented school of herbals and open up the much larger vista of natural history, by stimulating people's curiosity for the West. The time was favourable for Gennai to put this plan into practice, as Sugita Genpaku (1733-1817), his best friend and colleague in both herbal studies and Dutch learning, recalls in his famous memoirs *Rangaku kotohajime* ('The Dawn of Dutch Studies in Japan,' completed 1815): "About this time [i.e. 1760's], people somehow began to be enamoured of anything Dutch. They would treasure imported vessels and other curious things. A dilettante, if worthy of the name, never failed to have a collection, large or small, of things Dutch. This was especially true at the time when the ex-lord of Sagara Tanuma Okitsugu, (1719-1788) held control over the government as a powerful councillor to the Shogun, and the people were extravagant and gay [...]. People flocked every spring to the inn where the Dutch party was staying." [7]

杉田玄白

田沼意次

Apart from "rose water" there were indeed almost forty kinds of "barbarian" products mentioned or discussed in *Butsurui hinshitsu*. Forty among a total of three hundred sixty items amounts to a very high percentage. It shows clearly the author's predilection for overseas products and constitutes a distinctive feature of the book among the multitude of other herbals in eighteenth century Japan. These foreign products were collected and examined, as we have noted above, in the course of five exhibitions, which he helped to organise, or organised himself in Edo between 1757 and 1762. [8] In one case even a Dutch book was turned into a material for the explanation of a Western product. Following is the entry on "Prussian blue", the European pigment for oil painting which Gennai himself had presented at the fifth exhibition of 1762:

> "Berein brau [Berlijns blauw]. The Red-Haired [Hollanders] brought this to Japan. Similar to *hensei* [Chinese blue], but lighter in substance, yet deeper and brighter in colour than *hensei*.
> In my personal collection I have an illustrated book of Dutch flora that features several thousands varieties, each represented true to life in shape and colour. The blue pigment used in these illustrations appears to be Prussian blue. The colour is truly exquisite [...]." [9]

The book of Dutch flora referred to here is assumed to be a volume of Emanuel Sweerts's *Florilegium*. The large, beautiful book, the first to enter Gennai's collection of Western books, is supposed to have been presented to him by a Dutch surgeon named Bauer, as a token of gratitude, shortly after their meeting at the Nagasakiya inn in Edo in the spring of 1761, where Gennai had shown and presented to him a piece

vij. BOECK. 329
Saffraen-cruyt met de bloemen.

ILL 1

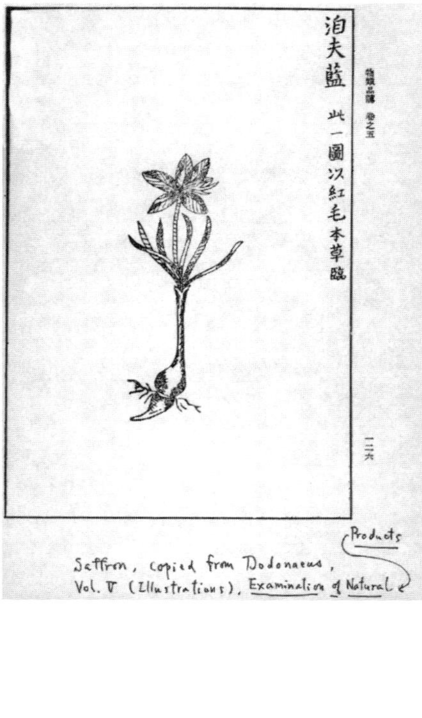
泊夫藍
此一圖以紅毛本草臨
物類品隲 卷之五
一二六
Saffron, copied from Dodonaeus, Vol. V (Illustrations), Examination of Natural Products

ILL 2

of Japanese "slangensteen" ('snake stone' or 'dragon's bone'). At any rate, this was a very peculiar way of using a precious book: not to identify a flower but simply to refer to the colour, probably hand-painted, of its illustrations. It was similar to Gennai's reference to Dodonæus' book, which had not yet come into his possession at the time of compiling *Butsurui hinshitsu*. In the entry on saffron, for instance, Gennai criticises Li Shizhen's description as erroneous and mentions Dodonæus: "in recent times a Dutch scholar named Dodonæus published a herbal in which he gave an extremely detailed illustration of saffron [...]. See the picture in the annex to the present book." [10]

We do not understand why he selected here only saffron from that vast volume of Dodonæus: he may have come across the illustration of saffron when he was hastily turning the pages of the book during a visit to his teacher, Tamura Gen'yu, who already owned a copy. In any case, these references to Dutch books tell us how important the illustrations were for these men of insatiable curiosity, who understood no word of any European language, and who could at best decode a few letters of the alphabet.

Sugita Genpaku, Gennai's close colleague, had a similar experience a few years before he seriously embarked upon the translation of Kulmus' *Ontleedkundige tafelen* (commonly known in Japanese as *Tâheru anatomia*) in 1771. In his memoirs, Genpaku recalled that, ターヘル・アナトミア when in 1765 Yoshio Kozaemon, the official Dutch interpreter from Nagasaki, proudly showed young Genpaku his new acquisition, Laurens Heister's *Heelkundige onderwyzingen* ('Instructions in Surgery,' 1741), at the Nagasakiya inn, "the illustrations in the book looked markedly different from those in Japanese or Chinese books." Thus, though not knowing a word of Dutch, "just viewing their exquisite precision, I felt as if being enlightened. So, I borrowed the book for some time as I wanted to copy at least the pictures [...]."

The appeal of the illustrations in these scientific books was not limited to a small circle of intellectuals in eighteenth century Edo. According to George Sarton, it was much the same in Europe, the home of Dodonæus. Sarton writes in a chapter on natural history in the Renaissance: "There is no doubt that the illustrations contributed very much to the popularity of these early herbals. Some readers in that age, as in our own, preferred looking at pictures to reading the text; this was especially true when the herbals were written in Latin, which put them out of reach of all but the learned doctors. Love of plants and interest in herbs and roots was not by any means confined to scholars; even women might want to know more about them and to be able to recognise them in the fields. Hence the pictures were very welcome. This is shown by the publication of books with a minimum of text and a maximum of pictures."[11]

Only looking at, referring to or sporadically copying illustrations in Western books, however, did not satisfy the scholars of Edo. They were quite naturally eager to translate them or at least to have them translated for them into Japanese. Again Sugita Genpaku recalls those years of groping in the dark in his memoirs *Rangaku kotohajime*: "So often when I met Hiraga Gennai and others, we would say to each other: 'The more we become aware of Dutch learning, the more strongly we are impressed by its empirical spirit. It would be a great benefit, if we would translate their books into Japanese. It is a pity that so far no one has tried it. Really, we must somehow find a way [...]. Even just one volume of them, if rendered into Japanese, would be of immense value to the nation!"[12]

It was in the early winter of 1770 that Hiraga Gennai finally left Edo for his second visit to Nagasaki, with the intention of realising his dream of translating into Japanese a book of European natural history with the help of interpreters in that open port. He carried in his travel bag the heavy volume of Dodonæus. But he was too much of a restless and mercurial mind: he tried his hand at many new pursuits and experimentations, including a Western-style oil painting, but he does not seem to have made any steady progress in the study of Dutch grammar during this one year stay in Nagasaki. When in 1772 he carried back his Dodonæus to Edo almost as intact as it had been at his departure, the translation of Kulmus' *Ontleedkundige tafelen* by the group of Sugita Genpaku, Maeno Ryôtaku and Nakagawa Jun'an (前野良沢、中川淳庵) was already well underway. The Dutch books of natural history by Dodonæus, Jonston, Sweerts and others, fascinated and enthralled the inquisitive mind of the ex-samurai Hiraga Gennai, and led him to dream of writing at least ten volumes, covering the wealth of flora, fauna and minerals of the Japanese archipelago, a feat that would rank him with his European predecessors. His heroic ambitions and high-spirited actions strongly encouraged many scholars, writers and artists of his time to go beyond the received intellectual and cultural boundaries of Tokugawa Japan and opened the road to a new, modern style

D'AMBOINSCHE

RARITEITKAMER,

Behelzende eene BESCHRYVINGE van allerhande
zoo weeke als harde

SCHAALVISSCHEN,

te weete raare

KRABBEN, KREEFTEN,

en diergelyke Zeedieren,

als mede allerhande

HOORNTJES en SCHULPEN,

die men in d'Amboinſche Zee vindt:

Daar benevens zommige

MINERAALEN, GESTEENTEN,

en ſoorten van AARDE, die in d'Amboinſche, en zommige omleggende Eilanden gevonden worden.

Verdeelt in drie Boeken,

En met nodige PRINTVERBEELDINGEN, alle naar 't leven getekent; voorzien

Beſchreven door

GEORGIUS EVERHARDUS RUMPHIUS,

van Hanauw, Koopman en Raad in Amboina, mitsgaders Lid van het Kyzerlyke kweekſchool der onderzoekers van de Natuurkunde in 't Duitſche Roomſche Ryk opgerecht onder den naam van

PLINIUS INDICUS.

T' AMSTERDAM,

By JAN ROMAN DE JONGE, Boekverkoper, 1741.

ILL 3

of natural history as well as to the development of Dutch-European studies in Japan. But in 1780, at the age of 52, the protagonist himself, exhausted, was destined to be drawn into a deadly whirlpool he had created himself.[13]

ILLUSTRATIONS

ILL 1 Saffron (Saffraen-cruyt met de bloemen) from Dodonæus' *Cruydt-Boeck*, 1618 edition, p. 329. Collection Katholieke Universiteit Leuven, Central Library, Tabularium, B 3998.

ILL 2 Saffron plant with flower, as copied from Dodonæus' herbal in *Butsurui hinshitsu* (1763), maki 5: illustrations (sanbutsu zue) (Irita Seizô, ed., *Hiraga Gennai zenshû* (original edition of 1932), vol. I, p. 126)

ILL 3 Title page of Georgius Rumphius, *D'Amboinsche rariteitkamer, behelzende eene beschryvinge van aller hande zoo weeke als harde schaalvisschen, te weeten rare krabben, kreeften, en diergelijke zeedieren, als mede allerhande hoorntjes en schulpen, die men in de Amboinsche zee vindt: Daar beneven zommige mineraalen, gesteenten, en soorten van aarde, die in d'Amboinsche, en zommige omleggende Eilanden gevonden worden* (Amsterdam, Jan Roman de Jonge, 1741). Collection Université catholique de Louvain, Bibliothèque générale et de sciences humaines.

ILL 4 Frontispice of Georgius Rumphius, *D'Amboinsche rariteitkamer ...* , (Amsterdam, François Halma, 1705, reprinted in 1741 edition by Jan Roman de Jonge). Collection Université catholique de Louvain, Bibliothèque générale et de sciences humaines.

ILL 5 Title page of Johannis Swammerdam, *Historia Insectorum Generalis, ofte Algemeene Verhandeling van de Bloedeloose Dierkens* (Utrecht: Meinardus van Dreunen, 1669). Collection Katholieke Universiteit Leuven, Central Library, Tabularium, BTAB A83299.

ILL 6 Picture of insect: gatefold in Johannis Swammerdam, *Historia Insectorum Generalis, ofte Algemeene Verhandeling van de Bloedeloose Dierkens* (Utrecht: Meinardus van Dreunen, 1669) between pages 136 and 137. Collection Katholieke Universiteit Leuven, Central Library, Tabularium, BTAB A83299.

ILL 7 Title page and frontispice of Noël Pluche, *Le Spectacle de la Nature ou Entretiens sur les Particularités de l'Histoire Naturelle*, Tome premier, Paris, les Frères Estienne, MDCCLIV (1754). Collection Université catholique de Louvain, Bibliothèque générale et de sciences humaines.

ILL 8 Frontispice of Joh. Jonstoni, *Thaumatographia Naturalis, in decem classes distincta*. Amstelodami, Apud Joannem Janssonium à Waesberge, et Elyzeum Weyerstraet, MDCLXV (1665) Collection Université catholique de Louvain, Bibliothèque générale et de sciences humaines.

ILL 9 Title page of Joh. Jonstoni, *Thaumatographia Naturalis, in decem classes distincta*. Amstelodami, Apud Joannem Janssonium à Waesberge, et Elyzeum Weyerstraet, MDCLXV (1665). Collection Université catholique de Louvain, Bibliothèque générale et de sciences humaines.

ILL 10 Joh. Jonstoni, *Thaumatographia Naturalis, in decem classes distincta*. Amstelodami, Apud Joannem Jans sonium à Waesberge, et Elyzeum Weyerstraet, MDCLXV (1665): classis septima in qu quadrupedum admiranda. Collection Université catholique de Louvain, Bibliothèque générale et de sciences humaines.

ILL 11 Portrait of Hiraga Gennai (1728-80) by Kimura Mokurô, councillor to the Lord of Takamatsu, included in Kimura's book on Gesaku writers Gesakusha kôhoi (1845), reproduced in Irita Seizô, ed., *Hiraga Gennai zenshû* (original edition of 1932)

ILL 12 Portrait of Hiraga Gennai (1728-80). This portrait was included in an autograph manuscript entitled *Sentetsu zôden*, attributed to Katsuragawa Hoshû. Before the war it was in the possession of the library of the Imperial University of Tokyo, but is since presumably lost. The portrait is alleged to have been a first-hand copy of an original attributed to Morishima Chûryô (alias Katsuragawa Hosan), a disciple of Gennai. It was reproduced in Irita Seizô, ed., *Hiraga Gennai zenshû* (original edition of 1932).

REFERENCES

1 Haga Tôru. *Hiraga Gennai* 「平賀源内」. Tokyo: Asahi Shinbunsha, 1981; paperback edition, 1989.

2 Irita Seizô, ed. *Hiraga Gennai zenshû* 「平賀源内全集」 ('The Complete Works of Hiraga Gennai'). Reprint edition. Meicho Kankôkai, 1970. All quotations of Gennai are from this edition.

3 Sugita Genpaku. *Dawn of Western Science in Japan*. trans. Matsumoto Ryôzô and Kiyooka Eiichi. Tokyo: Hokuseidô Press, 1969.
4 Sugita Genpaku, *Rangaku kotohajime* 「蘭学事始」. Ed. Ogata Tomio. Tokyo: Iwanami bunko, 1959; re-edition, 1982.

NOTES

1 Full title: *D'Amboinsche rariteitkamer, behelzende eene beschryvinge van allerhande zoo weeke als harde schaalvisschen, te weeten rare krabben, kreeften, en diergelijke zeedieren, als mede allerhande hoorntjes en schulpen, ... die men in de Amboinsche zee vindt: daar beneven zommige mineraalen, gesteenten, en soorten van aarde, die in d'Amboinsche, en zommige omleggende eilanden gevonden worden.*

2 Full title: *De historia piscium libri quatuor, jussu & sumptibus Societatis regiae londinensis editi. In quibus non tantum de piscibus in genere agitur, sed & species omnes, tum ab aliis traditae, tum novae & nondum editae bene multae, naturea ductum servante methodo dispositae, accurate describuntur. Earumque effigies, quotquot haberi potuere, vel ad vivum delineatae, vel ad optima exemplaria impressa; artifici manu elegantissime in aes incisae, ad descriptiones illustrandas exhibentur. Cum Appendice historias & observationes in supplementum operis collatas complectente. Totum opus recognovit, coaptavit, supplevit, librum etiam primum & secundum intergros adjecit Johannes Raius e Societate regia.*

3 Isaac Bruckner (1686-1762) was a cartographer, engraver and globe maker, appointed geographer to the court of Louis XV of France. He is the author of a *Nouvel Atlas de Marine* (Berlin, 1749). A Dutch translation was published in The Hague.

4 For a detailed analysis about the identification of these book titles see Haga Tôru, *Hiraga Gennai* (Tokyo: Asahi Shinbunsha, 1981), 302 FF.

5 Haga Tôru, *Hiraga Gennai* (Tokyo: Asahi Shinbunsha, 1981), 316.

6 *Butsurui hinshitsu, maki* no. 1, Irita Seizô, ed., *Hiraga Gennai zenshû*, vol. I, 11-12.

7 Translation by Ryôzô Matsumoto, in: Sugita Genpaku, *Dawn of Western Science in Japan*, trans. Matsumoto Ryôzô and Kiyooka Eiichi (Tokyo: Hokuseidô Press, 1969), 16-17.

8 Described in detail in chapter 8 of Haga Tôru, *Hiraga Gennai* (Tokyo: Asahi Shinbunsha, 1981), 105-128. In his own account of the 1762 exhibition, entitled *Tôto yakuhin-kai*, he makes explicit reference to Dodonæus' *Cruydt-Boeck*.

9 *Butsurui hinshitsu, maki* no.2, Irita Seizô, ed., *Hiraga Gennai zenshû*, vol. I, 35.

10 *Butsurui hinshitsu, maki* no. 3, Irita Seizô, ed., *Hiraga Gennai zenshû*, vol. I, 61.

11 George Sarton, *Six Wings: Men of Science in the Renaissance* (Meridian Books, The World Publishing Company, 1966), 135.

12 Translation by Ryôzô Matsumoto, o.c., 25-26.

13 For fuller details of Hiraga Gennai's biography, see Hiraga Gennai nenpu, appended to Haga Tôru, *Hiraga Gennai* (Tokyo: Asahi Shinbunsha, 1981).

'T AMSTERDAM, Gedrukt by FRANÇOIS HALMA Boekverkoper 1705.

ILL 4

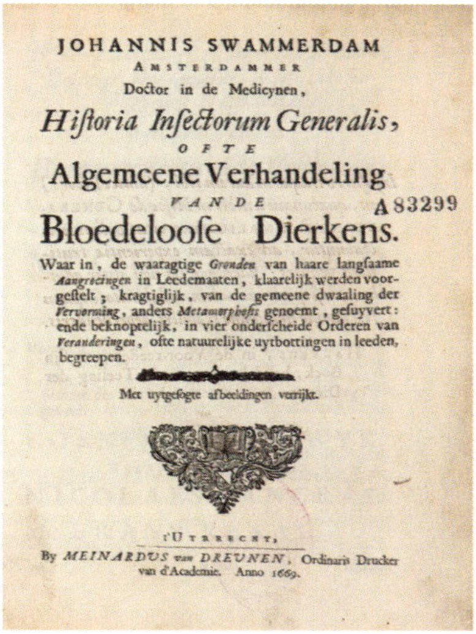
JOHANNIS SWAMMERDAM
AMSTERDAMMER
Doctor in de Medicynen,
Historia Insectorum Generalis,
OFTE
Algemeene Verhandeling
VAN DE
Bloedeloose Dierkens.
Waar in, de waaragtige *Gronden* van haare langsaame *Aangroeingen* in Leedemaaten, klaarelijk werden voorgestelt; kragtiglijk, van de gemeene dwaaling der *Vervorming*, anders *Metamorphosis* genoemt, gesuyvert: ende beknoptelijk, in vier onderscheide Orderen van *Veranderingen*, ofte natuurelijke uytbottingen in leeden, begreepen.
Met uytgesogte afbeeldingen verrijkt.
t'UTRECHT,
By *MEINARDUS* van *DREUNEN*, Ordinaris Drucker van d'Academie. Anno 1669.

ILL 5

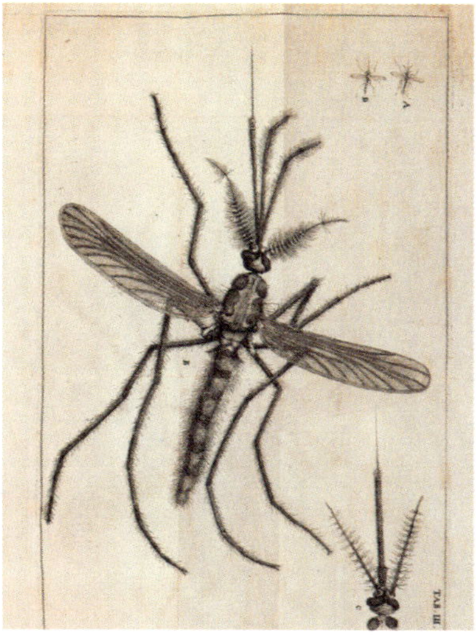

ILL 6

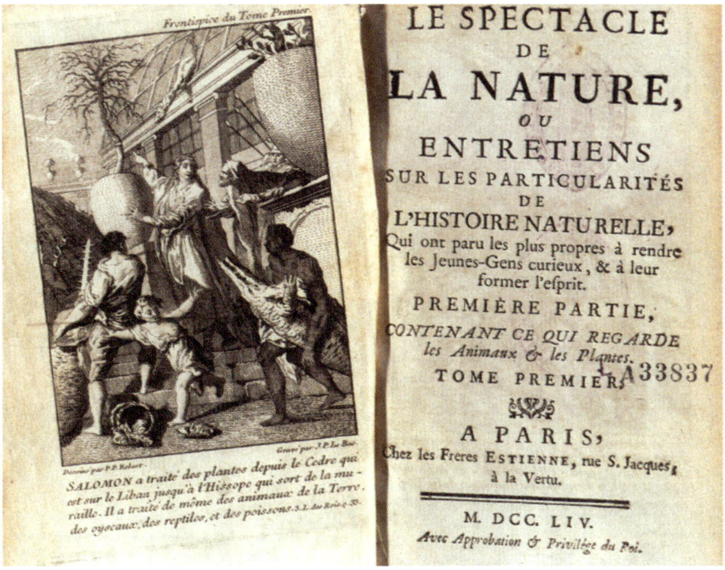
Frontispice du Tome Premier.
Dessiné par P.P. Robert. Gravé par J.P. Le Bas.
SALOMON a traité des plantes depuis le Cedre qui est sur le Liban jusqu'à l'Hissope qui sort de la muraille. Il a traité de même des animaux de la Terre, des oyseaux, des reptiles, et des poissons. 3. L. des Rois. 4. 33.

LE SPECTACLE
DE
LA NATURE,
OU
ENTRETIENS
SUR LES PARTICULARITÉS
DE
L'HISTOIRE NATURELLE,
Qui ont paru les plus propres à rendre les Jeunes-Gens curieux, & à leur former l'esprit.
PREMIÈRE PARTIE,
CONTENANT CE QUI REGARDE les Animaux & les Plantes.
TOME PREMIER.
A PARIS,
Chez les Freres ESTIENNE, rue S. Jacques, à la Vertu.
M. DCC. LIV.
Avec Approbation & Privilége du Roi.

ILL 7

ILL 8

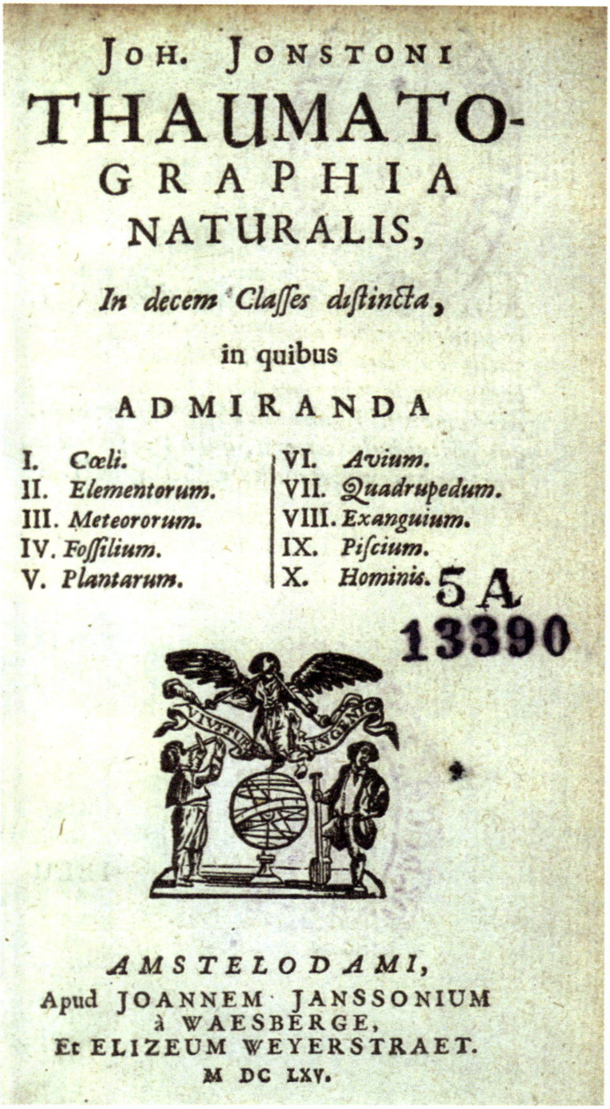

JOH. JONSTONI

THAUMATO-
GRAPHIA
NATURALIS,

In decem Classes distincta,

in quibus

ADMIRANDA

I. *Cœli.*	VI. *Avium.*
II. *Elementorum.*	VII. *Quadrupedum.*
III. *Meteororum.*	VIII. *Exanguium.*
IV. *Fossilium.*	IX. *Piscium.*
V. *Plantarum.*	X. *Hominis.*

AMSTELODAMI,
Apud JOANNEM JANSSONIUM
à WAESBERGE,
Et ELIZEUM WEYERSTRAET.
M DC LXV.

ILL 9

THAUMATO-GRAPHIÆ NATURALIS

CLASSIS SEPTIMA.

In quâ

QUADRUPEDUM ADMIRANDA.

Seneca l. 3. de Ira. cap. 30.

Frivolis turbamur & inanibus. Taurum color rubicundus excitat, ad umbram aſpis exurgit. Vrſos leoneſque mappa proritat. Omnia quæ natura fera ac rabida ſunt, conſternantur ad vana. Idem inquietis & ſtolidis ingeniis evenit: rerum ſuſpicione feriuntur.

ILL 10

ILL 11

ILL 12

XII

THE DEVELOPMENT OF JAPANESE BOTANICAL INTEREST AND DODONÆUS' ROLE: FROM PHARMACOPOEIA TO BOTANY AND HORTICULTURE

Yôzaburô Shirahata
International Research Center for Japanese Studies, Kyoto

徳川時代
蘭学
本草学者
松平定信

Dodonæus' *Cruydt-Boeck*, in particular in its Dutch editions of 1618 and 1644, was one of the great Western books that exerted a strong influence on the Japanese intellectuals of the Tokugawa period (*Tokugawa jidai*, 1600-1868). It can be said without exaggeration that Dodonæus' book opened the way to *Rangaku*, which literally means 'Dutch studies', but generally denotes the Western scientific inquiry of nature, an approach which was quite new to the Japanese scholars of the time. Especially the illustrations in the book made a deep impression on the intelligentsia, notably the *Honzôgakusha* or scholars of phytology. For more than a century, scholars would repeatedly attempt to read or translate the book and even two and a half centuries after the publication of its first edition, Matsudaira Sadanobu (erstwhile chief councillor to the Shogun, fl. 1758-1829), undertook the project of making a complete translation into Japanese. The question as to how deeply Dodonæus' book has influenced Japanese pharmacopoeia and phytology, has been debated from diverse angles, but we still do not have an accurate grasp of the extent to which it has stimulated the development of actual botany in Japan. In this essay, I would like to investigate the role Dodonæus' book has played in stimulating interest in Japanese herbs or plants.

1 The "discovery" of Dodonæus

青木昆陽

The eighth Shogun Tokugawa Yoshimune attached much more importance to practical or utilitarian studies than his predecessors and actively promoted the introduction of Western scientific knowledge and technology. This was no doubt inspired by his ambition to foster an indigenous enterprise and production. He relaxed the ban on Chinese translations of Western books provided they did not deal with the practice of Christianity. He also ordered two scholars, Aoki Kon'yô and Noro Genjô, to study the Dutch language. This new attitude of the Shogun was the initial impetus of what subsequently would develop into *Rangaku* or "Dutch Studies".

野呂元丈

It all began in the spring of the second year of *Kyôhô* (1717). Yoshimune had inherited the Shogunate just the year before. It is assumed that he ordered for two Dutch books, the *Cruydt-Boeck* by Dodonæus, and the *Naeukeurige beschryving van de natuur der viervoetige dieren, vissen en bloedloze water-dieren, vogelen, kronkel-dieren, slangen en draken* (Amsterdam 1660) by Jan Jonston [John Johnston], to be taken out from the Momijiyama bunko, the Shogunal library, or at least one of these.[1] According to the diary of *Opperhoofd* Aouwer, who had an audience with Yoshimune in that year, Yoshimune had Johnston's book shown to him, and asked him some questions about it. These were extraordinary questions to be asked during a routine meeting between the Shogun and the Dutch officials from Deshima at Edo castle (*Edo-jô*), because ordinarily the exchanges during such audience never concerned concrete topics. It was unusual for the Shogun to ask specific questions about a book through an interpreter and since he did not understand a word of Dutch, his interest must have been aroused by the illustrations it contained. The questions were a demonstration of his strong intellectual curiosity and his deep interest in the West.

享保
紅葉山文庫
出島
江戸城

In that year, medical officers of the Edo Bakufu too visited the Nagasakiya inn, where the Dutch delegation was staying and again asked questions about Johnston's book. Kurisaki Dôyû and three other doctors showed Johnston's book to the surgeon of the delegation, Willem Wagemans, and asked him about the names and habitats of the animals that were illustrated in the book. Namura Yazaemon, the chief interpreter (*ô-tsûji*), was reportedly on hand to do the interpretation. This query cannot have been unrelated to the Shogun's interest. Thus in the second year of *Kyôhô* Johnston's book moved all of a sudden into the limelight, fifty-four years after its introduction into Japan.

長崎屋
栗崎道有
名村八左衛門
大通詞

It was indeed in the third year of *Kanbun* (1663) that the book *Naeukeurige beschryving* had been introduced into Japan. In that year *Opperhoofd* Hendrik Indijk took it as one of the gifts to be presented to the Shogun on the occasion of the annual audience at Edo castle. Dodonæus' *Cruydt-Boeck* had already been presented to the Shogun in the third month of 1659 (second year of *Manji*) by *Opperhoofd* Zach-

寛文
万治

arias Wagenaer. However, the Bakufu official in charge of the audience at that time was unable to appreciate the value of the book of Dodonæus properly, and he reportedly gave orders that another book with larger and finer illustrations be presented. It was to satisfy this Japanese request, that Johnston's book was presented four years later.

Dodonæus' book is a large volume of about 1.500 pages, containing more than 2.000 illustrations, but the illustrations are woodcuts and relatively small, being arranged one or two to a page, whereas the illustrations in Johnston's book are all copperplates of full-page size, and of a more luxurious execution than Dodonæus' illustrations. The Bakufu official seems to have been satisfied with the Johnston book. Since, however, there was nobody on hand who could read the Dutch text, both books were stored away in the Momijiyama bunko, without ever seeing the light of day again.

It was thanks to Yoshimune's curiosity that these 'forgotten' Dutch books were eventually rescued from oblivion and it were no doubt the illustrations that aroused the Shogunal interest in the books. The illustrations in the two books differ from one another in three respects:

1 Subject matter: illustrations of animals versus of plants.
2 Medium: copper prints versus woodcuts.
3 Size: full-page size versus small size.

The Japanese are likely to have been more attracted by the animals, the copper prints and the large size. There is no doubt that Dodonæus' book also contained many plants that were new and strange to the Japanese, but the unknown animals in Johnston's book must have struck their fancy more powerfully than the plants. Especially people who were not specialists were likely to be fascinated by illustrations of unknown creatures. Moreover, the Japanese were already familiar with the technique of woodcuts, while the technique of copper prints was new to them. Moreover, copper prints allow for visual effects that woodcuts cannot approximate, and this too cannot have failed to strike the fancy of the Japanese. In addition, Johnston's illustrations were of a more powerful expression and larger in size. As a result, Johnston's book was the first to arouse interest and elicit questions among the Japanese. Strangely enough, however, no attempt was made at that time to translate Johnston's book into Japanese. Not until the 寛保 first year of *Kanpô* (1741), twenty-five years after Yoshimune had first shown interest in the book, did Noro Genjô try to understand its contents. One of the reasons why this book was left unstudied for such a long time, is that, even if they were given explanations by the Dutch about the contents of the book, there was not enough intellectual interest among the Japanese of the time. Even Noro Genjô, who finally manifested an interest in the book, contented himself with a few questions, and without any further ado went on to focus his questions on Dodonæus' book. We can think of several reasons for this shift, but it is important to note that he was essentially a scholar of *Honzôgaku*. In Johnston's book, there were few descriptions about herbs and their medicinal effects, whereas Dodonæus' book abounded with explanations about medicinal herbs and their effects. After only a few questions about Johnston's book, Noro shifted his attention to Dodonæus' book and continued asking questions for ten years over in an effort to grasp and understand its contents.

本草学

II Understanding Dodonæus' book

Noro Genjô (1693-1761) was born in Hatano, Seta-mura, Taki-gun, in Ise province. He went to Kyoto to study Confucianism under the guidance of Namikawa Tenmin, an authority of the Kogaku school of a stature equal to Itô Tôgai, and to study *Honzôgaku* with Inô Jakusui. In 1720 (fifth year of *Kyôhô*) he was persuaded by Niwa Seihaku [Shôhaku], a man from the same region and a fellow-pupil of Inô Jakusui, to go to Edo with him and take up a post as official herb collector for the Bakufu. In 1724 (ninth year of *Kyôhô*) he was given a residential site by the Bakufu. In 1739 (fourth year of *Genbun*) he became a physician in ordinary (*Omemie*) to the Shogun, and in 1741 (first year of Kanpô) he was appointed by Yoshimune as *Oranda honzô goyô* ('official in charge of Dutch pharmacopoeia') and started to study Johnston's book in earnest. In the third month of that year Noro paid a visit to the Dutch delegation that was just in Edo for the yearly audience with the Shogun. He was received by *Opperhoofd* Van der Waeijen, Secretary Briel and Surgeon Musculus. With the book of Johnston before him and with the help of the chief interpreter Yoshio Tôzaburô, he asked them for clarification about a few matters. Later Noro wrote a book, based on the answers he had received on this occasion, entitled *Oranda kinjû chûgyo no zu wage* ('Illustrations of Dutch Animals, Insects and Fishes, Explained in Japanese').

波多野勢多村多気郡
伊勢国
並河天民、古学派
伊藤東涯、稲生若水
丹羽正伯
元文
御目見
吉雄藤三郎
阿蘭陀禽獣虫魚図和解

This book was very simple indeed: most entries were hardly anything more than a list of names of animals, insects or fishes. For example, the

馬 entry for "horse" (*uma*) simply mentioned "pâruto" (*katakana* trans-
野馬 literation for Dutch "paard"), while the entry for *yaba* ('wild horse')
had the word "*eruto pâruto*" (*katakana* for Dutch "*wild paard*").
These were simply transcriptions of the Dutch equivalents as he had
heard them pronounced by his Dutch informants. An exceptional case
was the entry about the elephant, because it contained more detailed
information. It gives "*oorihare*" as its Dutch equivalent (i.e. "*olifant*"),
解 while in the explanation *ge* it says: "the length of the elephant, meas-
丈、尺 ured from the tip of its nose, can reach one *jô* and nine *shaku*, and in
説 the commentary *setsu* section, the author goes on to report that "these
animals exist in great number in countries visited by the Dutch. However, in none of these countries do they regard their meat as edible. Neither have the Dutch heard that their skins, bones or dung are used as medicine. However the tusks, called "*ihôruto*" (in *katakana*, Dutch "*ivoor*"), is used for medical purposes".

In the preface to his book Noro explains about *ge* and *setsu*: *ge* refers to the Japanese translation of what is written in the body of the text, whereas *setsu* refers to oral explanations the Dutch gave in reply to his questions. In other words, Noro Genjô selected from Johnston's book the plates that he understood or that struck his fancy, and the accompanying text he had a Dutchman explain to him. The record of what he had thus collected is the *Oranda kinjû chûgyo no zu wage*. The number of selected items was only eighty-one, and the explanations are quite simple, as already stated.

Reading through the preface of Noro Genjô's book, we can understand what he felt while he was in the process of "translating" this book. He writes: "This book (i.e. Johnston's book) was not meant as a herbal to identify plants with medicinal properties. Therefore it does not give any description about the medicinal effects of plants, but only about their shape or appearance. Even (the Dutch) said that they could not understand the contents properly because the explanations included so many Latin words."

Consequently, not only had the Dutch he had asked for explanation, not enough knowledge of Latin to satisfy his queries, but he must also have felt that, contrary to what he had expected, the Johnston book hardly contained any descriptions of medicinal properties of herbs. The fact that he did not write any sequel to the *Oranda kinjû chûgyo no zu wage* suggests that he subsequently did not submit any further queries to the Dutch about the book. He must have lost interest in it and stopped his attempts at translating it, although there still remained many items. After that time, nobody showed interest in Johnston's book.

The following year, the chief interpreter Yoshio Tôzaburô submitted queries about eleven plants to the Dutch delegation, including the roses, grapes, rice and maize. On the basis of this exchange he compiled the book *Shin Yû Oranda honzô* ('Dutch Herbal, [Compiled] in the Year of *Shin Yû*'). The following year saw the compilation of another book titled *Oranda honzô* ('Dutch Herbal'), a collaborative effort of the chief interpreter Nakayama Zenzaemon and the assistant interpreter Shige Shichirôzaemon who took care of the Japanese translation (*wage*) of two plants, while the translations of the remaining plants were the work of Noro Genjô. The work of compiling Japanese versions of Dutch herbal descriptions with Noro Genjô as the key person, went on for about ten years and continued until 1750 (the third year of *Kan'en*). Yet in the end no more than 119 herbs were dealt with, which reminds us of how arduous an enterprise translating from Dutch into Japanese was at that time.

辛酉阿蘭陀本草
阿蘭陀本草
中山善左衛門
茂七郎左衛門
和解
寛延

III Matsudaira Sadanobu's interest in botany, horticulture and Dodonæus.

It was Matsudaira Sadanobu who planned to translate Dodonæus' book completely. He ordered Ishii Shôsuke, a former interpreter in Nagasaki, to translate the book. After Ishii's death, Yoshida Seikyô, a clansman of the Tayasu clan residing in Edo, succeeded him and completed the translation in 1823 (sixth year of *Bunsei*). Unfortunately the 170 volumes of the manuscript translation, together with part of the block copies that had already been completed, as well as all engraved blocks were almost all lost in the great fire that hit Edo in the third month of 1829 (sixth year of *Bunsei*). The collection that Waseda University Library presently holds is believed to be a part of the manuscript translation that Matsudaira Sadanobu had ordered.

石井庄助
吉田正恭
田安家
文政

The title of the book was to have been *Ensei Dodoneusu Sômoku-fu* ('The Western Herbal of Dodonæus Illustrated'), and if the book had been published it would have been the crowning achievement of all the efforts that had been invested in *Rangaku* since its rise. Yet, more than 200 years had already passed since the original had been published, so even if this translation had been published, it would hardly have been

遠西ドドネウス草木譜

more than the presentation to a Japanese readership of a major classic in the field of natural history and botany.

Ironically, in the same year that the manuscript translation of

泰西本草名疏 Dodonæus' book went up in flames, *Taisei honzô meiso* ('The Nomen-
伊藤圭介 clature of the Western Herbals') by Itô Keisuke (1803-1901) was published. This book presents in alphabetical order the Latin names of the plants in Thunberg's *Flora Japonica*, followed by their Japanese equivalent. Thunberg did evidently not adopt the taxonomic system of Dodonæus, but rather that of his master Linnaeus. By adopting this system it was introducing the methodology of modern botany in Japan. The focus was no longer on the collection of information about plants, as embodied in natural history, but on the research of plants as embodied in botany.

Let us take a look at some of the characteristic features of *Ensei Dodoneusu Sômoku-fu*. While in the original the illustrations are inserted into the corresponding passage describing the plant, in the *Ensei Sômoku-fu* each illustration is isolated and enlarged to full-page size. On the next page, the Latin name is mentioned, sometimes followed by the German and/or French, and English name. Then follow the Japanese and Chinese names. Next come a general description, a notice on the medicinal properties and therapeutic value, and sometimes the compiler (supposed to be Yoshida Seikyô) has added supplementary comments on the curative virtues.

As to the overall structure of the book, the translation closely follows the descriptions and the illustrations are faithfully reproduced, but the order of items has been rearranged, so as to fit the traditional Japanese make-up of herbals. Yoshida Seikyô, who was the chief compiler, and had a rich knowledge of Japanese herbal science, evidently supplemented the translation with data about the medical virtues of plants, in an effort to present the book as a genuine herbal (*Honzôsho*). When he saw the original Dodonæus with its 1.500 pages, Yoshida must have felt that the book was intended as a comprehensive study about plants, and therefore worthy of a full translation into Japanese.

Why then did Matsudaira Sadanobu plan to publish a complete translation of Dodonæus' *Cruydt-boeck*? First of all, he had always had
天明 a strong interest in the natural world. In 1786 (sixth year of *Tenmei*),
老中 one year before he was appointed *Rôjû* ('senior councillor'), for exam-

ple, he had ordered to transfer honeybees from Kishû Wakayama to Edo, to breed them with a view to honey production, in three places in Edo: the Fukiage garden and Ninomaru garden in Edo Castle and the garden in the Edo residence of the Kishi clan. He was also greatly interested in gardening. In 1792 (fourth year of *Kansei*), he made the grounds which the Bakufu had granted him, into his suburban residence (*shimo-yashiki*), and laid out a garden, which he called *Yokuon-en* ('garden of my lord's favour'). This garden became the scene for his experiments in gardening, cultivation of plants and horticulture. After retiring as *Rôjû* in 1812 (ninth year of *Bunka*), he made the *Yokuon-en* his permanent residence. Styling himself *Raku-ô* ('happy old man'), *Fûgetsu-ô* ('old man of wind and moon'), *Kyokuhô* ('peak of the rising sun') or *Kagetsu Shujin* ('master of flowers and moon'), he indulged himself in a life of poetical composition, essay writing, gardening and horticultural pursuits.

紀州和歌山
吹上の庭、二之丸の庭
寛政
下屋敷
浴恩園
文化
楽翁
風月翁、旭峯
花月主人

The *Yokuon-en* boasted a layout known as *kaiyû-shiki* ('strolling garden'), a style much favoured by *daimyô* and high-ranking *bushi* during the Edo period. Moreover, because the garden was by the seaside, it had an ideal location for *shio-iri* ('drawing in the tide'), which meant that seawater could run into the pond in the middle of the garden through a small artificial waterway. This provided an ideal location for social gatherings and poetry parties with friends, but at the same time served as the testing ground for his experiments in gardening and horticulture, as well as a place for exhibiting samples.

回遊式
大名、武士
潮入り

Consequently the garden served as a kind of botanical garden, where all kinds of varieties could be admired. One area of the garden was called *Chigusa no sono* ('garden of a thousand grasses') and was even counted as one of the fifty-one most beautiful gardens of the time. Here were grown a variety of horticultural species including lotus, sasanqua and peach. In particular rare varieties of lotus were being collected and grown, including 'asahi', 'tennyo', 'goji-in' and 'tôzan-tenjiku-hasu'. There were also collections of peaches of both varieties, the ones appreciated for their flowers and the ones grown for their fruits. Moreover, he also laid out a section for herbs and medical plants, called *Go-yakuen* as well as an orchard called *karin*. All this is proof that from the mid-Edo period onwards, the burgeoning interest in herbs and natural history among the *bushi* or ruling class was reflected in the layout of the *daimyô* gardens and had an impact on the landscape architecture. It played a significant role in their culture and lifestyle.

千草の園
朝日、天女、護持院、唐山天竺蓮
御薬園
菓林

Matsudaira Sadanobu bred several kinds of plants, but had a particular fancy for the lotus. In his book *Seikô-fu* 清香譜, which he compiled after his retirement from public life, he reportedly illustrated over ninety kinds of lotus. Unfortunately this book was lost during the Second World War, so that we cannot know its exact contents. However, another book entitled this one *Seikô gafu* 清香画譜, a collection of illustrations of lotuses, is still extant in Tenri University Library, and is believed to be related to Matsudaira's missing book. Judging from this *Seikô gafu*, Matsudaira's *Seikô-fu* appears to have been an album of beautifully coloured illustrations of a number of lotus varieties. Beside each illustration the date of the drawing is given, and the shape and inflorescence of the flower is described. It can be said to be a scientific pictorial based on actual observation. It is probably the record of Matsudaira's observation of the lotuses in the *Yokuon-en*.

Originally the lotus flower is the flower of the Buddha. It was only treated as a prop in Buddhist ritual. But thanks to Matsudaira Sadanobu, the lotus became an object of observation and appreciation, of which numerous variations were bred. It was the elegant shape and delicate fragrance of the lotus that he found attractive. *Seikô*, meaning "pure fragrance" denotes the lotus flower. Matsudaira's predilection for the lotus may be understood as indicative of a typically Japanese appreciation of plants.

Matsudaira's interest in Dodonæus' *Cruydt-boeck* was an extension of his interest for the natural world and his love for gardening and horticulture. However, his interest in plants is different from *Honzô-gaku*, the traditional study of herbs with medicinal properties. His interest in plants was free from a concern for medical effects, and was informed by a scientific and horticultural attitude that focuses on the plants themselves, their shapes and their characteristics. Furthermore, his interest in plants was an interest in their beauty, for aesthetics have always informed Japanese interest in plants. His botanical interest as it is embodied in *Seikô-fu* is an aesthetic interest in the broad sense of the term, and that is why its expression has taken the form of a pictorial. Dodonæus' book stimulated Japanese interest in plants on many levels. It was assimilated in a typically Japanese way, and aroused interest in horticulture, heightened the aesthetic appreciation of plants and strengthened the urge to depict them.

REFERENCES

1 Arai Hakuseki 新井白石, Matsumura Akira 松村明, Matsunaga Teitoku 松永貞徳, and Odaka Toshirô 小高敏郎. *Taion-ki. Oritakushiba no ki, rantô kotohajime* 「戴恩記、折たく柴の記、蘭東事始」. Tokyo: Iwanami shoten, 1964.

2 Clusius, Carolus and Dodoens, Rembert. *Ensei Dodoneusu Sômokufu* 「遠西独度涅烏斯草木譜」. Edited by Sugimoto Tsutomu 杉本つとむ. 5 vols. Tokyo: Waseda Daigaku shuppan-bu, 1997-1998.

3 Itazawa Takeo 板沢武雄. *Nichi-Ran bunka kôshô-shi no kenkyû* 「日蘭文化交渉史の研究」. Tokyo: Yoshikawa Hirobumi kan, 1959.

4 Katagiri Kazuo 片桐一男. "Oranda tsûji Seiga Kurô (Arai Shôjûrô / Morihira Uemon) hoi" 「阿蘭陀通詞西雅九郎（荒井庄十郎＝森平右衛門）補遺」. *Rangaku shiryô kenkyûkai "kenkyû hôkoku"* 「蘭学資料研究会「研究報告」」 247 (1971).

5 Matsushima Hiroshi 松島博. *Kinsei Ise ni okeru honsô gakusha no kenkyû* 「近世伊勢における本草学者の研究」. Tokyo: Kôdansha, 1974.

6 Nichi-Ran Gakkai eds. 日蘭学会編, *Yôgaku-shi jiten* 「洋学史事典」. Tokyo: Yûshôdô, 1984.

7 Okamura Chibiki 岡村千曳. "Dodoneusu Cruiydt-Boeck no hôyaku – Ensei Sômokufu — ni tsuite" 「ドドネウスCRUIYDT-BOECKの邦訳＝遠西草木譜＝について」. *Rangaku shiryô kenkyûkai "kenkyû hôkoku"* 「蘭学資料研究会「研究報告」」 31 (1958).

8 Okamura Chibiki 岡村千曳. "Dodoneusu Cruydt-Boeck no hôyaku ni tsuite" 「ドドネウスCruydt-Boeckの邦訳について」. *Igaku no ayumi* 「医学の歩み」 28(8) (1959).

9 Ônishi Gen'ichi 大西源一. *Noro Genjô-den* 「野呂元丈伝」 (1915).

10 Ôtsuki Nyoden 大槻如電. *Nihon Yôgaku hennen-shi* 「日本洋学編年史」. Edited by Satô Eishichi 佐藤栄七増訂. Tokyo: Kinsei-sha, 1965.

11 Sakurai Yûkichi 櫻井佑吉 ed. *Noro Genjô-den shiryô* 「野呂元丈伝資料」. Matsusaka: Mie-ken shidankai, 1916.

12 Sugimoto Tsutomu 杉本つとむ. "Nihon saisho no honkakuteki Yôroppa hakubutsu-sho – *Ensei Dodoneusu Sômokufu* to sono hon'yaku" 「日本最初の本格的ヨーロッパ博物書 – 「遠西濁度涅烏斯草木譜』とその翻訳」. *Shokubutsu to bunka* 「植物と文化」 (15).

13 Sugita Genpaku 杉田玄白. *Rangaku kotohajime* 「蘭学事始」. Edited by Ogata Tomio 緒方富雄. Tokyo: Iwanami Bunko, 1959.

NOTES

1 See Sugita Genpaku, *Rangaku kotohajime*, 1959.

ILLUSTRATIONS

ILL 1 general view of the *Yokuon-en* (from *Seikô gafu*)
ILL 2 orchard with pavilion (from *Seikô gafu*)
ILL 3 Pavilion, shed and bonsai on display; in the background display of lotuses (from *Seikô gafu*)
ILL 4 orchard and herb garden (from *Seikô gafu*)
ILL 5 *garden of a thousand grasses*: in the foreground flowering peach trees (from *Seikô gafu*)
ILL 6 *garden of a thousand grasses*: a variety of lotuses and a bed sasanqua's (from *Seikô gafu*)
ILL 7 beach, bank of eulalia's and enclosed flower garden (from *Seikô gafu*)

ILL 1

ILL 2

ILL 3

ILL 4

ILL 5

ILL 6

ILL 7

PART IV

THE RANGAKU CONTEXT

DODONÆUS IN JAPANESE: DESHIMA SURGEONS AS MEDIATORS IN THE EARLY INTRODUCTION OF WESTERN NATURAL HISTORY

Harmen Beukers
Faculty of Medicine, Department Metamedica, Leiden University

The publication of *Kaitai shinsho* in 1774 is considered to be a turning point in the intellectual history of Japan: this translation of a Dutch anatomical atlas into Japanese was to be canonised as the most complete Japanese anatomy book of the Tokugawa period. The choice of the book is interesting since it underlines the visual argument as an evidence-based selection criterion. The comparison of the actual dissected body with the drawings in Dutch anatomy books had convinced 杉田玄白 Sugita Genpaku, one of the authors of the *Kaitai shinsho*, of the correctness of the European representations or, what he called the 'real things'. In 1771, after having seen two Dutch anatomy books for the first time, *viz.* Johann Adam Kulmus' *Ontleedkundige tafelen* (1734) and Bartholin's *Anatomia: ofte ontledinghe des menschelicken lichaems* (1656, 1658), he wrote:

> "Of course, not a word in them we could read, but the structures of internal organs and the skeletal frames illustrated in them appeared very different from those we had seen in books or had heard of in the past. We concluded that these must have been drawn from the real things." [1]

骨ヶ原 After Sugita and his colleagues had attended a dissection in Kotsugahara, in March 1771, their main purpose became:

> "[...] to show somehow to the people that the real structure of the human body was different from the one described in Chinese books." [2]

Genpaku considered Kulmus' *Tafelen* to be the more valuable of the two, and decided to translate the explanatory text that went with the drawings. The considerable annotations in the footnotes, the historical passages and philosophico-religious discussions would remain untranslated.

The main difficulty in the translation process was of course finding adequate terms for anatomical structures and for European medical concepts. In this respect it is important to mention Sugita's colleague Maeno Ryôtaku, who was to contemporary standards rather well versed in the Dutch language, which earned him the nickname "Ranka" (he who has gone Dutch). Maeno had studied Dutch first under Aoki Kon'yô, and later – by order of Okudaira, *daimyô* of Nakatsu – in Nagasaki under the interpreters Nishi Zenzaburô, Yoshio Kôsaku and members of the Narabayashi family. The Nagasaki interpreters constituted an indispensable link on the Japanese side in the process of knowledge transfer – a process in which on the Dutch side the Deshima surgeons played an important role.

前野良沢

青木昆陽

奥平、中津

西善三郎、吉雄幸作

出島

1 A new lingua franca

The closure of Japan in the 1630's created a language problem in the contacts between Japan and the Dutch. Until then Portuguese had been the common language to communicate with foreigners. The early interpreters at Deshima were known to be Portuguese-speaking. Gen'emon, the interpreter of Inoue Masashige of Chikugo was, according to the *Dagh-register* (*Deshima diary*) of 1652, fluent in Portuguese. It seems that this language was still important in the early eighteenth century; even as late as September 1734 the *Opperhoofd* (Jap.: *kapitan*) refers to four interpreters who speak Portuguese. That language was the basis for the dictionaries published by the Jesuits in the sixteenth century. Their Latin-Japanese-Portuguese dictionary may even have been useful to the Japanese to understand certain aspects of western science. Even in the eighteenth century it was still considered a treasure by the interpreters who possessed a copy. Although natural sciences and mathematics had been taught at the Jesuit colleges in Japan, they failed however to make a permanent impact.

井上政重、筑後

甲比丹

After the expulsion of the Portuguese, the language barrier posed considerable problems in the contacts between the Dutch and the Japanese, not only in scholarly matters but also in trade. In the *Dagh-register* kept by the *Opperhoofd* we find references to the incapability of Japanese interpreters. In 1672, for instance, they had great difficulties in reading and translating documents presented by the *Opperhoofd*

長崎町奉行 to the Nagasaki magistrate (*Nagasaki-machi bugyô*). The incident provided the *Opperhoofd* with a welcome argument to stress the need that Dutchmen could be given the opportunity to learn Japanese. The argument was repeated in 1683, when the *Opperhoofd* discovered fraud in official translations by the interpreters. At any rate, one year later there seems to have been a black servant in Deshima, who, according to the diary, could speak some Japanese, and who for that reason joined the court journey to Edo. As we shall see later on, other Deshima inhabitants too turned out to be fluent in Japanese. The situation had hardly improved in the early eighteenth century. *Opperhoofd* Boockesteijn 大通事 complained to chief interpreter (*ô-tsûji*) Imamura Gen'emon about the 今村源右衛門 proficiency of newly appointed interpreters. He foresaw serious difficulties if more able persons were not appointed.[3] Imamura himself was praised for his mastery of the Dutch language, as was his colleague Kafuku Manjirô. On the other hand, chief interpreters like Nakayama Sôemon and Narabayashi Ryôemon were known to have a poor understanding of the language or spoke it deficiently.

Whatever the complaints of the *Opperhoofd* may have been, all the initial activities related to the introduction of western sciences and techniques depended on the official Nagasaki interpreters. They had personal contacts with the Dutch during their daily visits to Deshima, as well as during the compulsory annual journey of the *Opperhoofd* and two staff members to the court in Edo. The interpreters' poor knowledge of the Dutch language was on the one hand due to the lack of dictionaries and grammars and on the other hand to government restrictions. Moreover, as their principal duty was to serve trade, their emphasis was on language related to daily business. They had no primary intention of carrying on systematic scholarship. They even lacked academic training. The situation changed when in the 1720s Shogun Yoshimune sanctioned the study of Dutch by two scholars, the physi- 野呂元丈 cian Noro Genjô (1693-1761) and the Confucian scholar Aoki Kon'yô (1698-1769), the teacher of the aforementioned Maeno Ryôtaku. Both scholars made use of the Nagasaki interpreters to query the Deshima surgeons during their annual visit to Edo. The most important were Motoki Nidayu, Nishi Zenzaburô and Yoshio Kôsaku.

The role of the Nagasaki interpreters was essential to the compilation of a Dutch-Japanese dictionary. The first complete one, the so-called *Haruma-wage* or *Edo-Haruma*, was not published before 1796. It marked the completion of a work started by the interpreter Nishi

Zenzaburô (1718-1768). As a member of the Nishi family of hereditary interpreters, he had learned Dutch in Deshima from 1722 on. Zenzaburô was finally appointed chief interpreter in 1754. He accompanied the Dutch missions to Edo as many as six times. He tried to compile a Dutch-Japanese dictionary, using Pierre Marin's Dutch-French dictionary as a reference, but he did only get through the letter 'B'. Aoki Kon'yô too published a list of 721 Dutch words in *Oranda moji ryakkô* ('Primer of the Dutch Language,' 1744). It was in fact a revision of earlier lists compiled by the interpreters. These lists gave the Dutch words only in *kana*, whereas Kon'yô's list featured the Dutch words in roman letters as well as their pronunciation in *kana*.

It is difficult to evaluate the contributions of the official interpreters in their entirety. The remnants of their early works are only fragmentary. At any rate, even before the publication of *Kaitai shinsho*, there existed unpublished, and for a general scholarly public inaccessible translations of Dutch medical books. The interpreter Motoki Ryôi 本木良意 (1628-1697) translated the Dutch edition of Johann Remmelin's *Pinax microcosmographicus*, a set of anatomical tables. His *Oranda zenku naigai bungô-zu* 和蘭全躯内外分合図 ('Dutch Chart of the Inner and Outer Parts of the Body') became only known to the medical profession after its publication by Suzuki Sôden 鈴木宗伝 in 1772. His younger colleague Narabayashi Chinzan 楢林鎮山 (1648-1711) translated six parts of Ambroise Paré's book on surgery (first Dutch edition of the complete works: 1592)[4] under the title *Kôi geka sôden* 紅夷外科宗伝 ('School Tradition of Red Barbarian Surgery,' 1706). It circulated in manuscript form as a 'secret manual' for *kômô* 紅毛 ('red-hair') surgeons. Chinzan was a member of a hereditary interpreters' family and studied under various surgeons at Deshima, especially Willem Hoffman (Deshima 1671-1675). He passed through the successive interpreters' ranks before he was finally appointed physician in ordinary to the Bakufu in 1691.

II Early interest in Dutch medicine

It is not self-evident that Japanese physicians valued western medicine higher than their own. As the aforementioned translations show, only a specific part of western medicine aroused curiosity, *viz.* surgery, in this case especially the treatment of wounds, contusions and fractures, and the surgical instruments. Narabayashi's translations also included anatomy, since Paré considered the description of the body structures introduced by Andreas Vesalius (1543) as fundamental to the surgical practice. The translation by Narabayashi Chinzan thus included books on wounds, plasters and anatomy. His translations of the parts on distillation and apothecary are evidence of an interest in western *materia medica*. Since the simple medicines were mainly of vegetable

origin, it is not surprising to find a parallel interest in western botany.

The interest for 'Dutch' surgery arose during the 1650's, when its efficacy was demonstrated in the treatment of high-ranking officials by the Deshima surgeon Caspar Schamberger (Deshima 1649-1651).[5] During his stay in Edo in 1650, he treated successfully the senior councillor, the "*rijksraad Mino-samma*" Inaba Masanori (1623-1696), *daimyô* of
小田原、美濃 Odawara and lord of Mino, who suffered from gout. It stimulated Inaba's interest for western medicine. The *Dagh-register* of 1668 reports a *Spiegel der anathomie* and the *Hortus Eystettensis* by Basilus Besler as books ordered by Inaba. In the 1680's the councillor probably also ordered Dodonæus' *Cruijdeboeck*, but when the "*geschilderde kruydeboek van Dodoneus*" arrived in November 1682, Inaba Masanori had
正往 just renounced his worldly possessions in favour of his son Masayuki. According to *Opperhoofd* Andreas Cleyer's diary, "Mino-samma" did no longer want to receive presents. The book was therefore requested
松平加賀様 by "Matzendairo Cange-zamma", i.e. Matsudaira Kaga Sama or
前田綱紀 Maeda Tsunanori (1643-1724), according to Cleyer, "second in the empire", nephew of the Shogun.[6] The book is now in the Library of Kanazawa University, erstwhile of the Kaga clan. It is interesting
伊奈波神社 to note that Inaba probably possessed another copy, since the Inaba shrine in Yodo still owns a *Cruijdeboeck* by Dodonæus.

Another example of a high-ranking official was the governor of Nagasaki, Inoue Masashige (1585-1661), in Dutch sources known as
筑後殿 Sickingodonne, i.e. 'Chikugo-dono', a champion of the Tokugawa system. His strong interest in useful western sciences and technology was probably related to his own health problems: haemorrhoids, bladder stones and catarrh. In a request dated to the year 1652 he ordered an anatomy book and a botany book in Portuguese, since, as we have mentioned earlier, his interpreter mastered only that language. An invoice dated later that year in Batavia mentions:[7] "een Herbarius van Dodoneus affgeseth voor d' Hr Sickingodonne 120 — een Historia Naturalis voor idem 52 —". The first book is evident; the second is probably Plinius' [Pliny] *Historia naturalis* (1635). The packing list of 1655 describes the 'Herbarium van Dodonæus' for Inoue as a book in folio with silver mounting and gilded edges.[8] In 1659 the *Opper-*

hoofd Zacharias Wagenaer presented another copy to the Bakufu. Four years later, *Opperhoofd* Hendrik Indijck brought as a gift for the Shogun a "book on the nature of animals" (probably Jan Jonston [John Johnston]'s *Naeukeurige beschryving van de natuur der viervoetige dieren, vissen en bloedloze water-dieren, vogelen, kronkel-dieren, slangen en draken*, Amsterdam, 1660) and for the councillors a 'book of the description of animals and fishes' (possibly Plinius' *Boecken, handelende van de natuere*). The books mentioned are finely illustrated books in folio, more suitable as presents to high officials than for the bookshelves of ship's surgeons.

These high-ranking officials became deeply interested in western medicine and botany. They sent their physicians to Nagasaki to be instructed by the Deshima surgeon. Both Inaba and Inoue had great influence on the trading conditions – Inaba was a "bugbear and spiritual father of the valuation trade of the VOC in 1672." Therefore, their requests were granted by the Dutch East India Company as much as possible, in the first instance of course by the Deshima surgeons if the requests regarded their knowledge and skills. In this period, the surgeons occupied an important, if not the foremost position in the introduction of western knowledge. That position was further strengthened during Tokugawa Yoshimune's reign from 1716-1745. He adopted a policy of importing western books on practical subjects like medicine, botany and astronomy.

III Deshima surgeons

The historiography of medicine and sciences in Japan is strongly dominated by the traditional view that attributes essential roles in the introduction of western sciences to four Deshima doctors: Willem ten Rhijne, Engelbert Kaempfer, Carl Peter Thunberg and Philipp Franz von Siebold. This emphasis is mainly a consequence of the fact that they published eyewitness reports about things Japanese. One cannot deny their importance in the transfer of information from Japan to Europe, but that alone does not justify a comparable status in the opposite direction, particularly not, if one realises that – with the exception of von Siebold – they only stayed for a short period at Deshima. Ten Rhijne was in Japan from 1674 to 1676, Kaempfer from 1690 to 1692 and Thunberg from 1775 to 1776.

The situation was different for the Deshima surgeons in the eighteenth century. Most of them spent at least four years on the island.

Their careers differed from the aforementioned university trained physicians. Generally speaking surgeons belonged to the class of artisans. In most Dutch towns the training of surgeons and their certification was regulated by a surgeons' guild. Apprentices lived in a surgeon's home and acquired knowledge informally. In larger towns the guild provided an additional course in anatomy, sometimes even in botany. More than any other cities, Amsterdam and Middelburg rigged out ships for the VOC. Their chambers assembled the crew and consequently also enlisted the ship's surgeons. As early as 1610 the chamber Zeeland, and probably not much later that of Amsterdam too introduced an examination for ship's surgeons.

The traditional division of labour between medical doctors and surgeons did not apply, since the former were virtually absent in the merchant fleet. As a consequence ship's surgeons also treated internal diseases such as scurvy, beri-beri, dysentery and typhoid fever. They even had to prepare medicines themselves. Thus the training of ship's surgeons was not limited to the traditional teachings of anatomy and surgery, but also included basic internal medicine and pharmacy. For that reason from the early eighteenth century on Amsterdam provided a winter course for lower-grade ship's surgeons. The Middelburg examiners compiled textbooks for prospective ship's surgeons. An example is Johannes Verbrugge's popular textbook *Het nieuw-hervormde examen van land- en zee-chirurgie*. Between 1708 and 1768 it went through thirteen reprints. All this made that ship's surgeons, compared to their colleagues ashore, had a higher remuneration and a more attractive position: on board they served as junior officers and in Asia they often became independent medical men.

There was still another possibility to become a ship's surgeon. Larger ships usually had in addition to the senior surgeon (*oppermeester or opperchirurgijn*) a second surgeon (*ondermeester or onderchirurgijn*) or a surgeon's mate (*derde meester*). If the senior surgeon died during the voyage to the Cape, personnel applying for the vacancy (not only the lower-rank surgeons) could be examined by a committee of senior ship's surgeons present at the Cape, under the presidency of the senior surgeon of the hospital at the Cape. In Asia it was the senior surgeon residing in Batavia who was responsible for the examination and appointment of surgeons.

For the introduction of western medicine in Japan it was advantageous that the medical practice at Deshima was in the hands of surgeons. Their education was practice-oriented and less influenced by purely academic discussions. Their handbooks had the same practical orientation; they were concise, written in the vernacular, avoiding Latin as much as possible.

iv Prolonged stay[9]

Three senior surgeons are known to have had some proficiency in Japanese: Wagemans, Musculus and Ketelaar. Only the latter one practised relatively briefly in Japan. Willem Ketelaar, from Vlissingen, arrived in the Dutch East Indies as second surgeon in December 1715. He was appointed senior surgeon in Japan in October 1722. Three years later, on 20 October 1725, Japanese officials ordered *Opperhoofd* Thedens, without stating the reason, to send Ketelaar and the provisional assistant back to Batavia. After a few years of service on different ships Ketelaar returned back home to Zeeland in June 1728.

The two other senior surgeons mentioned belong to a small group that stayed rather long in Japan [cf. table 1]. For instance, the chief surgeon Willem Wagemans, sent by the VOC chamber Delft, went to Deshima in the same year that he arrived in Batavia. With two intermissions he practised there until he was expelled on account of his proficiency in the Japanese language. His second surgeon from 1707 on was Hendrik de Vogel from Amsterdam. He died at Deshima on 11 January 1719 and was buried there by his compatriots the next day after an autopsy by the servants of the "*bongiosen*" i.e. the *bugyô*, the magistrate entrusted with inspection duties. In November of the preceding year De Vogel had drawn up his last will and testament. Since he was only a second surgeon, nothing is to be found about him in the *Daghregisters*, except for the sad story of the suicide of his slave Pedro van Bengalen, who committed suicide after being accused of the theft of his deceased master's money.[10]

The career of David Drinkman from Meurs is rather exceptional. He arrived in Batavia on the ship "Valckenisse" (from the VOC chamber Zeeland) as second surgeon in 1723. When he arrived in Nagasaki, in October 1725, he was senior surgeon. In 1729 he started a career in the trade and was finally appointed *Opperhoofd* at Deshima. He resigned from VOC service on 20 November 1736 as senior surgeon and *Opperhoofd*.

The careers of Philip Pieter Musculus and Doede Everts ran for a long time together. Both arrived in Nagasaki in July 1739. At his official installation in October of that year *Opperhoofd* Thomas van Rhee had in fact three medical men available: senior surgeon Philip Pieter Musculus, second surgeon Eelke Riskes and third surgeon Doede Everts. The Amsterdammer Riskes came as second surgeon on the ship Batavia to the Dutch East Indies, but in Deshima he was enrolled as second carpenter. He kept that position until he left Japan in 1745.

Musculus, from Claar-Oostersteijn in Germany, arrived in Batavia as senior surgeon for the VOC chamber Enkhuizen in December 1736. He reached the harbour of Nagasaki on 21 July 1739 on the ship "Popkensburg", and would leave Japan on 28 October 1747. After Wagemans he has the second longest stay at Deshima and made eight court journeys. Within a year after his return to Batavia he died in the hospital there on 24 September 1748. Musculus' stay in Japan coincided with the latter part of Tokugawa Yoshimune's rule. He personally experienced the re-orientation of Japanese scholars to western sources. The court journey of 1746 gives some insight into the actual state of Dutch learning. *Opperhoofd* Jan Louis de Win reports on 21 April 1746 about the visit to the palace:

> "We [*Opperhoofd* de Win and surgeon Musculus] were given three scrolls of paper. On each of them we had to write Dutch words with red ochre. We wrote *CRAANVOGEL* [crane], *SPARREBOOM* [spruce-fir] and *SCHILDPAD* [tortoise]. When we had finished, a slide in the screen was pushed open, leaving a space measuring two hands. A very old gentleman accompanied by the gentleman of fifteen to sixteen years previously mentioned observed us from a distance of three tatami. Two sheets of paper were given to Musculus, who had to write the words *BERG* [mountain] and *BAMBOES* [bamboo]. Thereupon the slide was closed and we returned to the *vertoefzaal* [...] Awa gave Musculus 12 sheets of paper on which he has to write Dutch words, which will be given to the crown prince."[11]

These sheets of papers with Dutch words and their transcription in 大槻如電 kana were still known to Ôtsuki Nyoden.[12] Musculus left Japan two years after Yoshimune's retirement as Shogun.

Doede Everts from Groningen went to the East Indies as third surgeon for the Amsterdam chamber in 1737. He arrived in Batavia in

January 1738. He arrived in Nagasaki together with Musculus. After two years he was promoted second surgeon, which meant an increase in salary from 14 to 24 guilders. In July 1745 the ship "De Vrijheidt" and two other ships arrived in Nagasaki carrying the new *Opperhoofd*. Everts left Deshima in 1745, and served fourteen months on "De Vrijheidt". In August 1747, he returned to Deshima as senior surgeon on the ship "Batavier", having practised on board of that ship for ten months. With one year intermission he was again active in Japan and made four court journeys. He could not make his fifth trip to Edo, because of the illness he contracted in February 1753. Although the next year it was planned that Everts would join *Opperhoofd* Hendrik van Homoed and the assistant Johannes Reijnouts on the court journey, he fell ill again in January and had to stay back at Deshima. In November 1753 he left Japan definitively. He died on his way back to Holland on 25 July 1756.

v Japanese interest in Dodonæus' Cruijdeboeck

A strong impetus for the study of western sciences came from Shogun Yoshimune. After the death of Tokugawa Ietsugu in 1716, Yoshimune was chosen as successor. He refused three times before he finally submitted himself to the decision of the family council. After coming to power he adopted a policy known as the *Kyôhô no kaikaku* ('*Kyôhô* Reforms'), which included the encouragement of the study of relevant western sciences. The official sponsorship of western learning coincided with a large-scale agricultural reform: the opening of new agricultural areas, the introduction of new products (sugarcane, sweet potatoes, wax trees) and new methods (sugar refining) and the encouragement of large-scale animal husbandry. Moreover, in 1721,[13] the Shogun established the Koishikawa Yakuen, a garden where medicinal plants were grown so that Japan would no longer be dependent on imports. Therefore, Yoshimune was interested in practical information from the West. From 1717 on, the year of the first court journey of the Dutch during his Shogunate, he showed great interest in querying the foreigners through his court officials, particularly on botany. According to Numata[14] (quoting the *Shomotsukata*), Yoshimune occasionally had Dutch books brought to him. One day in March 1724 chamberlain Arima Ujinori wanted to borrow a "Dutch botany book", and after a search he was told the following day that the book was not in the library. Later it was found out that the book was in the Shogun's private chambers.

徳川家継

享保の改革

小石川薬園

沼田

The new interest began indeed in 1717. In that and the following year the Dutch received queries about animals during their stay in Edo. Opperhoofd Jan Aouwer reports on 11 April 1717:

> "Kizaemon [chief interpreter Nakayama Kizaemon] brought me a book that had been taken from the Curiosity Cabinet of the Shogun. He asked me if I could translate the names of the animals in the book into Japanese. I said it would be possible but it would take a long time because the names of thousands of animals are in it. The book by Johnston was a folio volume in red Russian leather with gilt stripes. The book was still in an excellent condition although it had been presented by *Opperhoofd* Indijck in 1663." [15]

On 30 March 1718 *Opperhoofd* Christiaen van Vrijeberghe wrote in his diary:

> "I translated the names of several animals and their places of origin from Johnston's book *Beschrijvinge van de natuur der dieren* that was shown to Aouwer last year." [16]

Starting in 1731, more pertinent questions concerning dried herbs collected in Japan were asked by Niwa Seihaku [Shôhaku] (丹羽正伯) or Sawa Safak (1691-1756), described in Dutch sources as 'imperial botanist', 'Shogunal botanist' or 'imperial apothecary'. Seihaku studied medicine and botany under the famous botanist Inô Jakusui (稲生若水) of Kyoto. From 1720 on he collected, at the order of the Bakufu, botanical specimens throughout Japan. Two years later he was appointed physician in ordinary to the Shogun. In 1731 and 1732 he asked the senior surgeon Hendrik Thompson to help him identifying herbs.

From the 1720's on the diary kept during the court journey, frequently refers to "*Kiodaij*", i.e. Fukami Kyûdayû or Fukami Arichika (1691-1773). Kyûdayû was the descendant of a Chinese immigrant who had come to Nagasaki in the early seventeenth century, to serve there as translator. Kyûdayû was librarian of the Shogunal library since 1734. In 1737 he showed a keen interest in Dodoens's work. On 25 March he paid the Dutch a visit at their inn in Edo and "asked several questions pertaining to the books on herbs by Rembertus Dodonæus". The following day he asked "more questions about Dodonæus". At the end of March and the beginning of April the queries were repeated. At the conclusion of the court visit in 1737 *Opperhoofd* Jan van der Cruijsse was asked to order from Batavia among other things "a book about herbs with drawings of the herbs and their effects" and "a book about all known animals with illustrations and also the description of the nature of the animals".

During Musculus' second court journey, in 1741, the queries concerning natural history were taken up again. On 22 April 1741 Kyûdayû paid a visit to the Dutch and showed "a book by R. Dodoneus entitled the *Cruijdt-boeck*". He asked Musculus questions about the effects of rosemary. A couple of days earlier a "court physician" had shown "a book by E. Swart" with drawings of flowers and trees to the *Opperhoofd* and asked him the names. The Opperhoofd succeeded with the help of the Latin, French, German and Dutch index. The book was probably E. Sweerts's, *Florilegium, tractatus de variis floribus et aliis Indicis plantis ad vivum delineatum* (1615 or 1631). Two weeks later the chief interpreter Yoshio Chûjirô (吉雄忠次郎), known in Dutch sources as "Tosabro", sought an explanation about what was written about the herbs in "Swart's" book. "Musculus tried to explain – the Opperhoofd wrote – but he did not understand much of it". Motoki Ryôei (本木良永) (1735-1794), grandson of the aforementioned Motoki Ryôi and translator of important books on heliocentricity, would later use Sweerts's book while compiling his *Honyaku Oranda honzô* (1771).

The aforementioned 'court physician' visited the inn of the Dutch again on 24 April and showed "a book written by J. Johnston". Tosabro translated several captions. The next day Tosabro asked the *Opperhoofd*:

> "[…] whether Musculus could join him in his room since the court physician wanted to check the captions translated by him yesterday. I [Opperhoofd Jacob van der Waeijen] told Musculus only to talk about the book and about nothing else." [17]

The following years, the "personal physician of the Shogun" was no longer interested in animals. When he visited the Dutch in the years 1742-1745 – sometimes accompanied by students – his questions concerned the effects of certain herbs. Musculus tried to explain these effects. In 1743 the physician showed his source book. *Opperhoofd* Van der Waeijen noted on 18 March:

> "The personal physician of the Shogun paid us a visit. He showed

us a book containing pictures of herbs and flowers and he also showed us dried herbs. He wanted the effect of the dried herbs to be explained to him. Musculus explained it."[18]

The next day the *Opperhoofd* reported:

"The troublesome personal physician of the Shogun asked me to summon Musculus to the room of [chief interpreter] Mangero. Afterwards Musculus told me that he had explained the effect of certain herbs, which the personal physician of the Shogun had shown to him."

The aforementioned personal physician of the Shogun was Noro Genjô (1693-1761). At the age of twenty, he went to Kyoto to study Confucianism, medicine and botany. The latter subject he studied with the botanist Inô Jakusui. In 1719 Noro became official botanist of the Bakufu and travelled together with the aforementioned Niwa Seihaku throughout Japan, collecting botanical specimens. He published a report of his trip to the north of Japan in 1722: *Hokuriku hobutsu* ('products of the North'). In 1739 he had an audience with Yoshimune 御目見医師 and was appointed *omemie ishi*, ('physician in ordinary'), who could practise in the inner Bakufu court. The same year he and Aoki Kon'yô received permission from the Shogun to learn Dutch. It is said that Noro acquired a fair knowledge of Dutch as a result of his annual contacts with Nagasaki interpreters, especially Yoshio Kôsaku, and Dutch surgeons, particularly Musculus and Everts. The results of the 1741 interview about Jonston's book (according to Noro's notes the 1660 阿蘭陀禽獣虫魚図和解 edition) were recorded in a manuscript entitled *Oranda kinjûchûgyo-zu wage* ('Japanese Translations of Dutch Illustrations of Birds, Animals, Insects and Fishes'). That Noro lost interest in the subject becomes clear from one of his remarks:

"This is not a botanical treatise treating medicinal herbs, and is therefore of no use whatsoever, giving only the outside shape."[19]

It is also evident from the *Opperhoofd*'s journal that from 1742 on Noro turned to the work of Dodonæus, which he considered to be more relevant to medicine. Through the interpreters he continued interviewing Musculus and Everts every year except one up to 1750, when he completed his notes on Dodonæus concerning 118 plants and their medici-阿蘭陀本草和解 nal use: *Oranda honzô wage* ('Dutch Botany Explained in Japanese').

After Yoshimune retired in 1745 in favour of his son, the official encouragement of Dutch studies ceased and turned into a policy of negligence for the remainder of the century. Accordingly, the references of the *Opperhoofden* to queries about natural sciences at the court, which had started in 1722, stopped in 1745. Medical and botanical studies of Dutch sources had to come from private initiatives. Natural history books were therefore no longer popular as official gifts to court officials. Especially the College of Interpreters now received Dodoens's *Cruijdeboeck* frequently: in 1757, 1760, 1761 (even three copies) and 1775. The *daimyô* of Satsuma was also eager to receive a copy; he requested a "Kruijtboek van Dodoneus" in 1772. Thunberg wrote in his *Voyage au Japon* (1796) that the Dutch gave the following books to the Japanese: Jonston's *Historia Naturalis*, Dodonæus' *Herbarium*, Woyt's *Gazophylacium*, the Dutch translation of the book of Heister and a beautiful copy of *Plantes* by Muntingh. Thunberg's own influence becomes clear in the course of time. In 1800 the College of Interpreters received for the first time more up-to-date natural history books: Linnaeus' *Natuurlyke historie of uitvoerige beschryving der dieren, planten en mineraalen*, etc., Martinet's *Katechismus der natuur* including the *Aanmerkingen*. This does not mean that Dodonæus' *Cruijdeboeck* was no longer of interest to Japanese scholars. In 1815 Udagawa Yôan published his *Ensei Dodoneusu buppin-kô meiso*, and fourteen years later Matsudaira Sadanobu published parts of *Ensei Dodoneusu sômoku-fu*, a translation initiated by Ishii Tsune'emon (Shôsuke, Tôkô) and completed and arranged by others.

遠西鐸度涅烏私物品考名疏
松平定信
遠西独度涅烏斯草木譜、石井常右衛門
庄助、当光

VI Concluding remarks

We have illustrated the introduction of western sciences with an example taken from natural history. The impetus for a strong interest in botany in the eighteenth century came from Yoshimune's need of practically useful knowledge to support his agricultural reform. It induced the Shogunal librarian Fukami Kyûdayû and the Shogunal botanists Noro Genjô and Niwa Seihaku to search foreign sources.

The exchange of knowledge was predicated on the availability of an intermediary and of mediators. The intermediary turned out to be Dodonæus' herbal, a voluminous book in folio. It is not an obvious choice; it is a detailed, complicated and highly professional text. Its illustrations, however, helped to overcome the language barrier. The visual information served as a guide to select relevant plant monographs for translation. The diaries kept by the Dutch *Opperhoofden* of Deshima give detailed information about the roles of the mediators, in particular, on the Dutch side the role of (now lesser known) surgeons such as Musculus and Everts. During their relatively long stay in Japan, they joined in many court journeys. In Edo they were used by the Bakufu officials as experts helping to select monographs for translation, to explain the effects of certain herbs, and to check in a later stage the translations made by the interpreters. This example shows that the surgeons played an important, if not crucial, role in the transfer of western knowledge to Japan, at least during the reign of Tokugawa Yoshimune.

TABLE I

SURGEON	ARRIVAL	DEPARTURE
Willem Wagemans		
chief surgeon	1698	1699
	1700	1701
	1706	1717
Hendrik de Vogel		
Second surgeon	1707	1719
David Drinkman		
Senior surgeon	1725	1729
(junior) merchant	1729	1733
Opperhoofd	1734	1735
Philip Pieter Musculus		
Senior surgeon	1739	1747
Doede Everts		
third surgeon	1739	1741
second surgeon	1741	1745
senior surgeon	1747	1748
	1749	1753

NOTES

1 Sugita Genpaku, *Dawn of Western Science in Japan*, trans. Matsumoto Ryôzô (Tokyo: Hokuseido Press, 1969), 24.
2 Sugita Genpaku, *Dawn of Western Science*, 43.
3 Algemeen Rijksarchief [ARA] Den Haag, 1.04.21 Factorij Japan, inv. 142: 41 October 1731.
4 The Dutch version, acquired by Narabayashi Chinzan, was probably the 1649 Schipper Amsterdam edition. See Gabor Lukacs's contribution in the present volume.
5 Wolfgang Michel, *Von Leipzig nach Japan: Der Chirurg und Handelsmann Caspar Schamberger (1623-1706)* (München: Iudicium Verlag, 1999).
6 ARA, 1.04.21 Factorij Japan, inv. 96: 14 November 1682.
7 ARA, Kasteel Batavia inv. 776: 11 July 1652.
8 ARA, Kasteel Batavia inv. 797: 7 July, 1655.
9 The careers of the surgeons discussed here are mainly documented in ARA, 1.04.01 Archief V.O.C., inv. 5558, 6042, 12815 and 14172. Many of the details concerning their activities at Deshima can

easily be traced by consulting the English translation of the marginal notes of the Deshima Diaries: see Paul van der Velde and Rudolf Bachofner, *The Deshima Diaries: Marginalia* 1700 - 1740 (Tokyo: Japan-Netherlands Institute, 1992).

10 ARA, 1.04.21 Factorij Japan, inv. 130: 23 January 1719.

11 ARA, 1.04.21 Factorij Japan, inv. 155.

12 C.C. Krieger, *The Infiltration of European Civilization in Japan During the 18th Century* (Leiden: E.J. Brill, 1940), 33.

13 See Kasaya Kazuhiko's contribution in the present volume.

14 Jirô Numata, *Western Learning – A Short History of the Study of Western Science in Early Modern Japan* (Tokyo: Japan-Netherlands Institute, 1992), 37-39.

15 Van der Velde and Bachofner, *Deshima Diaries*, 209.

16 Ibid., 220.

17 ARA, 1.04.21 Factorij Japan, inv. 151.

18 Ibid., inv. 153.

19 Numata, *Western Learning*, 48; also Krieger, *The Infiltration*, 32.

THE EARLY PENETRATION OF FRENCH SCIENCE INTO JAPAN DURING THE TOKUGAWA PERIOD

Gabor Lukacs
Directeur de Recherche 1ère classe,
Centre National de la Recherche Scientifique, France

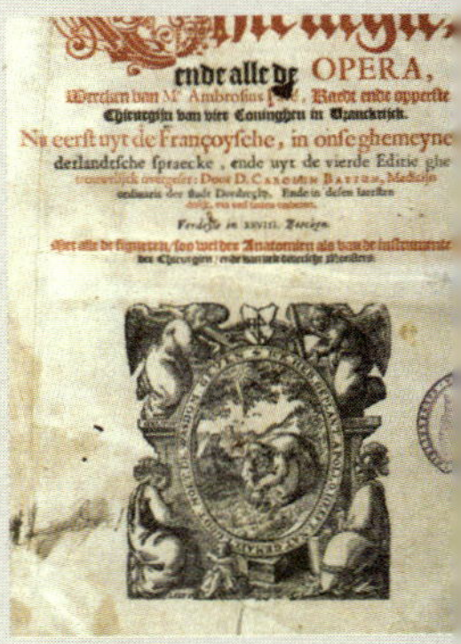

1 Introduction

The choice of my topic is inspired by the celebration in Japan of the year of France. Before going into the subject proper, I would like to list a few major events that occurred in the years '98' of the past centuries and had a bearing on the cultural and scientific relations between Japan and Europe. Japan's existence was made known to the European royal courts and intellectuals by Marco Polo's description. A prisoner in Genoa after returning from a seventeen-year stay in the Mongol Empire and China, Marco Polo dictated his manuscript in 1298, exactly seven hundred years ago. This text, written in a medieval form of rude French had a major influence on the representation of Japan in the mind and imagination of the educated Europeans. This impact is also reflected in the illustration of Japan on the first surviving globe of Martin Behaim (1492), preserved in the Germanisches Nationalmuseum, Nürnberg. In 1498 Vasco da Gama succeeded in reaching the west coast of India by sailing around Africa. Thus he opened the sea route to the Far East for the Portuguese navigators and became a pioneer in the establishment of relations between Europe and Japan.

豊臣秀吉

The year 1598 was rich in important events that influenced the relations of Japan with Europe. The great ruler Toyotomi Hideyoshi died that year. He had succeeded in uniting the country after a long period of civil war and became the first major opponent to the Christian religion in Japan. In the same year died the Spanish king Philip II, under whose rule the Spanish Empire extended its influence and territorial conquests to the Far East. Not many Japanese medical historians are aware of the fact that the top of the title page in the Plantin edition of Juan de Valverde's Anatomy *Vivae imagines partium corporis humanis aereis formis expressae* illustrates the coat of arms of the Spanish king. Equally in 1598, Pope Clement VIII issued an edict suppressing the exclusive right of missionary propagation of the Jesuitic Order in Japan. This initiative had disastrous consequences for the Japanese Christians.[1]

Serious conflicts marked the relationship between the Jesuits and the newly arriving Franciscans and Dominicans, leading ultimately to the expulsion from Japan of all the Catholic orders by 1640. In the same year 1598 the famous Padua professor of anatomy Johann Vesling was born. The illustrations of the Dutch version of his book entitled *Konstige ontleding der menschlijcken lichaems* (the Blasius translation of *Syntagma anatomicum*)[2] were the source of Yamawaki Tôyô's celebrated *Zôshi* which paved the way for the adoption of Western medicine and played an important role in the later rejection of the teachings of Chinese medicine in Japan.

山脇東洋
蔵志

The year 1798 is associated with two important events. In that year, Ôtsuki Gentaku completed the manuscript of *Chôtei kaitai shinsho*, considered the most complete Japanese anatomical treatise of the Tokugawa period (*Tokugawa jidai*, 1600-1868).[3] Udagawa Yôan, one of the most brilliant *Rangakusha* was born that year. Pioneer in the introduction of Western botany and chemistry into Japan, he was the author of the celebrated *Shokugaku keigen* and *Seimi kaisô*.[4]

大槻玄沢
重訂解体新書
宇田川榕庵

植学啓原、舎密開宗

II French scientific books in Japan and French books translated into Japanese.

Let us examine now the names of the French physicians, surgeons and scientists whose treatises were partially translated into Japanese during the Edo period (*Edo jidai*).[5] Table I lists the title of the original French editions. The Japanese translations were always based on the Dutch versions and most often the Japanese physicians and scientists were not even aware that the source of the translated works was initially a French treatise. The Dutch translation was not necessarily made from the first French edition. Similarly, the Japanese translation did not always use the first Dutch edition.[6]

Table I

1	Ambroise Paré (1510-1590), *Les Oeuvres ...* , Fo., Paris, 1575, ed. G. Buon.
2	Noël Chomel (1632-1712), *Dictionnaire Oeconomique ...* , Fo., Lyon, 1709, ed. imprimé pour l'auteur; with a supplement published in 1712.

JOAN VESLINGII
ONTLEEDINGE
des Menscheliken
Lichaems.
TOT AMSTERDAM,
Voor Gerrit Sweerman Boekverkoper inde Wolvestraet.

ILL 1

3	François Mauriceau (1637-1709), *Des maladies des femmes grosses et accouchées*, Qto, Paris, 1668, ed. J. Héviault.
4	Nicolas Lémery (1645-1715), *Traité universel des drogues simples*, Qto, Paris, 1698, ed. L. d'Houry.
5	Georges de la Faye (1710-1781), *Principes de chirurgie*, Oct., Paris, 1738.
6	Joseph Jérôme le François de Lalande (1732-1807), *Astronomie*, 2 vols. Qto, Paris, 1764, ed. Desaint and Saillant.
7	Pierre Lassus (1741-1807), *Pathologie chirurgicale*, 2 vols. Oct., Paris, 1805-1806, ed. Méquignon.
8	Laurent Antoine de Lavoisier (1743-1794), *Traité élémentaire de chimie*, 2 vols. Oct., Paris, 1789, ed. Cuchet.
9	Jean-Louis Baudelocque (1746-1810), *Principes sur l'art des accouchements ...* , Oct., Paris, 1781, ed. Méquignon l'ainé.
10	Antoine-François Fourcroy (1755-1809), *Philosophie chimique ou vérité fondamentale de la chimie moderne*, Oct., Paris, 1792.
11	Jacques-Pierre Maygrier (1771-1835), *Manuel de l'anatomiste*, Oct., Paris, 1811 (2nd ed.; the 1st ed. could not be found), ed. J. Merlin.
12	Anthelme B. Richerand (1779-1840), *Nouveaux éléments de physiologie*, Oct., Paris, 1801, ed. Richard, Caille et Ravier.
13	François Magendie (1783-1855), *Formulaire pour la préparation et l'emploie de plusieurs nouveaux médicaments*, 12mo, Paris, 1821, ed. Méquignon-Marvis.
14	Antoine Dugès (1797-1838), *Manuel d'obstrétrique*, Paris, 1826.
15	Jean-Pierre Girardin (1803-1884), *Leçons de chimie élémentaires, faites le dimanche à l'école municipale de Rouen*, 2 vols. 18mo, Paris, 1836-1837, ed. Rouvier.
16	Claude Bernard (1813-1878) et Charles Huette, *Précis iconographique de médecine opératoire et d'anatomie chirurgicale*, Oct., Paris, 1846, ed. Méquignon-Marvis.

It is of interest to note that only a small part of the works of Paré, Lémery and Lalande was printed in Japanese;[3] the translation of the other books remained unpublished, circulating among the specialists as manuscripts. Moreover, in the case of some of these authors it was not one of their major works, but a writing of minor importance that was brought to Japan and translated into Japanese. Thus no evidence is available that Baudelocque's *L'art des accouchements*, Bernard's *Introduction de la médecine expérimentale*, Magendie's *Physiology* and Maygrier's *Obstetrics* reached Japan during the Tokugawa period.

The number of French works that were translated during the Edo period was relatively high. This fact cannot merely be explained by their exceptionally important content. Another historically major event occurring in France at the end of the seventeenth century and totally independent from Japanese history, provides an adequate explanation. The French king Louis XIV desired to strengthen the unity of the country and curb the increasing influence of the Protestants. Therefore, he decided to repeal the Edict of Nantes, issued in 1598 by Henry IV, which guaranteed the Protestants the free practice of their religion. The consequence of this royal initiative was a major disaster for France, both economic and intellectual. Many highly educated Protestants, endowed with an enterprising spirit, fled the country. Some of them settled in the prosperous Netherlands, which was then at the height of its 'Golden Century'. They and their descendants distinguished themselves in various activities related to the printing industry and book publishing. TABLE II lists the names of some of the French refugees who achieved fame by translating and editing French books into Dutch and Dutch books about Japanese history into French.

TABLE II

1	J.A. de Chalmot, compiler of a monumental Dutch encyclopaedia in seven volumes[8] based on N. Chomel's *Dictionnaire Oeconomique*.
2	P. Desmaizeaux and F. Naud, translators into French of Kaempfer's *History of Japan*.
3	F. Halma, famous editor in the Netherlands of French books and an important Dutch-French dictionary. The latter was the source of the first Dutch-Japanese printed dictionary (*Edo-Haruma* 江戸ハルマ) published in 1796 by Inamura Sanpaku 稲村三伯.
4	P. Marin, author of an important Dutch-French dictionary, source of Nishi Zenzaburô's 西善三郎 attempted compilation of a Dutch-Japanese dictionary. Nishi's manuscript helped Maeno Ryôtaku 前野良沢 and his colleagues in the Japanese translation of Kulmus' *Ontleedkundige tafelen*.
5	P. Massuet, Amsterdam-based author of several books on history and medicine, French translator of Kulmus' *Ontleedkundige tafelen* and P. van Musschenbroek's treatise on Physics.
6	J. Morterre, famous editor of several French books in the Dutch language.
7	J. Neaulme, famous editor of several French and Dutch books in the eighteenth century (French translation of Kaempfer of 1729).
8	L. Renard, famous Amsterdam editor: a. *Atlas de la navigation et du commerce* ... , Amsterdam, 1715 (partially translated into Japanese and carrying for the first time Copernicus' name into the Japanese language); b. books on natural history.

In addition to the authors shown in TABLE I, the works of many French scientists were brought to Japan in the first half of the nineteenth century.[9] Although no evidence is available at the present time about their partial Japanese translation, some of these books were widely consulted by the increasing number of *Rangakusha* who had learnt to read the Dutch language. Some of the natural history and medical books were also appreciated for their beautiful illustrations. TABLE III gives a partial list of French scientific authors whose writings were available in Japan during the last decades of the Tokugawa regime (1600-1868). The precise influence of these books is difficult to assess.

TABLE III

1	M.J. Brisson (1723-1806), *Ornithologie* ... (F).
2	Comte de Buffon (1707-1788), *Histoire naturelle* ... (F).
3	J.G. Cloquet (1790-1883), *Anatomie de l'homme* (D).
4	G. Cuvier (1769-1832), *Leçons d'anatomie comparée* (F).
5	G. Desargues (1591-1661), *Manière universelle de M. Desargues pour pratiquer la perspective* (D).
6	H.-L. Duhamel du Monceau (1700-1782), *Éléments de l'architecture navale* (D).
7	J.-P. Girardin, *Manuel de chimie appliquée* (D).
8	J.-P. Girardin, *Leçons de chimie élémentaire appliquée aux arts industriels* (this title is likely but cannot be ascertained) (D).

9	J. Guillemeau (1550-1612), *Les oeuvres de chirurgie* (D).
10	L.B. Guyton de Morveau (1733-1816), *Traité des moyens de désinfecter l'air* (D).
11	An unidentified book on botany by A. de Jussieu (1686-1758) or A.L. de Jussieu (1748-1836) (G).
12	Comte de Lacépède (1756-1825), *Histoire naturelle des poissons* (F).
13	Comte de Lacépède, *Histoire naturelle des quadripèdes ovipares et des serpens* (G).
14	S.-F. Lacroix (1765-1843), *Éléments d'algèbre* ... (D).
15	P. de la Hire (1640-1719), *Tabulae Astronomicae* (D).
16	P. de la Hire, *Les éléments des sections coniques* ... (the title is likely but cannot be ascertained).
17	J.-B. de Lamarck (1744-1829), *Histoire naturelle des animaux sans vertèbres* (F).
18	J. Lieutaud (1703-1780), *Précis de la médecine pratique* (F).
19	A. Richard (1794-1852), *Botanique médicale* (D) and *Éléments d'histoire naturelle médicale* (F) (These works are two editions of essentially the same book).
20	N. Savart (1765-ca. 1826), *Cours élémentaire de fortification* ... (D).
21	O.-G. de Ségur (1779-1818), *Lettres élémentaires sur la chimie* (D).

These are the titles of French works for which evidence is available from the Dutch Archives[10] and other Japanese sources[11] proving their presence in Japan in the first half of the nineteenth century. The books may have been any of the French editions, marked (F), any of the corresponding Dutch, marked (D), or German, marked (G) translations. This list does not claim to be exhaustive.
(The presently available data from the archives are far from complete. The edition dates are often missing and the indicated ambiguous titles have to be clarified.)[12]

TABLE IV lists some of the French scientific books that had an impact on Japanese science through de Chalmot's encyclopaedia.[13]

TABLE IV

1	M.J. Brisson, *Ornithologie* ...
2	Comte de Buffon, *Histoire naturelle* ...
3	G. Rondelet (1507-1566), *Libri di Piscibus marinis*, Lyon, 1554-1555, M. Bonhomme, ed. (This book was a major source for Jonston's chapter on fishes in *Naeukerige Beschrijving van der Natuur* ... , Amsterdam, 1660, I, Schipper, ed. and it is many times cited in it).
4	E.L. Geoffroy (1725-1810), *Histoire abrégée des insectes qui se trouvent aux environs de Paris*, Paris, 1762, ed. Durand.
5	R.-A. Ferchault de Réaumur (1683-1757), *Mémoires pour servir à l'histoire des insectes*, Paris, 1734-1742, ed. Imprimerie Royale.

French scientific books having had some indirect impact on the science of the late Tokugawa period via their numerous citations in de Chalmot's encyclopaedia.

In view of the limited space at my disposal, I am not in a position to discuss in detail the impact of all the partially translated French books listed in TABLE I on Japanese scientists. Therefore, I have selected for discussion the work of the French surgeon Ambroise Paré. He was chronologically the first French scientist to exert a considerable influence in Japan.

But before going any further, let us first examine the situation of Japanese medicine at the time of the arrival of the first Europeans. It is generally accepted that at the end of the sixteenth century Western medicine was not more advanced than its Japanese counterpart, except in the fields of anatomy and surgery.[15] It is for example well known that for problems of internal medicine the Jesuit missionaries preferred to be treated by Japanese physicians rather than by the Western ship surgeons who were accompanying them on their long journey. The relatively underdeveloped state of Japanese anatomy and surgery has been rationalised in the light of Japanese medical history. From the seventh century on, Japan decided to fully adopt Chinese medicine. Under the influence of Confucian teachings dissections and bloodshed by surgical intervention were forbidden in Chinese medicine. As far as dissection is concerned, the situation was by the way not very different in Europe up to the end of the fifteenth century. Let us not forget that the prohibition of dissections was only lifted by Pope Sixtus IV (1414-84, r. 1471-84), the builder of the famous Sistine Chapel. A similar prohibition remained in effect in Japan up to the end of the eighteenth century.

In the field of surgery a revolution took place in Europe in the second half of the sixteenth century.[16] It was associated with Ambroise Paré (1510-1590), whose influence on Japanese surgery in the eighteenth century was immense.[17] A discussion of the accomplishments and innovations of Paré, physician in ordinary to four French kings, as well as of his tremendous impact on Japanese surgery, would actually require a special lecture if not a full symposium. Four of his discoveries had a tremendous impact on the surgery of the late sixteenth and early seventeenth centuries:

His accidental discovery, during the Turin campaign in 1536, of the gentle dressing composed of egg yolk, oil of roses and turpentine, for the treatment of firearm wounds instead of cautery. He demonstrated that gunshot wounds were not poisonous (first disclosed in *La méthode de traicter les playes*, 1545).

The application of blood vessel ligature in amputations (first disclosed in *Dix livres de chirurgie*, 1564).

The precise location of bullets and their extraction from the body by placing the soldier back in the position in which he received the shot, in order to determine the bullet's precise trajectory (Battle of Perpignan, 1542), (first disclosed in 1585; 4th enlarged edition of Paré's collected works entitled: *Oeuvres*).

Paré's revival of the podalic version in difficult deliveries re-popularised the procedure which had been described by Soranus of Ephesus, the leading authority on gynaecology and obstetrics of the Antique World (first disclosed in *Briefve collection*, 1549 and then in *Deux livres de chirurgie*, 1573).

It is well known that Paré did not follow the classical education of the physicians of his time and did not know Latin. For that reason, he published all his books in French.[18] Printing of the Fo. first edition of Paré's collected works was completed in 1575. The immense success of this edition should be understood in view of its outstanding scientific, artistic and typographic qualities and not least in view of its appropriate timing, responding as it did to an urgent need among the surgeons of late sixteenth century France. It contained 295 illustrations, most of them already featured in Paré's separate earlier works.[19]

The first foreign nation to discover Paré were the Dutch. The political conditions in the Low Countries, under Spanish rule during the sixteenth century, explain the particular attention paid to the war-experience of the French surgeon. The fourth enlarged edition of Paré's *Oeuvres* (1585) was translated into Dutch by the Ghent physician Carel Batten, who had had to leave his native town with the Protestant exodus. The book was published in Dordrecht in 1592. The translator produced a remarkable work complete with faithful reproductions of Paré's illustrations. Subsequently, twelve additional Dutch editions of Paré's work were printed in the Netherlands during the seventeenth century.[20] The unusual demand for and success of the Dutch versions after the establishment of peace in the Low Countries are quite surprising, especially when taking into account that some of the leading physicians in the Netherlands rejected Paré's invention of blood vessel ligature and kept using the old technique of cauterisation in amputations.

Medical historians agree that Paré's Dutch version, acquired by Narabayashi Chinzan (1648-1711) was the 1649 Schipper Amsterdam edition [ILL 1].[21] In the light of the Japanese civil war of the sixteenth century in which firearms originally introduced by the Portuguese, were widely used in the struggle for power, it is highly regrettable that Paré's 楢林鎮山

book did not reach Japan a hundred years earlier. By the time the Dutch version came into the hands of the Nagasaki interpreters, the treatise was considered in Europe as only of historical importance, except for the Book devoted to the Plague. However, at the end of the seventeenth century no advanced monograph on surgery was available in Japan.

Narabayashi Chinzan,[22] member of a hereditary interpreters' family, was brought up in close contact with the Dutch language. At the age of eighteen he received the title of assistant interpreter (*ko-tsûji*) and by the time he acquired Paré's book, his knowledge of Dutch had improved so much that he became a chief interpreter (*ô-tsûji*). Learning the Dutch language must have been extremely difficult in the Nagasaki of those days. Contacts with the few foreigners at Deshima were limited and no dictionary or Dutch grammar book was available. It appears that Narabayashi's acquaintance with the Dutch physician Willem Hoffmann was the starting point for his interest in medicine. However, in Nagasaki he may have met such prominent scholars as A. Cleyer (1634-1697), W. Ten Rhijne (1647-1700) and E. Kaempfer (1651-1716). The stimulating atmosphere in the contacts with the Dutch physicians eventually induced Narabayshi to give up his position as interpreter and devote his energy and time only to medicine. He was rewarded in 1691, when the Shogunate decided to offer him a position of official physician. In 1706 he completed the manuscript [ILL 2] that is considered to be a partial translation of Paré's Dutch version, but also contains medical knowledge originating from a treatise written in the seventeenth century by the Chinese physician Chin Jikko. Narabayashi's translation was the starting point of the long process that made it possible for Western medicine to be valued in Japan, culminating in its full adoption after the Meiji Restoration.

小通詞 大通詞 出島

The second major physician interested in Paré's surgical work was Arashiyama Hoan (1633-1693).[23]

嵐山甫安

His Paré manuscript has been preserved in the Tokyo University Library. According to Mestler, although undated, it was produced even before the Narabayashi manuscript. Just how Arashiyama had become proficient in the Dutch language remains unclear. However, his close contacts with the Dutch physicians at Deshima are attested to by the extant certificate, signed by N. de Roy and D. Bush, which recognises his aptitude in the art of surgery. Narabayashi's original manuscript entitled *Geka sôden* did not survive. However, several old copies are known to have existed before the Second World War. Doe reproduced[24] an illustration from Fujikawa's copy (1 volume, 36 leaves of text with 22 leaves of coloured drawings) and mentioned the existence of another copy at Nagasaki Medical College.

外科宗伝

According to Mestler[24] and Ôtori,[26] one complete set of *Geka sôden*, preserved at the Library of Tokyo University and prefaced by the famous botanist and Confucian teacher Kaibara Ekiken (1630-1714), who referred to it as a new treatise and not a translation, is made up of six parts:[27]

貝原益軒

1 Shikake sho 仕掛之書 (on pathology and treatment of diseases)
2 Kinsô sho 金瘡書 (on treatments of wounds)
3 Kôyaku sho 膏薬書 (on treatments using unguents)
4 Abura no sho 油之書 (on treatments using oils)
5 Yushuyô shozu 油取様書図 (on the removal of oils)[28]
6 Kinsô tetsuboku-zu 金瘡跌撲図 (illustrations of the treatments of wounds).

Four other translators are known to have been involved in translating Paré's writings: Nishi Gentetsu (1681-1760), a certain Akitaka, Irako Mitsuaki (1737-1798?) and Ôtsuki Gentaku (1757-1827).[29] In addition to this list, three important surgeons, Yoshio Kôgyû or Kôsaku (1724-1800), Hanaoka Seishû (1760-1835) and Koshimura Norimoto used some illustrations derived from Paré's book in their own treatises.[30] Nishi Gentetsu's translation (1735) may be entitled either *Kinsô tetsuboku ryôji no sho* or *Kinsô fûboku ryôji no sho*. Irako Mitsuaki's printed translation (1769) is entitled *Geka kunmô zu-i*. Full identification of the contents of Irako Mitsuaki's title page with Paré's work has not been successful yet. Irako's name in ILL 3 is shortened to one character (rô), the fourth and fifth Chinese characters of the second column read Mitsuaki (Kôken). Furthermore, the title page reveals that the translation involved two of Irako Mitsuaki's disciples whose identification appears difficult at present. The work seems to have been carried out under the direction or with the advice of Mitsuaki. The contents on the title page of *Geka kunmô zu-i* seem to have been inspired by chapters III and IV, which are entitled respectively "Les signes & ingements des playes" and "Du prognostic des playes", as written on page 286 of book IX of the first French edition of 1575. However, it is difficult to recognise Paré's original writings in Irako's text. In addition to his printed book, Irako Mitsuaki drew a picture scroll in one roll, known as *Irako shi kinsô fûboku chiryô no maki*, whose illustrations were those of Narabayashi's *Geka sôden*.[31] These figures represent sur-

西玄哲
伊良子光顯
吉雄耕牛、幸作
華岡青洲
金瘡跌撲療治之書
外科訓蒙図彙
郎
光顯

gical operations as performed by the Dutch. However, no explanatory text accompanied the picture scroll. Concerning Nishi Gentetsu's translation, the various comments agree that this most precise manuscript, quite faithful to Paré's Dutch text, was almost identical to the second part of Narabayashi's work.

An interesting mystery surrounds the origin of a unique printed two-volume book entitled *Geka kunmô zu-i mokuroku*,[32] preserved in the Bibliotheca Walleriana at Upsala University.

This book, also derived from Paré and edited in 1764 or 1767, was reported to contain drawings different from those of Arashiyama, Narabayashi and Nishi [ill 4 & 5]. The author of this book was probably a physician called Akitaka but this statement requires further confirmation.

ill 4 is definitely derived and copied from Narabayashi after modification of Paré's woodcut (1575 edition, Book X., p. 392). However, the elaborate illustration of the triploid and its surgical application [ill 5], although bearing some resemblance to Paré's woodcut (1575 edition, Book IX., p. 304), was derived from Scultetus' (1595-1645) *Armamentarium chirurgicum*, one of the most influential surgical treatises of the seventeenth century (second edition, Hagae, 1656, Plates nos. 3 and 32). Scultetus described in detail the precise use of his triploid and cited Paré as having used instruments for the same treatments. An enlarged Dutch edition of Scultetus' book, with illustrations, published in Amsterdam in 1672, must have reached Japan.

iii Various aspects of Paré's influence on Japanese medicine

Japanese authors write Paré's name in various ways, either phoneticised in Chinese characters or in *katakana*:

an fu ro shi u ・ pa a re 俺 貌 魯 止 烏・巴 亜 暦	by Ôtsuki Gentaku in *Kyûen tekiei*
an bu ru 安 武 児	by Sugita Genpaku 杉田玄白 in *Kaitai shinsho* 解体新書
an bu ru shi su ・ pa a re 安 勃 盧 悉 斯・八 穵 列	by Ôtsuki Gentaku in *Chôtei Kaitai shinsho* 重訂解体新書
a n bu ro a su ・ pa re ア ン ブ ロ ア ス・パ レ	by Sakai Shizu[32]

Paré's considerable influence on the evolution of Japanese medicine during the eighteenth century covers a wide range of areas. His teachings were transmitted to Japan through the various Dutch editions of his book. The various translated manuscripts were copied and widely circulated among Japanese surgeons. Mestler mentions Paré's treatise among those few books that "made the greatest impression on the Japanese historical mind".[34] Evidence is available to demonstrate Paré's impact in Japan beyond surgery, in such specialities as anatomy[35] and obstetrics.[36] In addition, physicians well known to the Japanese such as Blankaart (1650-1704), Kulmus, Palfyn (Palfijn) and Ruysch also propagated Paré's surgery through the numerous quotations in their own works, which were brought to Nagasaki by the agents of the Dutch East India Company. In the process of the compilation of *Kaitai shinsho*, a number of Western books were consulted by Sugita Genpaku and his colleagues, both for the comprehensive presentation of the text as well as for the appropriate anatomical illustrations to be entered.[37] Paré's Dutch treatise, in the possession of Maeno Ryôtaku, was amongst these books. However, the precise influence of Paré's anatomy on *Kaitai shinsho* has not yet been demonstrated.

Yoshio Kôgyû, the famous Nagasaki surgeon, Dutch interpreter and prolific translator, used to give a certificate of proficiency to his students on long paper rolls. At least one such certificate, dated 1790, has survived. It is decorated with a drawing that imitates Paré's technique of healing a dislocated shoulder.[38] Late in the eighteenth century, Katsuragawa Hoshû edited Ôtsuki Gentaku's translation,[39] entitled *Kyûen tekiei*, originating from Paré's *Génération de l'Homme* (Book XXIII in the first and XXIV in the enlarged fourth edition of his collected works). A fragment of Paré's Dutch version was brought to Nagasaki in 1724 at Shogun Yoshimune's order with a view to translating its index into Japanese. Thus the great Shogun must also have been acquainted with Paré's treatise.[40]

桂川甫周
九畹摘英
吉宗

iv Ambroise Paré in twentieth century Japan

This short section is based on the special lecture[41] given by the famous Japanese specialist of Paré, Professor Ohmura (Ômura) Toshirô, on the occasion of the symposium held at Laval in 1990 to commemorate the four hundredth anniversary of the death of the great Renaissance surgeon. In his presentation, Professor Ohmura expressed the gratitude of the Japanese people to Ambroise Paré and informed the audience about

ILL 1A

ILL 1B

ILL 2A

ILL 2B

ILL 2B

ILL 2B

the visit of a team of the Japanese Television to Laval to pay a tribute to the French surgeon. Professor Morioka, surgeon to the Emperor, was also of the party that made the pilgrimage to Paré's native city. According to Professor Ohmura, the Japanese people were well acquainted

with Paré's famous motto, printed for the first time in the fourth edition of his *Oeuvres*: "Je le pansay, Dieu le guarit" ('I dressed him, and God healed him'). However, most Japanese did not know that Paré was the original author of these words.

In view of the historical importance of Paré's work for Japanese medicine, Professor Ohmura translated into modern Japanese both *The Book of Fractures* and *The Book of Luxation or Dislocation* and published them during the 1980s. He also informed the participants of the symposium that a picture of Paré's famous statue by David d'Angers on the main square of Laval, illustrated the cover of the November 1990 issue of the *Japanese Surgical Review*. Moreover, a Japanese committee was established for the celebration of the four hundredth anniversary of Paré's death and about four hundred participants gathered on that occasion. In addition, in 1991 a lantern was donated by the committee and placed in a garden in Laval to express the Japanese people's gratitude to the French surgeon.[42] Most books written in Japan about eighteenth century medicine mention the date of translation of the Narabayashi manuscript, 1706, as a landmark in Japanese medical history.

v A French painting related to sixteenth century Japanese history

The question also arises as to how Japan was represented in the European mind and more specifically in the French aristocracy's imagination at the beginning of Japan's seclusion period. In this regard I cannot resist the temptation to illustrate the only oil painting in the Louvre Museum pertaining to Japanese history.

The canvas [ill 6] was painted by Nicolas Poussin (1594-1665), the most famous French artist of the seventeenth century. Its subject represents the miraculous recovery by St. Francis Xavier (1506-1552) of the
鹿児島 dead daughter of an inhabitant of the city of Kagoshima. The upper half of the painting (not shown on ill 6) displays Christ in glory. Francis Xavier's companion Juan Fernandez Oviedo (1526-1567), probably the first Westerner to speak the Japanese language to some degree, shown from the back, is looking up to Christ expressing his gratefulness

and admiration. Reflecting his modesty, Fernandez always remained a coadjutor brother, the lowest rank in the Society of Jesus.

The history of this painting, commissioned for the high altar of the Paris Jesuit Church of the Faubourg Saint-Germain (now rue Bonaparte) is well documented.[43] The church was demolished in the eighteenth century and the painting subsequently entered the Louvre. King Louis XIII was annoyed by the undisputed reign of Simon Vouet (1590-1649) over the Paris artistic world. As a result, to counterpoise Vouet's influence he asked Richelieu (1585-1642) to have Poussin return from Rome, where the painter had become highly successful under the protection of powerful patrons. Reluctantly, but yielding to the pressure, Poussin returned to Paris in 1640. He did not appreciate the atmosphere of the French capital and its cold and rainy climate. Nor was he happy with the new artistic commissions of decorative character. Taking advantage of Richelieu's and the King's death (1643) Poussin did not hesitate to return to Rome.

In the seventeenth century, the unveiling of a major altarpiece was considered an important event. The King, the royal court and those occupying a rank of some consequence in high society must have visited the church. On Poussin's painting they discovered the unusual physiognomy of Japanese men. However, the Japanese women were painted with Western features and the recovering young girl was lying in a European-style bed, unknown to the Japanese. If Poussin was familiar with the features of Japanese men, it was the result of his long stay in Rome. Poussin had not seen the face of any Japanese women. The first Japanese delegation was staying in Rome from February to August 1585. Although the Italian artists manifested little interest in representing the features of the Far Eastern envoys, the physiognomy of some of them was immortalised in works of art. One of the walls of the Vatican Library is decorated with a fresco showing the Japanese emissaries during a procession.[44] The Louvre owns a beautiful drawing by Federico Zuccaro (1540/43-1609) showing from the back one of the members of the Japanese embassy. During their short visit to Venice, Tintoretto (1518-1594) painted the portrait of one of the delegates and the canvas decorated the wall of the Senate's room for some time. Evidence is available that paintings on screen illustrating Japanese everyday life had already been imported to Rome and Lisbon at the end of the sixteenth century. The Louvain artist van Winghe made drawings of one such folding screen representing the gate and tower of Azuchi castle.[45] Several Western artists and book illustrators equally tried to represent Chinese noblemen as well as scenes of everyday life in that Asian country. However, surprisingly, on most of these illustrations[46] it is impossible to recognise the specific facial features of the Japanese and Chinese. They are represented as hardly distinguishable from Westerners.

The legendarised figure of St. Francis Xavier, his life and alleged miracles inspired several works of art during the Counter-Reformation.[47] Rubens painted a huge altarpiece for the Antwerp Jesuit Church (ca. 1617).[48] His disciple Erasmus Quellenius (1607-1678) was commissioned to decorate the Mechelen Jesuit Church with several paintings depicting the Saint's sermons, conversions and miracles. Giovanni B. Gaulli (called *il Baccicio*), (1639-1709) painted Francis Xavier's solitary death on the small island of Shangchuan close to the Chinese coast (S. Andrea al Quirinale, Rome). Carlo Maratta (1625-1713) illustrated the moment when the Portuguese were about to bury the Saint's body (Gesu, Rome). However, in none of these works one can find any detail or person reminiscent of Asia. Jacques Stella (1596-1657) painted Francis Xavier lying on a mat while being visited in a dream by an angel. The heavenly messenger is holding a map in front of his eyes showing the countries where the great missionary will accomplish his future activities. Although Japan is on the map, the angel is pointing to the southern part of China. No Japanese city is indicated on the map.[49] It is interesting to note that in spite of the importance the Jesuits attached to art and its power to convey a message, they failed to make a substantial contribution to the visual representation of the continent they aspired to convert to the Christian faith. For instance, Matteo Ricci (1552-1610) teaching astronomy to the Chinese, never became a subject for painters of the Counter-Reformation. Even at the end of the seventeenth century, when Father Andrea Pozzo (1642-1709) was involved in the decoration of the Sant' Ignazio di Loyola Church in Rome and transformed the entire ceiling into a vast celebration of the Jesuit mission, the symbol of Asia still followed Cesare Ripa's obsolete *Iconologia* (1611). The latter did not have anything to do with the Chinese or Japanese world.[50]

Interestingly, a Japanese art historian, Kimura Saburô, has demonstrated recently that the subject of the altarpiece was a fabrication by the propaganda of the Catholic Church of the Counter-Reformation.[51] None of the numerous letters of Francis Xavier or his companions, sent from Japan to Rome, mention the depicted or, any other miraculous events. However, the Jesuit historian Luis Frois described in his manuscript of 1584, that Luis de Almeida, a Jesuit priest, had visited the city of Kagoshima and healed an eighteen year old gravely ill girl. In the Japanese chronicle *Kakken uwai* dated 1583, i.e. about thirty years after St. Francis Xavier's death, the visit of a European priest to Kagoshima is likewise related.[52]

The spirit of the Counter-Reformation and the personality and missionary activity of St. Francis Xavier stimulated the publication of several biographies of this illustrious companion of the founder of the Jesuitic Order. The biography published by Orazio Torsellino in 1594 reproduced the narrative of Luis Frois about the recovery of the Japanese girl in Kagoshima. However, Luis de Almeida's name was substituted by Francis Xavier's. Kimura Saburô has convincingly demonstrated[53] that Poussin's source of inspiration was the French translation of Torsellino's book (1608).

The name substitution is easy to explain in the light of the evolution of the Jesuit Order's history at the end of the sixteenth century. As a consequence of the growing influence of Jewish converts within the Jesuit Order, the decision was taken in 1593 by the General of the Order, Claudio Acquaviva, to henceforth reject their application for admission. This rule was to be strictly observed for several centuries.[54] Although Luis de Almeida's merits were recognised as the first Jesuit physician to introduce Western medicine into Japan, he was the descendant of an old Portuguese Jewish family. Anti-Jewish sentiment combined with the renown of Francis-Xavier must have resulted in the substitution of his name.

NOTES

1 Charles R. Boxer, *The Christian Century in Japan*, (1549-1650) (Berkeley/Los Angeles: University of California Press, 1967).
2 Saburô Miyashita, "A Bibliography of the Dutch Medical Books Translated into Japanese," *Archives internationales d'histoire des sciences* 25 (1975): 16; Grant K. Goodman, *The Dutch Impact on Japan*, (1640-1853) (Leiden: Brill, 1986).
3 Teizô Ogawa, "The Beginnings of Anatomy in Japan," *Okajima's folia anatomica Japonica* 52 (1975): 59-72.
4 Tôgo Tsukahara, *Affinity and Shinwa Ryoku, Introduction of Western Chemical Concepts in Early Nineteenth Century Japan* (Amsterdam: Gieben, 1993).
5 Miyashita, o.c.
6 Ibid.
7 Miyashita, o.c., 42.
8 Second edition: Noël Chomel & Jacques A. de Chalmot, *Algemeen Huishoudelijk-, Natuur-, Zedekundigen- en Konst-Woordenboek* (Leiden/Leeuwarden: de Chalmot), first edition: 1768-1777; second edition: 1778-1786.

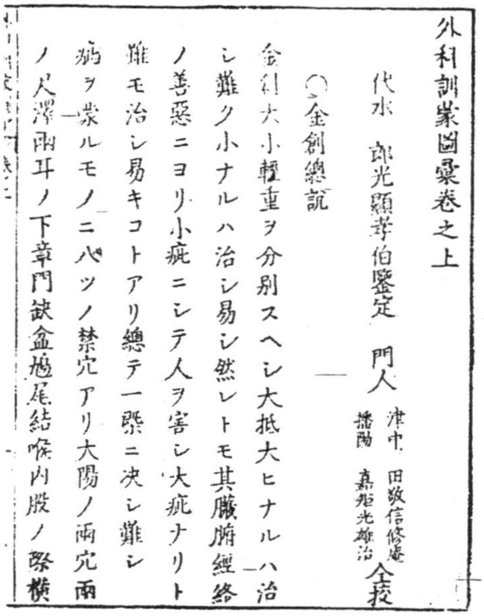

外科訓蒙図彙巻之上

伏水　郎光顯孝伯鑒定　門人　津中　田敬信修庵　播陽　嘉矩光雄治　仝校

○金創總説

金創大小輕重ヲ分別スヘシ大抵大ヒナルハ治シ難ク小ナルハ治シ易シ然レトモ其臓腑經絡ノ善惡ニヨリ小疵ニシテ人ヲ害シ大疵ナリト雖モ治シ易キコトアリ總テ一槩ニ決シ難シ病ヲ祟ルモノニ八ツノ禁穴アリ大陽ノ兩穴兩ノ人澤兩耳ノ下章門缺盆旭尾結喉内股ノ際横

ILL 3

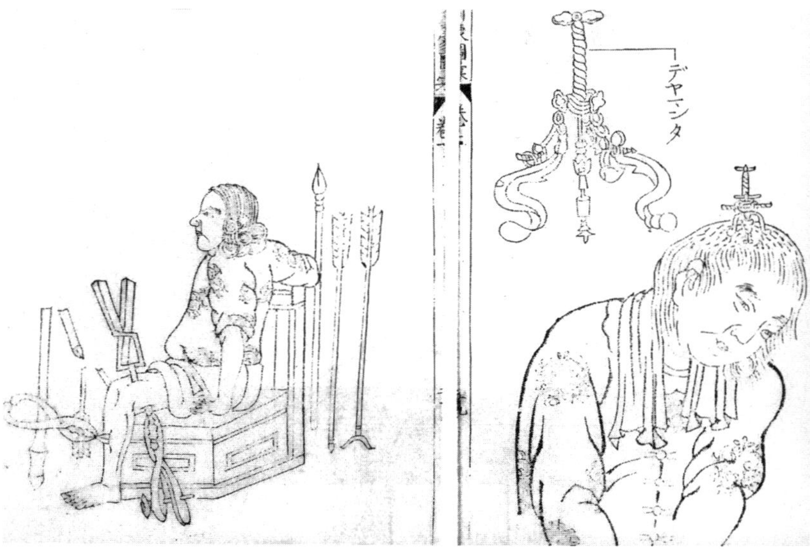

ILL 4

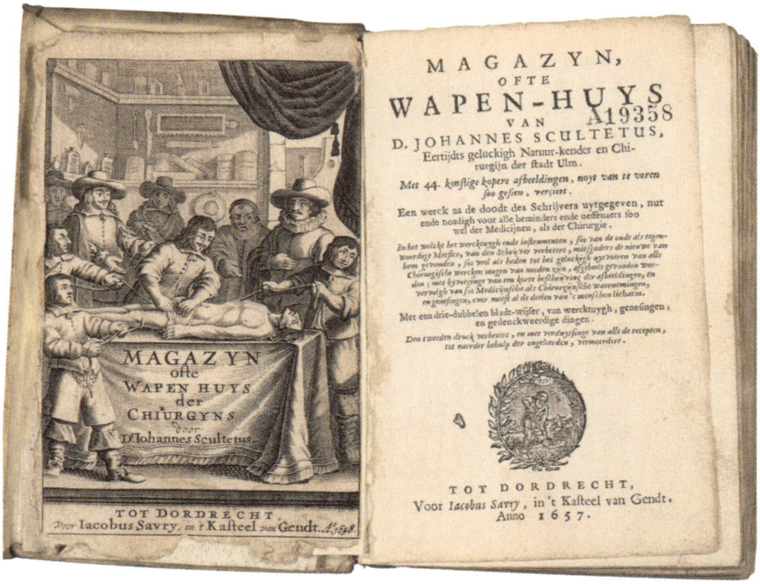
MAGAZYN ofte WAPEN HUYS der CHIRURGYNS door Dr. Iohannes Scultetus
TOT DORDRECHT, Voor Iacobus Savry, in 't Kasteel van Gendt
MAGAZYN,
OFTE
WAPEN-HUYS
VAN
D. JOHANNES SCULTETUS,
Eertijdts geluckigh Natuur-kender en Chirurgijn der stadt Ulm.
Met 44. konstige kopere afbeeldingen, noyt van te voren soo gesien, verciert.
Een werck na de doodt des Schrijvers uytgegeven, nut ende noodigh voor alle beminders ende oeffenaers soo wel der Medicijnen, als der Chirurgie.
Met een drie-dubbelen bladt-wijser, van wercktuygh, geneſingen, en gedenckweerdige dingen.
TOT DORDRECHT,
Voor Iacobus Savry, in 't Kasteel van Gendt.
Anno 1657.

ILL 5

ILL 6

9 James Mac Lean, "The Introduction of Books and Scientific Instruments into Japan, (1712-1854)," *Japanese Studies in the History of Science* 13 (1974): 9-68; Tsuruichi Hayashi, "A List of Dutch Books on Mathematical Sciences, Imported from Holland to Japan before the Restoration of 1868," *Nieuw Archief voor Wiskunde* (1907): 232-237; Suketoshi Yajima, "Dutch Books on Science and Technology Brought to Japan in XVIII and XIX Centuries," *Japanese Studies in the History of Science* (1973): 76-79.
10 Mac Lean, o.c.
11 Hayashi, o.c.; Yajima, o.c.
12 Mac Lean, o.c.
13 Chomel & de Chalmot, o.c.
14 Ibid.
15 Goodman, *The Dutch Impact on Japan*, (1640-1853).
16 Paule Dumaître, *Ambroise Paré, chirurgien de quatre rois de France* (Paris: Perrin, 1986).
17 Gordon E. Mestler, *Bulletin of the Medical Library Association* 42, no. 3 (1954): 287-327; 44, no. 2 (1956): 125-159; 45, no. 2 (1957): 164-219; Toshiro Ohmura, "Voyage de la chirurgie d'Ambroise Paré de France au Japon," *Actes du colloque international Ambroise Paré et son temps* (1990): 245-256.
18 Janet Doe, *A Bibliography of the Works of Ambroise Paré* (Chicago: Chicago University Press, 1937).
19 Ibid.; Dumaître, o.c.
20 Doe, o.c.
21 Ibid.; Ohmura, o.c.; Mestler, o.c.
22 Goodman, o.c.
23 Ibid.
24 Doe, o.c.
25 Mestler, o.c.
26 Ranzaburô Ôtori, "The Acceptance of Western Medicine in Japan," *Monumenta Nipponica* 19, 3-4 (1964): 20-40.
27 A copy of Narabayashi's work from a French private collection was exhibited in Paris in 1972.
28 In fact to be read properly as *Abura toriyô no shozu*.
29 Miyashita, o.c., 42.
30 Mestler, o.c.
31 *Catalogue of the Historical Writings of the Japanese Society of Medical History* (Kyoto: Benridô, 1959); *Catalogue of the Books Exhibited at the "Yôgaku Kotohajime Ten", Earlier Phases of Western Civilization in Japan* (Tokyo, 1954).
32 Mestler, o.c.
33 Sakai Shizu 酒井シヅ, "Kaitai shinsho to chôtei kaitai shinsho"「解体新書と重訂解体新書」, in *Ôtsuki Gentaku no kenkyû*「大槻玄沢の研究」, ed. Yôgaku-shi kenkyûkai 洋学史研究会 (Kyoto: Shibunkaku, 1991).
34 Mestler, o.c.
35 Mieko Macé, "L'anatomie occidentale et l'expérience clinique dans la médecine japonaise du XVIe au XVIIIe siècle," *Nombres, astres, plantes et viscères* (1994): 135-179.
36 Miyashita, o.c., 42.
37 Macé, o.c.
38 Ohmura, o.c.
39 Miyashita, o.c.
40 Goodman, o.c.
41 Ohmura, o.c.
42 Ibid.
43 Alain Mérot, *Nicolas Poussin* (Paris: Hazan, 1990).
44 Donald F. Lach, *Asia in Making Europe* (Chicago/London: The University of Chicago Press, 1977).
45 Ibid.
46 C. Vecellio, *Habiti antichi e moderni* (Venice, 1598) and T. De Bry, *Indiae Orientalis* (Frankfurt, 1598).
47 Emile Mâle, *L'art religieux après le Concile de Trente* (Paris, 1932).
48 The painting was transferred to Vienna in 1776 after the Jesuit Order was abolished in the Southern Netherlands. The canvas is exhibited in the Kunsthistorisches Museum.
49 Department of Prints, Bibliothèque Nationale, Paris, Da 20 folio, no. 82C 109 157.
50 Lach, o.c.
51 Saburô Kimura, "La source écrite du Miracle de Saint François-Xavier de Poussin," *Revue du Louvre* (1988): 394-398.

52 Ibid.
53 Ibid.
54 Jacques Proust, *L'Europe au prisme du Japon* (Paris: Michel, 1997).

ILLUSTRATIONS

ILL 1 Johan Vesling, *Konstige Ontleding des Menschelijcken Lichaems; In 't Latijn beschreven Door den Edelen en Hoogh geleerden Heer Joannes Veslingius, ... vertaelt door Gerardus Blasius, Der Medicijnen Doctor, en Practizijn t'Amsterdam* (Amsterdam: Gerret Sweerman, 1661). Collection Katholieke Universiteit Leuven, Central Library, Tabularium 7A2729.

ILL 1A title page of the 1636 edition of Paré's book. *De Chirurgie ende alle de Opera, ofte Wercken van Mr. Ambrosius Paré, Raedt ende opperste Chirurgijn van vier Coninghen in Vranckrijck. Nu eerst uyt de Françoysche, in onze ghemeyne Nederlandtsche Spraecke, ende uyt de vierde Editie gehetrouwelijck overgest: Door D. Carolum Battum, Medicijn ordinarius der Stadt Dordrecht. Ende in desen laetsten druck, van veel fauten verbetert. Verdeylt in XXVIII Boecken.* Tot Amsterdam, gedruct by Cornelis van Breugel, MDCXXXVI (1636). Collection Université catholique de Louvain, Bibliothèque générale et de sciences humaines.

ILL 1B the title page of the Schipper edition of Paré's book, Amsterdam 1649, Library of the University of Amsterdam.

ILL 2 the repair of a dislocated shoulder from Narabayashi Chinzan's *Kôi geka sôden* (1706)

ILL 2B Drawings in Ambroise Paré, *De Chirurgie, ende Opera van alle de Wercken/ van Mr. Ambrosius Paré, Raedt ende opperste Chirurgijn van vier Koningen in Vranckrijck Uyt de Françoysche in de Nederlandtsche Sprake, uyt de vierde Editie ghetrouwelijck overgheset door D. Carolum Battum, Medicijn ordinaris der Stadt Dordrecht* (Vlissingen: Jacob Pick, 1655), pp.453-454. Collection Katholieke Universiteit Leuven, Central Library, Tabularium, B 12361. These pictures served as the basis for the drawings in *Kôi geka sôden* (ILL 2).

ILL 3 the title page of *Geka kunmô zu-i* (1769), as reproduced by Doe from Fujikawa's copy in her bibliographical work published in 1937.

ILL 4 drawings from the two different volumes of *Geka kunmô zu-i mokuroku* by Akitaka? (1764 or 1767), Bibliotheca Walleriana, Upsala University, Sweden.

ILL 5 Title page and frontispice of Scultetus, *Magazyn, ofte wapen-huys van D. Johannes Scultetus, Eertijdts geluckigh Natuur-kender en Chirurgijn der Stadt Ulm* (Dordrecht: Voor Iacobus Savry, in 't Kasteel van Gendt, 1657). Collection Katholieke Universiteit Leuven, Central Library Tabularium, A19358.

ILL 6 the lower half of the *Miraculous Recovery by St. Francis Xavier of the Dead Daughter of an Inhabitant of the City of Kagoshima*, oil on canvas by Nicolas Poussin, 1641, Musée du Louvre, Paris.

XV

"RERUM MEMORABILIUM THESAUROS," A TREASURY OF MEMORABLE THINGS – CARL PETER THUNBERG'S OBSERVATIONS DURING HIS YEAR IN JAPAN, 1775-1776

Catharina Blomberg
Faculty of Oriental Studies, Stockholm University

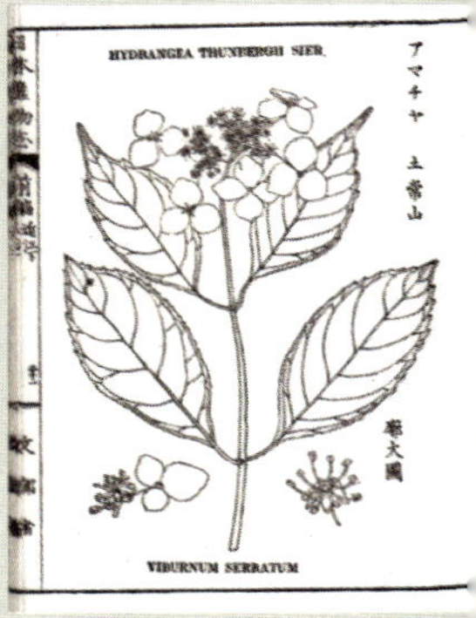

"Each and every Traveller thinks that he ought to become an Author and relate something remarkable to his Countrymen, despite the fact that he himself has had such scant information that he has neither understood nor coherently been able to comprehend what he has heard and seen, much less been able to transmit to others some intelligible idea about it. And this sole reason has produced more incomprehensible books than one can easily imagine." [1]

Carl Peter Thunberg's caustic remark in the foreword to his *Travels in Europe, Africa, Asia, Undertaken in the Years* 1770-1779, offers an interesting insight into his character. He was a born scholar, whose curiosity and powers of observation never seem to have flagged for an instant, even under the most trying circumstances, and he endeavoured to report the facts, as he understood them, accurately and without exaggeration. The bulk of Thunberg's voluminous *oeuvre* consists of scholarly papers published in the *Acta* of various academies and learned societies in Sweden and throughout Europe during his long career. His first major work, the *Flora Japonica*, was published in Leipzig in 1784, and this was followed by the account of his travels, published in four volumes in Upsala between 1788 and 1793. His *Flora Capensis* was published in Upsala between 1807 and 1813.[2] The long list of papers published in the *Acta* of the Royal Academy of Sciences, Stockholm, seventy-two in all, spans nearly five decades, from 1773 to 1821, as do the two hundred and ninety-three dissertations, written by Thunberg and defended in a public disputation by one of his students, from 1780 to 1828, a truly remarkable achievement even for a man whose scholarly career can be said to have lasted from his first preliminary treatise in 1767 to his death in 1828.

Carl Peter Thunberg was born in Jönköping, in the Swedish province of Småland, on November 11th 1743. His father, Johan Thunberg, was a bookkeeper at Hörle iron-works as well as a tradesman, and died early. His mother, Margaretha, né Starkman, the daughter of the alderman of the guild of pistol-makers in Jönköping, continued her late husband's trade in order to support her two small sons, and later married the merchant Gabriel Forsberg. Thunberg's parents intended him to go into trade, but at the age of twelve he was showing such scholarly promise that he was encouraged to continue his schooling, supplemented by private tuition under his teacher Håkan Sjögren, later dean of Växjö Cathedral. In 1761 he matriculated at Upsala University, where he sat for the then customary examinations in Theology, Law and Philosophy before specialising in Medicine, where he quickly became one of the star pupils of the famous botanist Carl von Linné (1707-1778), internationally better known as Linnaeus, his name before he was ennobled. Linnaeus, who had revolutionised botany with his sexual system of classification and was internationally renowned as one of the leading scholars of his day, arranged extensive foreign travels for a number of promising Swedish scholars through his friends and colleagues in Europe, e.g. the famous Dutch botanist Johannes Burmannus (Burmann, 1706-1779) and his son and successor Nicolaus Laurens (1734-1793), and Sir Joseph Banks (1743-1820), who had taken part in Captain James Cook's first voyage, 1768-1771, accompanied by Linnaeus' pupil Daniel Solander (1735-1782).

Thunberg presented his doctoral thesis *De Ischiade* (On Sciatica) in 1770, and in the autumn of that year travelled through Denmark and Holland to Paris, where he spent eight months on a small travelling scholarship from Upsala University in order to perfect his knowledge of Natural History, Anatomy and Surgery. He arrived in Paris on December 1st 1770, found his lodgings, recommended by Linnaeus, in the house of a tobacconist, M. Berth, in the *Rue de la Harpe* near the Sorbonne, and applied himself with vigour to his studies. It is quite typical of his seemingly boundless energy that he managed to find the time to visit both the Charité and the Hôtel Dieu hospitals within the first twenty-four hours of his stay. His studies were a combination of theory and practice, and he attended the lectures and demonstrations of such luminaries as the professor of anatomy and surgery Raphael Sabatier (1732-1811) and the botanist Bernard de Jussieu (1699-1777). Some of Thunberg's remarks on life in pre-revolutionary France are quite illuminating. He noted with approval that free medical care for all, regardless of social standing, was provided at the Hôtel Dieu, and when he paid a study visit to the workshop of the enameller Roux, who was famous for making false eyes, and who charged from 1 to 25 *louis-d'or* for them, he discovered that Roux used to distribute them free of charge to the poor once a month. Before leaving Paris he bought a set of the latest surgical instruments, which five years later, together with others acquired in Amsterdam, created a great deal of interest among the Shogunal physicians in Edo. Already when Thunberg passed through Amsterdam on his way to Paris, Burmannus the Younger had suggested that he ought to visit Japan, and upon his return to Holland in the autumn of 1771, it was arranged, through the *bona officia* of the Burmanni, that he was to travel to the Cape of Good Hope with the Dutch East India Company and spend some years there in order to examine the flora, sending specimens back to Holland, and also to perfect his command of the Dutch language in preparation for his journey to Japan.[3]

On December 30th 1771 Thunberg sailed as ship's doctor on board the *Schoonzigt*, and less than a week into the voyage an unfortunate incident occurred, which came very close to costing him his life. The ship's chaplain, who was in charge of the provisions and the menu, had by a strange oversight given the cook a quantity of lead white mixed with ordinary flour, and the resulting pancakes, eaten by the officers, the cook and his staff caused varying degrees of violent sickness. The first of Thunberg's published scholarly papers described this alarming event in minute and relentless detail, and it is further proof of his scholarly attitude that he managed to record his terrifying symptoms while being desperately ill for the best part of three weeks.[4]

Having arrived in Cape Town on April 16th 1772 Thunberg remained in the country until the beginning of March 1775. During his three years in Africa Thunberg made three long field trips, each lasting between three and five months, throughout the length and breadth of the Cape colony, enduring all manner of adventures and calamities. On one occasion his party was attacked by a wild buffalo, which gored two horses, and on his thirtieth birthday he and his mount sank into a hollow made by a hippopotamus near the bank of a river they were crossing. Between these trips Thunberg was busy drying seeds, mounting pressed plants, packing dried insects and stuffed birds and planting live saplings and bulbs in boxes in order to send them to Leiden and Amsterdam as well as to Linnaeus in Upsala. He also made brief excursions with new-found friends and colleagues, e.g. the Scots botanist Francis Masson (1741-1805), who was collecting plants for Kew

Gardens in London, and the French naturalist and explorer Pierre Sonnerat (1745-1814), with whom he climbed Table Mountain, noting with some amusement that Sonnerat managed to wear out three pairs of "dainty and thin French shoes" on this particular occasion. Masson and Thunberg corresponded in later life, and there is a letter dated the Cape of Good Hope, May 29th 1793, in which Masson refers to mutual acquaintances: "Many of your old friends still exist and have often enquired for you, viz. De Witt of Rode Zant [...]" The latter appears to have been a particular friend, and Thunberg had often visited his estate, consisting of vineyards and orchards. Thunberg studied and made notes on everything new and interesting, and apart from discovering several hundred hitherto unknown plants as well as animals and insects, he even managed to compile a brief vocabulary of the native "Hottentot" language, complete with click-sounds, an exercise he was to repeat in Java as well as in Japan. Idleness was obviously anathema to him, and on one occasion, when he was waiting for new funds in Cape Town, he expressed his utter distaste for his condition, however temporary, as an "otiosus Spectator", an idle onlooker.

On March 2nd 1775 Thunberg sailed for Batavia in the *Loo*, arriving on May 18th. Almost immediately he was appointed chief surgeon on the *Stavenisse*, which was bound for Japan, and engaged to remain in the country for a year as doctor at the Dutch Factory (*Oranda shôkan*), accompanying the *Opperhoofd* ('overseer', Jap.: *kapitan*) on the journey to Edo. Before the *Stavenisse* weighed anchor on June 21st Thunberg had ordered a number of new suits of clothing, "some silk, some cloth with froggings and other finery, in order to be able with dignity to show myself among the curious Japanese, who view the Europeans with more attention than ever the Naturalist can examine any rare Animal."[5] Remarkably he also managed to compile a small Malay vocabulary and phrase book, chiefly dealing with domestic matters, or indeed catastrophes, e.g. "Is it that late, why have you not yet cleaned the house?" and "Whose fault is it that I have not yet had any coffee?" The Dutch officers customarily brought a number of slaves to Japan as domestics, and Thunberg probably needed a smattering of Malay in order to communicate with his personal servant. The many fortifications on the canals in Batavia made him reflect on the situation of the Europeans in a strange land: "These defence measures, necessary for a people which is scant in number in a country encroached upon, make the City far from pretty, and leave a thinking Philosopher with less than agreeable notions about the manner in which the many delica-

オランダ商館
甲比丹

cies, expensive fabrics and furniture the European enjoys with so much pleasure are procured from these distant places."[6]

出島 Things could not have been more different in Japan, where the Europeans were confined to the small island of Deshima in the harbour of Nagasaki. "A European, who has to remain here, is as though dead and buried in a corner of the globe. [...] One's will is completely infirm and inert, since for the European there is no other will here but that of the Japanese, which must be strictly observed in every detail."[7]
Thunberg's account of his stay in Japan occupies a substantial part of his *Travels*, and his year in this remote country, which barely tolerated the presence of a few carefully guarded foreigners on its soil, was clearly the high point of his extensive travels. The *Stavenisse* dropped anchor in the harbour of Nagasaki on August 13th 1775, after a rough passage, and Thunberg noted the many time-consuming preparations which had to be undertaken before the officers and their personal luggage were allowed ashore and the unloading of the cargo could finally begin. The muster-roll of the entire crew, including the slaves, recorded names and ages only, as everyone was considered to be of Dutch nationality, although there were Swedes, Danes, Germans, Portuguese and Spaniards among the crew. Three years earlier an abandoned ship, the *Burg*, which contrary to Company orders had not been fired, had drifted ashore and been meticulously inspected by the Japanese authorities. Important quantities of contraband had been discovered, and as a consequence the rules concerning smuggling were enforced with increased stringency. Even the ship's captain and the head of the Factory were now subjected to careful searches when moving between the ship and Deshima, and the passes which the Europeans were obliged to carry at all times, contained not only their names but a list of the belongings about their persons on each occasion, including such items as pocket watches. Thunberg recorded the prolonged and rigorous examination of the ship, its cargo and those who had sailed in her, the ceremonial visits by Japanese officials, who were treated to cakes and liqueurs, served in crystal glasses from cut-glass decanters, while seated under a specially erected awning on the deck, and the constant need to apply for permission to move about, even when he, the ship's doctor, had to go aboard to tend to those who had fallen ill or met with accidents while unloading the cargo. Bedding was ripped open and the feathers stirred about, tubs of butter or jam were pierced with iron rods, a square hole was cut in the middle of each cheese, and random samples were taken from the eggs brought

in from Batavia. Despite certain mitigating circumstances, e.g. the fact that "Customs duties are quite unknown [...] A particularly fortunate advantage, which few other countries possess!"[8], Thunberg regarded the inflexible Japanese bureaucracy as tedious and annoying. European coins could not be taken into the country, and the really dangerous contraband was religious works, especially those illustrated with engravings. There were few restrictions concerning reading matter for personal use, however, and Thunberg noted that his books in Latin, French, Swedish and German were only cursorily examined by the interpreters, who were unable to read them. The officers, including Thunberg, were also allowed to keep their rapiers, although other weapons were not permitted.

Apparently undaunted by the many restrictions imposed on the Europeans at Deshima Thunberg seems to have organised his scholarly work from the beginning of his stay. Two days after the arrival of the *Stavenisse* he began examining the fresh fodder which was brought three times a day for the cattle, pigs, sheep, goats and deer which the Dutch kept at Deshima, "and sought out the rare plants which were to be found, in order to dry them for the Herbaria of Europe and which I did not myself have the liberty to gather in the surrounding fields in a country, where our pigeons, more at liberty, were less suspected, captive and guarded than the Europeans, zealous for trade and profit, who had arrived here after many mortal dangers so far from their own domicile".[9] He investigated the tiny community of Deshima thoroughly, describing the half-timbered houses where the Dutch Factory staff lived, the Japanese guard houses, the interpreters' large building, known as the Collegium, and the way in which the trade was organised, as well as the main goods for import and export. Among the more exclusive and sought after commodities was "Unicornu", chiefly rhinoceros horn but also narwhal, and Thunberg had himself bought a quantity, for which he received such a good price that he was able to pay off the advance he had received from the East India Company and finance his "beloved science" during his stay in the country. "The Japanese have an exaggerated idea, about its medicinal usefulness and power to prolong life, strengthen vitality, improve the memory and cure all ailments",[10] he said about "Unicornu", explaining that the Japanese interest in it had been recently discovered by chance, when a retired head of the Dutch Factory had sent a narwhal horn as a gift to one of his friends among the interpreters.

Thunberg was able to observe the Chinese who were trading in Nagasaki, and who were subject to the same restrictions as the Dutch, partly due to the fact, he stated, that they had attempted to bring into Japan religious works printed by the missionaries in China. The Chinese were however allowed to have a temple of their own in the city, which they could visit freely, and they were also permitted to own Japanese currency for their daily expenses. Although similar in colour and looks to the Japanese, Thunberg noted, the Chinese differed in their clothing as well as in their language. In his first published account of Japan, a paper on Japanese coinage presented by Thunberg when he was elected a member of the Royal Swedish Academy of Sciences in 1779, one of his opening remarks concerned the dress of the Japanese, who like "Asiatic Peoples in general, do not allow themselves to be governed by fashion like the Europeans. [...] The Japanese [...] surpass their neighbours in this constancy. Already from time immemorial they have used and still use today, without any change, their own justifiably so called national costume, which consists of one or more full-length dressing-gowns, which are tied around the waist with a belt." [11] This interest in the attire of the Japanese mirrors the utilitarian aspect of Thunberg's observations. The examination and collection of plants may have been his main task, but any other detail from the daily life, customs, manners, religion and political organisation of the Japanese which might prove useful in Europe was duly filed for reference. It is worth noting in this context that the Swedish king, Gustaf III, who attempted to improve his country's economy by means of sumptuary laws, had promulgated the general use of a so-called "Swedish costume" in 1778.

During September and October, when the *Stavenisse* was being loaded and made ready for the return journey to Batavia, Thunberg had to remain on board to tend to the sick, who were suffering from an epidemic of stomach ailments, chiefly diarrhoeas, which ravaged Nagasaki. As far as the sailors were concerned Thunberg blamed this on the climate, with hot days and cold nights, and in the city itself he
柿 considered a glut of ripe persimmons, *kaki*, to be a contributing factor. The ship had been moved out to the small island known as Papenberg in Nagasaki harbour, and Thunberg was able to go ashore on this and other islands, some of them uninhabited, and walk about freely collecting plants. "If one then comes to a village, which usually is quite large, an incredible number of people and children come thronging and screaming in order to observe the in their opinion peculiar Europeans. They are especially amused by our big round eyes and there-

fore always shout *Hollanda O-Me.*"[12] Among the medicinal plants discovered during these excursions Thunberg mentioned smilax (*Smilax china*), the root of which was considered to purify the blood, the only cure for venereal diseases known to the Japanese doctors. He claimed that the interpreters were "heartily pleased" to discover that this plant, imported from China at great expense, grew locally. He also found wild figs (*Ficus pumila* and *erecta*), *Fagara piperita*, which Thunberg called "pepper bush" and which was used in food instead of pepper but also made into a poultice with rice flour as a cure for boils and rheumatic joints, and nettles (*Urtica japonica* and *nivea*) from which the Japanese made ropes, some of them thick enough to be used on ships. From September 1st 1775 and for the duration of his stay he took temperature readings four times a day, and made meteorological observations.

The *Stavenisse* sailed in November, leaving Thunberg and thirteen other Europeans, as well as a few slaves, in what seemed like house arrest on Deshima. He complained that "The way of life of the Europeans is in most things the same as elsewhere in the East Indies, luxurious and disorderly,"[13] and described the unvarying evening routine of a constitutional along the island's two streets, followed by a visit to the "Chief", i.e. the *Opperhoofd* of the East India Company Factory or another officer. The officers enjoyed free board and lodging, and thus had few living expenses, unless they gave lavish dinner parties or spent their money on "the Sex". Thunberg commented with disapproval on the arrangements at Deshima, where prostitutes could be engaged for periods from a day to a whole year, and on the brothels in Nagasaki which the Chinese and Dutch could visit. "The Christians, whom it would be befitting to be enlightened by Religion and Moral, should never degrade themselves to licentious commerce with the more pitiable than unfortunate daughters of the Country; but the Japanese themselves, as Heathens, do not consider Lewdness a Vice, least of all in such places which are protected by Laws and the Powers that be."[14]

He was no friend of the "poisonous tobacco-pipe", and found these evening gatherings unutterably tedious and a waste of valuable time. There was not a great deal of demand on his professional services, however, and he was able to devote much of his time to the plants and insects he collected. He also cultivated the interpreters, many of whom practised medicine in Nagasaki and who were keen to learn all they could from the foreign doctor. In his obituary of Thunberg the theologian Carl Adolph Agardh characterised him as cheerful, lively and kind, and another contemporary, the archaeologist Johan Henrik Schröder, mentioned his lack of affectation and his candid and amiable manner as well as his abstemious life-style.[15] The openness and lack of guile in his demeanour seems to have impressed the interpreters, and the fact that his interest in science eclipsed all else must have been apparent to everyone who met him. Thunberg appears to have developed a genuine friendship with some of the interpreters, and the fact that they secretly procured a number of Japanese coins and maps of Japan as well as city maps of Edo, 都 Miyako (present-day Kyoto) and Nagasaki for his collection is proof of the extent of their confidence in him, as it was an extremely perilous venture to break the laws of the Tokugawa *Bakufu*. The coin collection had to be smuggled on board hidden in his shoes, at great personal risk to Thunberg, and he took meticulous care to make extensive notes on the value, age and other characteristics of each coin, forming a representative selection of the current coinage of Japan, including some old pieces. Most of the senior interpreters spoke quite fluent Dutch, and although Thunberg was to say about the Japanese in general that "They ask for information about everything and often ask questions until the onset of boredom",[16] he himself would appear to have employed similar methods. A routine was established, where Thunberg taught the interpreters botany and medicine, as well as other sciences, and they in their turn brought him rare and unusual plants, which they had collected themselves or received from friends in the interior.

He finally received permission from the Governor of Nagasaki to collect plants in the countryside, but this was immediately revoked when the authorities, ever inclined to interpret legal precedents literally, discovered that Thunberg was of more senior rank than the Dutch assistant doctor who many years previously had been allowed to search for medicinal plants during an epidemic. At last, on February 7th 1776, having received renewed permission, he was able to walk in the countryside, although his joy was somewhat tempered by the fact that he was accompanied by a large retinue of interpreters, *banjos* (i. e. guards, 番衆 *banshû*) and servants. The Japanese officials were given a run for their money by the briskly striding Thunberg who complained that he was obliged to treat his exhausted attendants to refreshments at some wayside teahouse and that this proved quite costly. During the following month he availed himself of this opportunity once or twice a week, weather permitting, examining kitchen gardens as well as wild flowers and plants. There were many vegetables and herbs also commonly grown in Europe, e.g. beetroot, carrots, asparagus, onions, lettuce, beans, peas, aniseed and parsley. Sweet potatoes (*Convolvulus edulis*) were grown on the mountain slopes near the villages, and Thunberg preferred them to ordinary potatoes (*Solanum tuberosum*) which did not seem to thrive in Japan. He noted the many uses of bamboo, that boxwood was used to make ornamental combs, and that the tobacco plant was one of the few tangible relics left by the Portuguese. The sample of finely cut Japanese tobacco which he brought home to Sweden still retains a faint aroma, incidentally.

By mid-February preparations for the journey to Edo, or *Hofreis*, were well under way, and some provisions, including several cases of wine, liqueur and beer bottles, as well as household utensils and empty packing cases for goods to be brought back, were sent ahead by boat to Shimonoseki. The Japanese New Year fell on February 19th 1776, 下関 and Thunberg described the festivities as well as the Japanese method for the computation of time, and also the *e-fumi* ceremony, which took 絵踏み four days in the city of Nagasaki, where the entire population were gathered by district and made to trample on copper plaques bearing pictures of the crucified Christ or the Virgin and Child, in order to prove that they were not Christians. On March 4th the embassy left Deshima, a vast procession of about two hundred Japanese officials, interpreters and servants, and three Europeans, the *Opperhoofd*, Arend Willem Feith, his secretary Herman Köhler, and Carl Peter Thunberg as "Doctor of Legation". The remaining Dutch at Deshima, as well as a large number of Japanese, accompanied them to a temple on the outskirts of Nagasaki, "where we rested for a while and treated our merry party to Sakki (i.e. *sake*)".[17] Thunberg was very gratified to note 酒 that when the party continued on their way, all the Japanese officials and others in Nagasaki and at Deshima who had any dealings with the Dutch Factory "were standing in groups, according to their station and prestige, along the route we travelled, more than half a mile in length on both sides of the road, which was for us the most beautiful sight, as well as the greatest honour".[18] The three Europeans travelled in "norimon", the palanquin used by high-ranking *bushi*. Thunberg described 武士 the comforts of this conveyance in glowing terms, its velvet cushions, silk curtains and bamboo blinds, and the small shelves where he could place writing materials, a few books, one bottle each of red wine and beer, and a lacquer box containing a "double sandwich". Packhorses carried the personal luggage as well as an apparatus for making tea, from which boiling water was available at all times, although Thunberg and the other Europeans "seldom used this stomach-weakening beverage".[19]

Sheltered from the elements and carried along at a comfortable pace Thunberg was able to observe the countryside, flora and fauna, buildings and people of Japan, and he made careful notes of the names of the places he passed through as well as the most important local crafts and 嬉野 other characteristics. At Ureshino he inspected the baths constructed around the hot springs, where the water was thought to cure such ailments as venereal diseases, lameness, scabies and rheumatism, and he later published a paper on hot baths in Africa and Asia.[20] The white, 肥前 translucent porcelain produced in the province of Hizen he had already seen at Deshima, and he was now able to obtain further information about its manufacture. Despite the low esteem in which the Europeans 江戸参府 were held by the Japanese, the party undertaking the *Hofreis* (Jap.: *Edo sanpu*) were treated like a *daimyô* procession, and given comfortable lodgings and provided with an escort by the local *daimyô* when passing through his domain. When spending the night in a town or city they were confined to their lodgings, and on the rare occasions when they were allowed to walk about, as in the port city of Shimonoseki, Thunberg noted that the gates of the street where the brothels were situated were kept carefully closed. At Shimonoseki Thunberg saw laver, "called Awa Nori", being collected, and throughout the journey, when there was a better chance of tasting genuine Japanese food than at Deshima with its European fare, he noted interesting dishes and food-蕎麦、蜜柑 stuffs, from buckwheat noodles, *soba*, and *mikan*, mandarin oranges, 梅干 to *umeboshi*, pickled plums. The party embarked on a sizeable passenger vessel at Shimonoseki, and sailed through the Inland Sea, *Seto* 瀬戸内海、兵庫 *Naikai*, to *Hyôgo*, near present-day *Kôbe*. It was a cold journey, held up by adverse winds for weeks, but the energetic Thunberg used the enforced idleness to give lectures on medicine to the interpreters, question them on their country, its government, administration and society, and add material to the Japanese vocabulary and phrase-book which he had already started to compile. One of the senior interpreters owned a copy of the *Dictionarium Latino Lusitanicum ac Iaponicum*, printed 天草 by the Jesuit mission press at Amakusa in 1595. The title page was missing from the copy seen by Thunberg, a quarto volume of 906 pages, one corner of which had been singed, and its owner refused to part with this rarity at any price.[21] Thunberg published the vocabulary, comprising about 1.500 words, in a chapter of his *Travels*, and his brief grammar and phrase book under the title *Observationes in Linguam Japonicam* in 1792.[22]

The large and prosperous city of Osaka made a great impression on Thunberg, who dubbed it the Paris of Japan, "the most amusing place in the entire country [...] where thousands of pleasant pastimes are available".[23] There was no time for exploration, however, and before dawn the next morning the party pressed on towards Miyako (today known as Kyoto), the imperial capital, where they were to spend four days. Thunberg regretted that the Imperial Palace could only be viewed from a distance, and described the position of the "spiritual" Emperor, living in seclusion surrounded by his consorts and courtiers, " ... the former Ruler of the Country, nowadays a Pope who only possesses power in Matters of the Church, who is considered so holy that no Male may look upon him."[24] The actual political ruler of the country was the tenth Tokugawa Shogun, Ieharu (1737-1786, regnavit 1761-1786) whom Thunberg referred to as the "secular" Emperor. If Osaka resembled Paris because of its gaieties, Miyako was of old the centre of traditional arts and crafts of the utmost refinement, and the travellers were shown lacquer work, silks, and objects of gold, silver and copper. We are not told whether they bought anything, but Thunberg mentioned ordering lacquer ware at Hakone, and in his collection, now in the Museum of Ethnology, Stockholm, there are several lacquer objects, e.g. boxes, plates and an *inrô*, the medicine case worn by *bushi* males suspended from their sashes.

家治

箱根

印籠

Everywhere along the *Tôkaidô* (Eastern Sea Route) farmers were working in the fields, and the fruit trees were in full bloom. Thunberg had hoped to be able to find many rare and unknown plants along the road to Edo, but he discovered, to his half-amused annoyance, that the diligence and care of the Japanese husbandman was such that hardly a single weed could be discovered in the meticulously tended fields, not even by his eagle eye. An opportunity presented itself, however, when the party crossed the Hakone Mountains, a steep and difficult passage. Thunberg got out of his "norimon" and walked, and although he was not allowed to stray far from the road, he put his training from the mountainous Cape country to good use and ran uphill so fast that he had ample time to gather quite a few rare plants in his handkerchief before the worried and breathless *banjos* had time to catch up with him. The party arrived in Edo on April 27th, and were installed in their rather cramped lodgings overlooking an alley, where hordes of street urchins screamed and shouted as soon as the Europeans showed themselves at the windows, climbing the walls of the houses opposite to catch a glimpse of them.

東海道

The Europeans were not allowed to leave the house before the Shogunal audience had taken place, but a steady stream of visitors arrived. The first to appear were a group of physicians and astronomers, who were received with suitable ceremony. Their attention was focused on Thunberg, the only scientist of the party, and after a preliminary exchange of courtesies the astronomers began asking for information concerning eclipses, which they were unable to compute accurately. Thunberg answered their queries through the interpreters to the best of his ability, but found his conversation with the physicians considerably easier, as the interpreters had some understanding of medicine, and a wide range of subjects was covered, from cancer and fractures to piles and toothache. Two of the younger doctors became daily visitors, and Thunberg received them privately, giving lectures in physics, economy, 桂川甫周 botany, surgery and medicine until late at night. They were Katsuragawa Hoshû (1751-1809), official physician to the *Bakufu*, "the Emperor's Physician in Ordinary, very young, kind-hearted, brisk and active", 中川淳庵 and Nakagawa Jun'an (1739-1786), "Physician in Ordinary to one of 蘭学 the noblest Princes of the Country", leading exponents of *Rangaku*, "Western Learning", in Japan.[25] The embassy's arrival in Edo had been preceded by the rumour that this year's Dutch doctor was no ordinary 蘭学者 barber-surgeon, and the two *Rangakusha* made the most of his three-week stay, according to Thunberg. "Although I was often exhausted by their importunity, I cannot deny that I have spent many enjoyable and edifying hours in their company."[26] Among the foreign books used by his "beloved Disciples" Thunberg mentioned John Johnston's (1603-1675) *Historia Naturalis*, Rembertus Dodonæus' (1517-1585) *Herbarius* (*Cruijdt-Boeck*, 1618/1644), Johann Jacob Woyt's "Treasury", i.e, his *Gazophylacium Medico-physicum* (1741), and Lorenz Heister's (1683-1758) *Chirurgie* (1718), and he was also shown several Japanese botanical and zoological works, some of them very poorly illustrated, he complained. Katsuragawa and Nakagawa began applying their newly acquired knowledge straight away, and Thunberg, who had brought a quantity of sublimate of mercury from Holland, was able to introduce this new cure for venereal diseases, which were very common in Japan in his day. At their request Thunberg wrote out a diploma in Dutch, giving details of the lectures he had given and the progress they had made, before parting from them, and he kept up a correspondence with them for several years after his return to Sweden, exchanging seeds and specimens.

The Shogunal audience took place on the 18th, and Feith, Thunberg and Köhler, dressed up in silks and gold lace, with a wide black silk cloak over the shoulders and girded with their swords, were taken to Edo Castle in their "norimon". The audience involved a great deal of waiting in different buildings in the compound, and various daimyô came to greet them, asking for specimens of their writing on fans or pieces of paper. Only the *Opperhoofd* was brought into the presence chamber, where he prostrated himself to the standing Shogun in the traditional Japanese greeting of a superior. He then joined the others in the antechamber, where a number of courtiers came to see them. From the sudden silences among these when certain men entered, Thunberg concluded that the latter must be very high-ranking personages, and he later learned from the interpreters that the Shogun himself had come in to take a closer look at the foreigners. After a brief tour of the palace, including the now empty audience chamber, the party were taken by "norimon" to visit the residences of all the members of the *Rôjû*, Senior Council, "although the day was far advanced and our early breakfast had had time to subside to a considerable depth", and returned to their lodgings in the evening, faint and exhausted.[27]

老中

The return journey was less formal, and there was time for some diversions. In Miyako there was the customary sightseeing tour, which included the Hôkôji temple with its *Daibutsu*, a gigantic seated figure of Amida Buddha, and the Rengeôin or Sanjûsangendô, where a large seated figure of Kannon is flanked by one thousand standing Kannon sculptures.[28] Thunberg had gathered some information concerning the religions of Japan, and was aware of the basic features of Shinto as well as the Indian origins of Buddhism, and the sight of the *Daibutsu* made a great impression. "Daibuds Idol, which was almost in the centre of the Temple, could strike both terror and veneration; terror, as regards its size, which is hardly likely to have its equal, and veneration, as regards the reflections one is caused by it to make."[29] In Osaka, where the party spent two whole days, they were allowed more freedom than ever before, were shown around the city, and attended "Comedies" and "Dances", i.e. *Kabuki* performances. Thunberg found these "peculiar" and indeed "preposterous" and although the actors were quite skilful, "in a way", the theatre was rather small and confined. He was considerably more impressed by a botanical garden, "although without an Orangery", where he was able to buy several rare bushes and trees, such as Japanese maple and sago palm (*Cycas revoluta*, Jap.: *sago yashi*). These were planted in a large wooden box, protected by hoops interwoven with twine, and shipped via Nagasaki and Batavia to the *Hortus Medicus* in Amsterdam.[30] It took Thunberg a great deal of per-

法広寺、大仏
蓮華王院、三十三間堂
観音
神道
歌舞伎
沙穀椰子・西穀椰子

suasion to be allowed to view the process by which copper bars were cast in cold water, a method which was the secret behind the particularly strong and shiny colour of Japanese copper. Through the *bona officia* of his friends among the interpreters Thunberg received as a gift a box containing samples which illustrated every step of this process, and he was overjoyed to bring this back to his former professor Torbern Bergman (1735-1784), the originator of analytical chemistry.[31] After his return to Deshima on June 29th, Thunberg spent the summer surveying and arranging his collections, and was able to make several excursions in the countryside, but he could not be persuaded to remain in Japan for another year and left Deshima on November 23rd 1776 for the *Stavenisse*, which was anchored at Papenberg.

He arrived in Batavia on January 4th 1777, and spent six months in Java before sailing to Ceylon on July 5th on board the Mars. Thunberg stayed mainly in Colombo, where he met several countrymen of rather diverse backgrounds and fortunes, and made botanical excursions in the company of a Sinhalese physician. He was particularly interested in the cultivation, harvest and export of cinnamon, a spice, which of old plays an important part in Swedish cookery.[32] Another plant which attracted his utilitarian interest was breadfruit (*Artocarpus incisa*), which he tried to introduce in Europe, describing with enthusiasm no less than fifteen dishes using it as an ingredient, and commenting that it tasted not unlike cabbage. On February 6th 1778 he finally left Asia, sailing in the *Loo* for Cape Town, where the ship remained for a few weeks before continuing to Europe. The long-tailed monkey he had bought in Ceylon died from exposure, and his breadfruit saplings and other rare plants were ruined in a severe gale in the English Channel. On October 1st the *Loo* arrived at Texel, and the next two and a half months were spent in Holland, before Thunberg travelled to England in mid-December. In London he met Daniel Solander and was allowed to go through Sir Joseph Banks's extensive herbarium. It was not until the end of January 1779 that he at last set out for home, and it is quite typical of his energy and thirst for knowledge that when he had to wait in Stralsund in the then Swedish Pomerania for the packet-boat to Ystad in Sweden, he used the extra time to travel to Greifswald in order to visit the University and especially the library.

On March 14th 1779 he set foot on Swedish soil, and never again left the country, despite tempting offers of prestigious positions abroad. He settled in Upsala, where he had been appointed demonstrator in botany, and in 1784 married Brita Charlotta (1752-1813) né Ruda. Carl von Linné the Younger (1741-1783) was not of the same calibre as his famous father, and when Thunberg succeeded to the professorship in 1784, the Botanical Gardens and collections of Upsala University had suffered some damage, the most irreparable being the loss of Linnaeus' herbaria, which his widow had sold to England. With the same energy that had characterised his assiduous work during his travels Thunberg devoted the next forty years to the consolidation, improvement and augmentation of the scientific heritage of Linnaeus. Already during his stay in Paris in 1771 Thunberg had been presented to King Gustaf III, who was visiting the city as Crown Prince, and immediately upon his return he was received in private audience by the King and gave a report on Japan. In 1785 Thunberg was created a Knight of the order of the Vasa (he was elevated to Commander in 1815, the first scholar to be so honoured) and donated his collections, comprising 18 cabinets of herbaria, 13 of insects, 11 of shells and mussels, as well as fish, amphibians, birds and mammals, to Upsala University. Gustaf III was as great a patron of the sciences as of the arts, and upon a petition from Thunberg he donated a plot for the new Botanical Gardens, just below Upsala Castle, as well as a sum from his privy purse for the construction of an orangery and other buildings. The inauguration took place on the centenary of the birth of Linnaeus, May 25th 1807, and Thunberg summed up his views on nature in his speech: "Although it is so seldom felt, and seldom fulfilled, it is nevertheless one of our foremost duties, – to know Nature, – to realise the value of Nature's great and wonderful Chain, – to regard its splendour devoutly and with admiration, – to use its products wisely and with consideration." [33]

Thunberg spent the winters in Upsala, and the summers at his estate Tunaberg, not far from the city, travelling between his homes in an old-fashioned narrow barouche, jocularly referred to by his students as "the rattle-snake", [34] and only reluctantly and very occasionally did he venture as far as Stockholm, a distance of seventy kilometres. His nine years of travels had provided him with enough impressions and material to last him to the end of his long scholarly career, and he was the undisputed European authority, not only on the flora of Japan, but also on the country's social conditions. In an introductory salutation in Thunberg's first major work, the *Flora Japonica*, 1784, his friend and senior colleague Jonas Theodor Fagraeus (1729-1797) congratulated him on having succeeded in following where Engelbert Kaempfer (1651-1716) had lead the way: "Aperuit Kaempferus, officio noster, viam. Tibi soli contigit hunc adire Corinthum", ("Our Kaempfer opened the way for the task. You alone succeeded in reaching this Corinth"), he stated, and continued, rather prophetically, to refer to Thunberg's work as "rerum memorabilium thesauros" ("a treasury of memorable things"), unaware of the fact that this was only the beginning ... .[35]

NOTES

1 Carl Peter Thunberg, *Resa uti Europa, Africa, Asia, fårråttad åren* 1770-1779, Del I (Upsala, 1788), foreword. English translations of Thunberg's texts are by the present author.

2 Its full title was *Flora Capensis, Sistens Plantas Promontorii Bonae Spei Africes*. A second volume was published in Copenhagen in 1818, and a complete edition was first published in Stuttgart 1823.

3 In his autobiographical notes Thunberg gives a list of his patrons: a Mayor Temminck and three other gentlemen, van Der Deutz, van der Poll and ten Hoven, who with Professor Burmannus paid his expenses and provided him with letters of recommendation to the Governor General of the Cape, Tulbagh, and his successor designate Rheede van Oudshorn. Vide *Carl Peter Thunberg* (1743-1828), *Självbiografiska anteckningar med bibliografi, sammanställda av Lars Wallin, Scripta Minora, Bibliothecae Regiae Universitatis Upsaliensis*, vol. 6 (Stockholm: Almqvist & Wiksell International, 1993), 10.

4 Carl Peter Thunberg, "Händelse at Blyhvitt (Cerussa) af förseende blifvit brukadt i mat, Insänd från Goda Hoppets Udde af Carl Peter Thunberg, Med. Doctor" ("Occurrence that Lead White (Cerussa) by Mistake was used in Food, Submitted from the Cape of Good Hope by C.P.T., Doctor of Medicine"), *Kongl. Vetenskaps Academiens Handlingar För År* 1773 (*Transactions of the Royal Academy of Sciences of the Year* 1773).

5 Carl Peter Thunberg, *Resa* ... , Part II, 352.

6 Op. cit., 294.

7 Carl Peter Thunberg, *Resa* ... , Part III, 71.

8 Op. cit., 29-30.

9 Ibid., 25.

10 Ibid., 54.

11 Carl Peter Thunberg, "Inträdes-Tal om de Mynt-Sorter som i äldre och sednare tider blifvit slagne och varit gångbare uti Kejsaredömet Japan" ("Inaugural Speech on the Types of Coins which in earlier and later times have been struck and used as legal tender in the Empire of Japan"), Hållet för Kongl. Vetenskaps-Academien, Den 25 Aug. 1779 (Held in the Royal Academy of Sciences, August 25th, 1779) (Stockholm: Johan Georg Lange, 1779), 4.

12 Carl Peter Thunberg, *Resa* ... , Part III, 67. The Japanese phrase means "Dutch big-eyes".

13 Op. cit., p. 71.

14 Ibid., p. 84.

15 *Biographie öfver Carl Petter (sic!) Thunberg, Kongl. Vetenskaps Academiens Handlingar för År* 1828 (*Biography of C.P.T., Transactions of the Royal Academy of Sciences for the Year* 1828) (Stockholm: P.A. Norstedt och Söner, 1829); *Vita Caroli Petri Thunberg, Adumbrata a Johanne Henrico Schröder* (*The Life of C.P.T., Sketched by J.H.S.*) (Upsaliae: Palmblad & C., 1833).

16 Carl Peter Thunberg, "Tal om Japanska Nationen" ("Speech on the Japanese Nation"), Hållet för Kongl. Vetensk. Academien vid Praesidii Nedläggande, Den 3 Novemb. 1784 (Held at the Royal Academy of Sciences on the occasion of relinquishing the Presidency, November 3rd 1784) (Stockholm: Johan Georg Lange, 1784), 8.

17 Carl Peter Thunberg, *Resa* ... , Part III, 106-107. We may note that the retinue accompanying the embassy had increased manifold since the mid-17th century, when Thunberg's countryman Olof Eriksson Willman accompanied Adriaen van der Burgh on the *Hofreis* in 1652, stating that the four Europeans travelled with about sixty Japanese. Vide Catharina Blomberg, "Idolaters and Devil Worshippers -Religion in Olof Eriksson Willman's Travel Diary From Japan 1651-1652," in *Le Vase de béryl, Etudes sur le Japon et la Chine en hommage à Bernard Frank*, ed. Jacqueline Pigeot and H.O. Rotermund (Arles: Éditions Philippe Picquier, 1997), 363-372; and "Jammaboos and Mecanical Apples – Religion and Daily Life in Olof Eriks-

son Willman's Travel Diary 1651-1652", *Itinerario, European Journal of Overseas History*, vol. XXII, no. 2 (Leiden, 1998), 85-102.

18 Op. cit., 107. The old Swedish mile was 36.000 feet, i.e. 10.688,5 metres.

19 Ibid., 111.

20 Carl Peter Thunberg, "Några Varma Bad uti Africa och Asien" ("Some Hot Baths in Africa and Asia"), *Kungliga Vetenskaps Academiens Nya Handlingar* (*New Transactions of the Royal Academy of Sciences*), vol. II (Stockholm: Johan Georg Lange, 1781). *Resa* ... , Part. III, 115.

21 This dictionary, the first ever printed in Japan, was based on a work by the Augustine friar Ambrogio Calepino (c. 1440-1510). Thunberg's countryman Willman mentioned being shown a copy of the same work, complete with title page, when visiting the *Ômetsuke*, 'Chief of Intelligence' Inoue Chikugo no Kami Masashige in Edo in 1652.

22 Carl Peter Thunberg, *Resa* ... , Part III, "Japanska Språket" (Chapter headed "The Japanese Language"), 294-353. *Observationes in Linguam Japonicam, a. Car. Petr. Thunberg, Nova Acta Regiae Societatis Scientiarum Upsaliensis (Observations on the Japanese Language, by C.P.T., New Transactions of the Royal Academy of Sciences)*, vol. V, (Upsaliae: Apud Joh. Edman, Direct. et Reg. Acad. Typogr., An. MDCCXCII), 258-273.

23 Carl Peter Thunberg, *Resa* ... Part III, 150.

24 Op. cit., p. 153.

25 Ibid., 199. Katsuragawa was the scion of a family of physicians, and Nakagawa was a *bushi* from Obama, not far from Nagasaki. For a discussion of the importance of these two leading *Rangakusha*, vide Grant K. Goodman, *Japan: The Dutch Experience* (London: The Athlon Press, 1986).

26 Ibid., 200.

27 Ibid., 218.

28 The Hôkôji was originally constructed in 1586 by Toyotomi Hideyoshi, and housed a wooden *Daibutsu* which was 19 metres tall. This was destroyed in an earthquake in 1596, and in 1612 Toyotomi Hideyori had a bronze image cast and constructed a new hall, which was seen by Olof Eriksson Willman and which burned down in 1662. Willman mentioned that the hall was 8 pillars in width and 12 pillars lengthwise, and Thunberg said that the temple he saw was constructed on 96 pillars, so an exact replica must have been made. The *Daibutsu* seen by Thunberg was made of wood, however, and since the destruction by fire of the last image in 1843 nothing remains of the *Daibutsuden* or Great Buddha Hall. Amida (Sanskr. Amitabha) is the Buddha of the Western Paradise. Kannon (Sanskr., Avalokitesvara) is the Bodhisattva of Mercy. Vide supra, note 17, my two papers dealing with Willman.

29 Carl Peter Thunberg, *Resa* ... , Part III, 246. For a discussion of Thunberg's views on Japanese religion see also Catharina Blomberg, "From the Horizon of the Enlightenment – Carl Peter Thunberg's Views on Religion in Japan," in *Florilegium Japonicum: Studies Presented to Olof G. Lidin on the Occasion of His 70th Birthday*, ed. Bjarke Frellesvig and Christian Morimoto Hermansen (Copenhagen: Akademisk Forlag, 1996), 29-40.

30 Op. cit., 248-251.

31 Ibid., 251-253.

32 His quite exhaustive paper, "Anmärkningar vid Canelen gjorde på Ceylon af Carl Peter Thunberg" ("Remarks on Cinnamon made in Ceylon by C.P.T.") was published in *Kongl. Vetenskaps Academiens Nya Handlingar*, Tom I. År 1780.

33 Carl Peter Thunberg, "Tal vid Invignings-Acten af den Nya Akademiska Trägården (sic!), Dess Orangerie och Samlings-Salar, med Kongl. Maj:ts Allernådigste Tillstånd Hållit i Upsala Uti Den Nya Botaniska Lårosalen D. 25 Maji 1807, Då Tillika Firades Framledne Archiaterns och Riddarens Carl von Linnés Hundra-Åriga Födelse-Dag, af Carl Peter Thunberg, Med. och Botan. Professor, Riddare af Kongl. Maj:ts Vasa-Orden" ("Speech at the Inauguration of the New Academic Garden, Its Orangery and Meeting-Rooms, with the Most Gracious Royal Permission Held in Upsala in the New Botanical Lecture Room, May 25th 1807, When the Late Physician in Ordinary and Knight Carl von Linnés One Hundredth Birthday was also Celebrated, by C.P.T., Professor of Medicine and Botany, Knight of the Royal Order of the Vasa") (Upsala: Joh. Fr. Edman, Kongl. Akad. Boktr., 1807).

34 J.H. Schröder, *Vita Caroli Petri Thunberg* (Upsala, 1835), 9.

35 Carl Peter Thunberg, *Flora Japonica* (Leipzig, 1784), salutation by Jonas Theodor Fagraeus, pp. IX and X. This is an erudite paraphrase of Horace, *Epistulae* I, 17, 36: "Non cuivis homini contigit adire Corinthum" (Not every man has the fortune of visiting Corinth), quoted here as a florid way of describing the completion of something hard to accomplish.

XVI

THE INFLUENCE OF HERMAN BOERHAAVE'S MECHANICAL CONCEPT OF THE HUMAN BODY IN NINETEENTH-CENTURY JAPAN

Frederik Cryns
Kyoto University

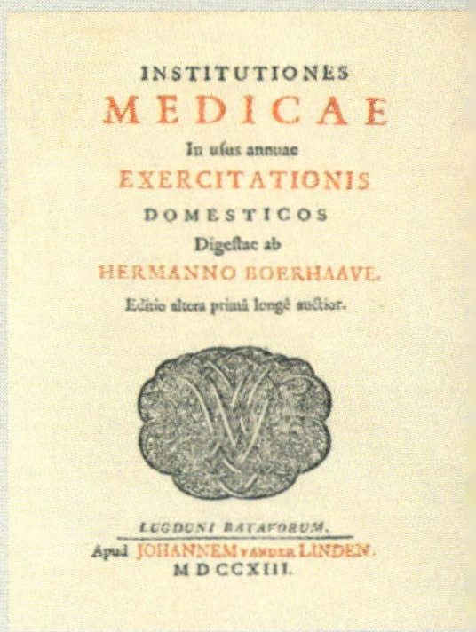

INSTITUTIONES
MEDICAE
In usus annuae
EXERCITATIONIS
DOMESTICOS
Digestae ab
HERMANNO BOERHAAVE.
Editio altera primâ longè auctior.

LUGDUNI BATAVORUM,
Apud JOHANNEM vander LINDEN.
MDCCXIII.

I INTRODUCTION

At the beginning of the eighteenth century the Dutch clinician and medical teacher Herman Boerhaave (1668-1738) succeeded in incorporating the various new scientific insights of the seventeenth century and the wisdom of Antiquity into a comprehensive medical system. His ideas exerted a profound influence throughout the whole of Europe. Two of his major works, the *Institutes* (1708)[1] and the *Aphorisms* (1709)[2] went through re-editions every year until the end of the eighteenth century. Although his system was based on immature science and was superseded by new discoveries in the nineteenth century, Boerhaave's greatest contribution was the foundation of a new scientific approach to medicine, in which empiricism was combined with the introduction of simple mechanical principles to explain all physiological processes in the human body. As can be learned from his *Oration on the Usefulness of the Mechanical Method in Medicine*, Boerhaave saw the human body basically as "composed in such a manner that its united parts are able to produce several motions of very different kinds which derive – fully in accordance with the law of mechanics – from the mass, the shape, and the firmness of the parts and from the way in which they are linked together."[3] For Boerhaave the human body was a mechanism, whose working could be fully understood by the science of mechanics. Disease was understood as a condition of the body in which one or more functions were prevented from working. The physician had to be a craftsman "who from his knowledge of the correct structure, discerns the defects of the parts and the ways and means of repairing them".[4]

Boerhaave's doctrines reached Japan through Dutch translations of his Latin works as well as the works of several authors who were directly or indirectly influenced by Boerhaave's Leyden school, such as de Gorter, van Swieten, Plenck, Huxham, Heister and many others. These works constituted the bulk of medical literature translated by the *Rangakusha*[5] between 1790 and 1830. However, most of the books imported into Japan were very practical, written as simple textbooks for the practitioner and not for a university audience. They did not contain much about the underlying system of medicine or the author's concept of the human body on which this system was based. His concept was in essence mechanical and alien to Japanese thought. However, in the course of the translation of Western medical works, the *Rangakusha* were at some point or another confronted with this underlying philosophical concept, no matter how poor the availability of explicit information on this subject was. It is indeed very difficult to separate scientific approach from the contemporary philosophical concept of nature. This is especially true in the eighteenth century, when the greater part of the human physiological and pathological system was still based upon speculation.

蘭学者

This paper endeavours to explore some characteristics of the way *Rangakusha* coped with this basically alien mechanical concept of the human body in Boerhaave's doctrines. For this purpose there are two sources which have to be examined because they are the only two among the Western works that were imported and translated in the Edo period, and that provide explicit information on the theoretical background of Boerhaave's system of medicine. Laurentius Heister's *Practicaal geneeskundig hand-boek* (Practical medical handbook, 1762)[6] contains a 154 page long dissertation on the mechanical method in medicine. Heister's work was translated by Udagawa Genshin (1769-1834)[7] as *Heisuteru naikasho* (n.d.).[8] Another work, Gerard van Swieten's *Verklaaring der korte stellingen van Herman Boerhaave* (Explanation of the maxims of Herman Boerhaave, 1760),[9] a detailed commentary on the *Aphorisms*, was translated by Udagawa's pupil Tsuboi Shindô (1795-1848)[10] as *Manbyô chijun* (1826).[11] Through an analysis of the translation of these two works, I would like to explore how Udagawa and Tsuboi approached Boerhaave's concept of the human body, in particular by focusing on the translation of the term "nature" in relation to the definition of the human body. There are however limitations to this method. It is very difficult to find out what moved the translator in his selection of Japanese equivalents. Yet I hope that a comparative study of the meanings of both the Western and Japanese terms will shed some light on the difficulties the *Rangakusha* encountered when

宇田川玄真

坪井信道

dealing with Western philosophical concepts and what solutions they had to offer.

II UDAGAWA GENSHIN'S TRANSLATION OF THE MECHANICAL METHOD IN MEDICINE

Heister's *Practicaal geneeskundig hand-boek* is the Dutch translation of the original Latin *Compendium medicinae practicae* (Amsterdam: Iansonio-Waesbergios, 1743) and was published by Jan Morterre in Amsterdam in 1762. The handbook proper is preceded by a dissertation entitled *Verhandeling van de voortreffelykheit en uitstekendheit der mechanisch-geneeskundige leerwyze* ('Dissertation on the Excellence and Superiority of the Mechanical-Medical Method'). The dissertation tries to demonstrate through a multitude of examples that the mechanical method is the only right method in the analysis of physiological and pathological processes and in the treatment of diseases. It is conceived as a critique of the Stahlian method. George Ernst Stahl (1660-1734) was a contemporary of Boerhaave and professor of chemistry at the University of Halle in Saxony. He tried to explain life and disease by the action of a "sensitive soul" or "anima" and applied his vitalistic views to the development of psychopathology and therapy. By showing the errors of the Stahlian method and explaining the answers of the mechanical scientists, Heister (1683-1758) tries to demonstrate the superiority of the mechanical method in medicine.

In Heister's dissertation, the fundamentals of the Stahlian doctrine are summarised as follows. The soul and nature of human beings are held to be identical and all medical problems can be reduced to the soul. The soul is the primal force of motion and builds up the body. It co-ordinates all physiological functions and processes and takes care that the body stays healthy. If we would replace the term "soul" by
気 "vital energy", we may find some resemblance with the notion of *ki* in traditional Chinese medicine. *Ki* is the term that describes the vitalizing substance that circulates through the body together with the blood and sustains the physiological activities of the body. Heister's most important argument for refutation of the Stahlian doctrine is that the soul does not remember anything of her physiological activities. This claim is of course predicated on the premise of a strict separation of matter and mind, resulting in a view of the body as exclusively inanimate. In other words he denies the possibility of the existence of a kind of vitalizing energy without cognition other than the soul. Heister makes a

distinction between the functions of the soul that occur as a result of the will and prior knowledge of the soul, and the functions of nature which have their origin in the structure of the human body, for example the function of the heart and the blood circulation. Still according to the same premise, the existence of such a vitalistic energy is refuted by the fact that the soul has no knowledge or memory of these physiological actions and that these actions cannot be altered by the will of the soul. Heister tries to prove his mechanistic view of the human body by means of an anatomical survey of the heart, the vessels, the stomach and the nervous system. These examples are partly based on Boerhaave's *Institutes*. Heister concludes that the human body is a hydraulic machine. The prime force of action is not assigned to the soul. There is no need for a soul. The human body is a mechanism that imitates and maintains its motion due to its own structure, a sort of *perpetuum mobile*. He uses the metaphor of the watermill. The only thing necessary for the watermill to get in motion is the flow of water. In the same way, the human body only needs the flow of blood.

Heister's *Practicaal geneeskundig hand-boek* was translated by Udagawa Genshin as *Heisuteru naikasho*. The manuscript in the Central Library of Kyoto University (Fujikawa Collection) also includes an abridged translation of Heister's dissertation on the mechanical method in medicine *Verhandeling van de voortreffelykheit en uitstekendheit der mechanisch-geneeskundige leerwyze*, entitled *Jinshin kyûri ijutsu-ron*. Udagawa shows not much interest in the actual debate between the mechanical and Stahlian method so that several parts concerning the critique on the Stahlian method have been omitted. The translation concentrates on the practical examples and the mechanical method as such.

歇伊私的児内科書

人身究理医術論

Udagawa translates the "functions of the soul" (werkingen der ziel) as *ishiki no sayô*. *Ishiki* is originally a Buddhist term. Contrary to the activities of the five senses, *ishiki* signifies the inner activities of cognition and thinking. In this sense it comes close to Heister's interpretation of the word "soul" as the cognitive agent. It is however quite far away from the original Stahlian meaning of vital energy. It is interesting to note that Udagawa did not choose the term ki to translate Stahl's concept of soul. In traditional Chinese medicine there are three qualities that are ascribed to a human being, namely *kei* (form) as the corporeal receptacle, *ki* as the vitalizing substance and *shin* as the mind.[12] The meaning of Stahl's soul, perceived as vital energy, is closer to *ki*, but Udagawa chose *ishiki*, the meaning of which lies in the category

意識ノ作用. 意識

形

神

of *shin*. Had he chosen *ki*, his readers could have construed the debate between Heister and Stahl as a debate between the "Western" mechanical view and the "Oriental" vitalistic view. With the present choice of terminology however, it is hard to make a link between the concept of soul in the Stahlian doctrine and the concept of *ki* in traditional Chinese medicine.

The "functions of nature" (werkingen der natuur) are translated as 自然ノ作用、自然 *shizen no sayô*. Udagawa translates "nature" in all cases as *shizen*, regardless of the various nuances in meaning the word has in the original text. Physics (natuurkunde) is translated as 自然術 *shizen-jutsu* and 自然ノ学術 *shizen no gakujutsu*. This is a literal translation, which does not convey the meaning of physics properly. As has been noted in earlier studies, the word *shizen* was only adopted to represent the Western meaning of "the natural world" or "nature as an object of scientific study" in the middle of the Meiji period. In the Edo period most *Rangakusha* used as equivalents for this Western concept words such as 天地 *tenchi* (heaven and earth), 万物、造化、造物 *banbutsu* (all things), *zôka* or *zôbutsu*. In his translation draft of de Gorter's *Gezuiverde geneeskonst* (1744), Udagawa Genshin's teacher 宇田川玄随 Udagawa Genzui (1755-1797) uses *shizen* as a translation for "natuur"[13] in the sentence "Although one does not know how nature functions here" (Alschoon men niet weet hoe de natuur hier in het werk gaat),[14] but in the printed version 西説内科撰要 *Seisetsu naika sen'yô* (completed in 1790), the above sentence is translated as 其造物者ノ為ス所其然ル所以ヲ知ルコト能ハズ, in which the interpretation of "natuur" as *shizen* is replaced by *zôbutsusha*.[15] Originally *Shizen* was only used as adjective or adverb in the meaning in which 自ら *onozukara* (naturally, as a matter of course, of its own accord etc.) is presently used. *Shizen* represented therefore not the natural world as an object, but had a more subjective connotation, implying that certain conditions or activities of a being emanated from an internal power. Phenomena originate from internal substances (*ki*) and *shizen* was used to express the spontaneous condition of the workings of these substances.

In the so-called *Edo-Haruma* Dutch-Japanese dictionary *shizen* appears as the first translation of "natuur", in a list which includes a number of other interpretations such as 性質, 形状 *seishitsu* (character) or *keijô* (shape, form).[16] The biggest difficulty the *Rangakusha* encountered with the word "natuur" was that it was used in Western works in two or more different meanings. Its more modern meaning was the aggregate of all natural things, i.e. the natural world. It was however also used to indicate the qualities by which something is what it is, the qualities within a being, which are the sources of its behaviour. Egbert Buys's *Nieuw en volkomen woordenboek van konsten en weetenschappen* ('New and Complete Dictionary for Arts and Sciences', 1769-1778), a popular reference book among the *Rangakusha*, gives eight different meanings for "natuur".[17] Udagawa however translates "nature" in all cases by *shizen*, regardless of the different meanings in which it is used. By the "functions of nature" Heister specifically means all functions of the human body, which are not controlled by the mind, i.e. the functions we put nowadays in the category of functions under the control of the autonomic nervous system.

The mechanical scientists perceived the body as a mechanism, which initiates and maintains its workings because of its own structure, like a closed circuit. The word "nature" signified this structure of the human body. During the seventeenth century there was a fierce debate about the meaning of "nature" as used by Hippocrates,[18] who allegedly stated that "nature cures diseases". The adherents of the Stahlian doctrine supported the view that "nature" in Hippocrates signified the soul. Heister however clearly states that "nature is not the soul, but the corporeal structure" (De Natuur is niet de Ziel. De Lichaamlyke Bouw is de Natuur).[19] Udagawa translates these two sentences as "*shizen to ishiki no kotonaru wo ron[zu]*" (自然ト意識ノ異ナルヲ論[ズ]) and "*jinshin zôritsu no shizen wo ron[zu]*" (人身造立ノ自然ヲ論[ズ]).[20] In the first sentence we can easily grasp Heister's meaning of the difference between soul and nature, but in the second sentence *shizen* is not understood as equal to the corporeal structure, but as the condition of the building process (*zôritsu*) of the human body (*jinshin*).

Heister tries to explain his statement that "nature refers to the structure of the body", through the device of an alleged objection by the Stahlians, which he describes as: "If nature were nothing else than the human body, which can exist of no other parts but material ones, then it would be impossible that, in a healthy condition, it would operate and manage everything so well". (Als de Natuur anders niet dan 't Lichaam was, 't welk uit geene andere dan stoffelyke deelen bestaan kan; zo ware het egter onmogelyk, dat het zelve in den gezonden staat, alles zo wel gepast verrigten en bestieren [...] zoude.)[21] The focus of the Stahlian objection is on the fact that, being a completely material thing, the corporeal structure cannot possibly execute immaterial functions such as the operation of its own struc-

ture. Udagawa however translates as follows: 吾輩「スタリアネン」意識ノ説ヲ廃シテヨリ教示スルニ蓋自然ト名クモノハ他ナシ人身中ニ在テ他ノ有形ノ諸具ト異ニシテスベテ平全ノ時ニ於テ諸物ヲ適従協合シ諸事ヲ弁別命令シ.[22] Here he misinterprets Heister's alleged objection of the Stahlians that the human body as such cannot exist of other parts than material ones (and that still according to the Stahlians, therefore an immaterial power is necessary to manage everything). Heister uses this objection to indicate that it is indeed possible for the human body to operate without an immaterial energy due to its structure. But Udagawa interprets this objection inversely. In his translation it seems that the Stahlians deny the existence of such energy and that Heister defines *shizen* as something in the human body, which is different from the material properties of the body such as the various organs. Udagawa emphasises this definition by adding an explanatory sentence at the end of the paragraph stating that "*shizen* supervises the transitions of the material and immaterial properties, conveys the phenomenalization of *ri* (the absolute principle of the universe) and is the fundamental source of the innate good capacities and fundamentally good nature, which harmonises all things" (故ニ自然ハ有形無形ヲ主化スル事ヲ主宰ニシテ象理ヲ通知シ諸物ヲ協和スルノ良能性善ノ大本也).[23] Udagawa uses *shizen* here as a noun and attributes to it the role of managing the properties of the human body. The term "phenomenalization of *ri*" shows some resem-
朱熹 blances with the Neo-Confucian metaphysics of Zhu Xi. All things
理 were thought to have their respective *ri* or fundamental principle and
気 their *ki*, a kind of substance which creates forms. Heister means however the opposite, namely that the body is created in such a way that it can function and cure defects in its structure (diseases) to some extent independently without the help of an immaterial power.

Heister makes this clear in the following words: "all changes occurring in the human body are being produced by the motion and the composition of its parts" (dat alle veranderingen, in het menschelyke lichaam voorvallende, door de Beweeging en de t'zamenstelling der Deelen voortgebragt worden) and further "it being very wisely arranged by the Creator in such a way that every part may retain the powers and capacities that have been allotted to it, according to the proportions and properties of its essence. From this then originate all operations and changes in this body, not by the order of the soul, nor in accordance with the direction of the soul, but due to a certain mechanical necessity." (Zynde het door den Schepper zeer wyslyk alzo geschikt, op dat ieder deel zyne bepaalde, hem toegelegde krachten en vermogens, naar evenredigheit en hoedanigheit van zyn weezen [...] behouden mogt. Hieruit ontstaan nu alle in 't zelve lichaam voortgaande verrigtingen en veranderingen, niet op het bevel, of volgens het bestier der ziele; maar door een zekere mechanische noodzakelykheit).[24] Udagawa translates the first part as follows: 故二人身諸般ノ変化スル諸具ノ輻輳造成シ運動ヲナス. This translation reveals that several parts (or tools) change several parts of the human body and that they congest, create and move, but not that the changes occurring in the human body are innate to the structure of the body. The second part, which is the key point of Heister's thesis, is translated as follows: 然レドモ造物者神通不測ノ妙用ヲ以テ製作シ又諸部順列配置ハ外ヨリ形象二因テ察ストイヘドモ其造成スルノ精力順整二流利快行シ開闔ノ機発スル事皆其天性自然ノ良能ヨリ発原スル事ヲ探索スベキ也.[25] Again, the idea of mechanical necessity is not reproduced faithfully. Moreover, Udagawa introduces the concept of *seiryoku*, one of the creative substances (*ki*) in traditional Chinese medicine, to explain the changes occurring in the body, while Heister precisely tries to prove that these changes are not produced by any kind of energy. Udagawa's interpretation ends up to be exactly the opposite of the original text.

精力

Udagawa's translation does not reveal clearly Heister's thesis that all motions originate from the structure of the body. This is most clearly displayed in the passages where he translates "a certain mechanical necessity" by "everything originates from these innate, "natural" capacities" (皆其天性自然ノ良能ヨリシテ発原スル事). Here again Heister's concept of the human body as an inanimate mechanism is replaced by the concept of an animated organism, driven by innate energies. This replacement is quite obvious, as seen in the light of the fact that European scientists were used to the sight of machines, such as clocks etc., in their daily life, whereas this was not the case in Japan. Natural philosophy, as it was developed by Newton, concentrated its efforts on the mathematical and mechanical explanation of the natural world. The European physicians had a grounding in this kind of mathematical knowledge, which they tried to apply to their own scientific field, whereas the Japanese physicians received a Confucian, philosophical education with no notion of mathematical concepts. Udagawa was not an exception, and so he tried to comprehend Heister's mechanical concepts within his own intellectual framework.

In some cases however, when the definition of the human body is expressed in clearer terms, Udagawa's translation comes closer to the meaning of the original. This is the case in the following example: "The structure and composition of our body is prepared by the Creator in no other way than as an ingenious hydraulic machine, or a clock" (dat de bouw en opstel van ons Lichaam anders niet, dan als eene konstige hydraulicque Machine of Water-werktuig, of als een uurwerk [...] door den Schepper toebereid is).[26] Udagawa stays close to the metaphor of the human body as machine by translating this sentence as: 人身ノ基礎構立スルモノハ他ナシ奇巧絶妙ナル運水ノ器自鳴鐘ノ如ク....真ニ造物者ノ製作感ズルニ勝ヘタリ.[27] 造物者 Note the translation of "Creator" as *zôbutsusha*. The reference to the "Creator of our mechanism" is not limited to Heister, but appears in almost all medical works of the period. 造物主、造物者 造物 Udagawa translates the term "Schepper" (Creator) in most cases as *zôbutsushu* or *zôbutsusha* and the term "schepping" (creation) as *zôbutsu*. The term appears in the writings of other *Rangakusha* too. *Zôbutsushu* was the same term employed by some Jesuits in China to represent the Creator in their Chinese writings. Giulio Aleni (1582-1649) used the 職方外紀 term in his *Zhifang waiji*, a work with a widespread circulation among the *Rangakusha*. The adoption of this term however does not imply that the *Rangakusha* understood its implications in a Western sense. The more common term used by the Jesuits and known to the 天主 *Rangakusha* to denote the Christian God specifically was *tenshu*. *Zôbutsu* is originally a Confucian term. It was used to refer to the earthly forms, which were created by the innate creative substance *ki*. The fundamental difference with the Christian notion is that these *zôbutsu* are conceived of as coming into existence due to an energy from within and not as an object created by a creator outside. When employing the term *zôbutsu*, the *Rangakusha* did not by the same token adopt the western view of nature as an object, functioning as a machine, put together and set in motion by the Creator as a craftsman would do, but used the terminology of their own philosophical system. As this term was regularly used in Confucian texts, its use did not arrest the attention of the reader.

人身固形ノ諸部 流動血液 人身中の器械 Heister's definitions of the solids (vaste deelen) and fluids (vloeibare deelen) are faithfully translated. The solids are defined as either vessels, through which the fluids flow, or certain machines. Udagawa's translation comes quite close to the idea of mechanical necessity when 各自己ノ主能一箇ノ作用アリテ he writes that every part has its own function and that these parts work together in sequence and form a unit to produce certain functions (相共ニ列次連属シテ又一物ト成テ一箇ノ作用ヲ為ス).[28] That the fluids work "according to the hydraulic or mechanical laws" (en dit alles wel op het naauwkeurigste volgens de wetten der Water-konstwerken [...] en naar het voorschrift der Mechanica)[29] is also quite correctly translated as 運水術及諸具巧術ノ法ニ従テ理合シ [...],[30] but when he adds that we have to perceive the working of the "natural" truth 自然真実ノ作用ヲ悟ベシ, Udagawa again curbs Heister's theories into his own intellectual framework.

We may therefore conclude that the mechanical concept of the human body is not represented clearly in Udagawa's translation, although the translation itself is quite literally. By reading only the translation without knowledge of the original text, it would have been difficult for his contemporaries to gain a clear notion of the theoretical foundations of the scientific approach of the European physicians towards the human body. The work is however important because there are sufficient indications that the theoretical concept on which the Dutch imported medical books were founded, was totally different from the Japanese traditional views. The fact that Boerhaave's name is mentioned several times in relation to the mechanical method, has perhaps contributed to the fact that Udagawa urged his pupil Tsuboi Shindô to start his translation of Boerhaave's *Aphorisms*.

III Tsuboi Shindô's introduction of Boerhaavian physiological theories

Heister's dissertation is the only work translated in the Edo period that provides such extensive information on the theoretical background of Boerhaave's theories. The *Rangakusha* Shingû Ryôtei (1787-1854) owned a copy of the Dutch version by Lové of the *Institutes*, Boerhaave's primary work in which his theories on the physiology of the human body are explained in detail, but no Japanese translation of this work has been found yet. The only work of Boerhaave that has been translated in the Edo period, was the *Aphorisms*. The work on which Tsuboi's translation is based, is *Verklaaring der Korte Stellingen van Herman Boerhaave over de Kennis en Geneezing der Ziektens* (Leyden: Joh. en Herm. Verbeek, 1763-1776), the Dutch translation of van Swieten's (1700-1772) *Commentaria in Boerhaave Aphorismos de Cognoscendibus et Curandis Morbis*. The *Aphorisms* is a general work on pathology and therapy. It consists of 1479 aphorisms, or short statements concerning practical medicine. Van Swieten's work is a commentary on each of these aphorisms, giving detailed explanations, numerous quotations from classic and contemporary medical literature and many clinical examples. Van Swieten based the commentaries on his lecture notes of Boerhaave's classes. We can assume that they capture the essence of Boerhaave's doctrines relatively faithfully. No detailed theoretical explanations of Boerhaave's mechanical concept of the human body can be found in van Swieten's commentary. The principles of mechanics have however a pervasive influence throughout the whole work in the many practical physiological and pathological examples it offers.

Tsuboi Shindô translated a considerable part of van Swieten's com-
蒲爾花歇萬病治準 mentary between 1823-26. The first part of his translation, *Buruhabe manbyô chijun* covers chapters 1 to 169, containing the introductory notes and general pathological explanation of the diseases of the fibres, the vessels, viscera and fluids. The second part, a translation of chap-
蒲爾花歇内病論 ters 558 to 718, entitled *Buruhabe naibyô-ron*, concerns the febrile dis-
萬病治準 eases. Both works are commonly referred to as *Manbyô chijun*. With this selection Tsuboi covers Boerhaave's general doctrines on internal diseases. The chapters 145 to 557, out of which Tsuboi translated only chapters 145-169, deal with surgery. The chapters following 718 deal with special pathology. Boerhaave's aphorisms are translated in classical Chinese, the commentaries of van Swieten in Japanese.

I would like to focus my analysis on some passages in this translation, which have an indirect link to the mechanical concept of the human body. Boerhaave defines disease as "all conditions of the human body, which prevent the vital, natural or animal functions." (Alle gesteldheid van het menschelijk lighaam, die de levendige, natuurlijke, of dierlijke werkingen verhindert, wordt Ziekte genoemd).[31] This sentence implies that Boerhaave had a functional view of health and disease. Proper working of these functions is called health, while disease is a condition in which these functions are impeded. Tsuboi translates as follows: 人身ノ三機ヲ活機、性機、神機ト謂フ. 一モ妨碍セ被ルル所有ル、是レヲ疾病ト謂フ.[32]
機 As the word *ki* means working or mechanism (*karakuri*), the translation stays very close to the original. This sentence in itself is however not enough to comprehend the human body as mechanical. At first sight, these functions can equally be interpreted as vitalistic. A more detailed analysis of the translation of the definition of the three functions is therefore necessary.

The "vital and natural functions" are, in accordance with Heister's definition, the functions that are not controlled by the soul. In his explanation about the vital functions, van Swieten only gives a tautological definition in the beginning. He defines vital functions as all functions without which there can be no life. In an example he refers to the phenomenon of drowning, observing that "only movement is lacking, if you can resuscitate it, life also returns" (de beweeging alleen ontbreekt, en kunt gy die weer opwekken, dan komt het leven ook weer).[33] Tsuboi does not translate this sentence literally, but interprets "movement" correctly as the blood circulation: "only the blood circulation is interrupted" (唯々血ノ運行休止スルノミナリ).[34] After this he gives his own interpretation of vital functions (*kakki*) as "one has to know that the 活機 vital functions are to be found in the blood circulation" (故ニ活機ハ血ノ運行ニ在リト知ルベシ). By this definition, Tsuboi displays a remarkable understanding of the mechanical theory. "Beweeging" is the translation of "motus" in the Latin edition of van Swieten's Commentaries.[35] The term "motus sanguinis" is used by William Harvey (1578-1657)[36] and Richard Lower (1631-1690)[37] to designate the circulation of the blood. The influence of these two anatomists is apparent throughout the *Aphorisms*. For Boerhaave the basic mechanism of life was the circulation of the blood. The heart was the motive power of this mechanism. Van Swieten writes that "there is still life however weak, as long as the heart opens and closes" (dat het minste leven nog bestaat in de werking van het gesloten en geopend hart),[38] which Tsuboi translates correctly as "it is evident that the primary existence of life is based on the circulation of the blood. The source of the circulation of the blood is the movement of the heart" (生命ノ存ズル事一ニ血ノ運行ニ由ル事昭昭タリ血ノ運行ノ起源ハ心ノ運動ニ在リ).[39]

The working of the blood circulation is viewed as a closed circuit with no beginning and no end. Here van Swieten faces the same difficulty as Heister, namely the ultimate origin of this movement. Heister tries to explain this by means of the concept of a *perpetuum mobile*. It was however known that the heart still continues its movement for some time, even without the help of the other parts of the human body. The existence of the autonomic nerve had not yet been discovered. Van Swieten therefore quotes the words of Galen: "movement is created into the heart" (Galenos heeft gezegd, dat de beweeging het hart is ingeschapen).[40] In other words he attributes this as yet inconceivable part of the human mechanism to the work of the Almighty Creator. Tsuboi translates as follows: 運動ハ心臓ノ稟受スル所ナリ.[41] The term *hinju* 稟受 employed for "ingeschapen" (created into) literally means "to receive from heaven" and is in this sense a correct translation. *Hinju* is however a purely Confucian term, which in my narrow view was not used by the Jesuits in China. It did not imply the existence of a Creator outside the human body but was only used in the meaning of "innate characteristics". Tsuboi consistently avoids terms used by the Jesuits and adopts Confucian terms to designate Western religious concepts. This stands in contrast to Udagawa's automatic use of *zôbutsushu*.

The second category of functions, which are not ruled by the soul, are the natural functions (natuurlijke werkingen). These are the digestive functions or as van Swieten states: "all those functions that provide the human body with such things as it has lost (= nutrition)".[42] The digestive system is an important pillar in Boerhaave's medical system. It was believed that the absorbed food was gradually converted in the digestive tract, including the vessels, the lungs and the blood itself. A good digestion depended on the action of the solids and the aid of some fluids. Although some chemical characteristics were attributed to it, the digestive system was mainly conceived of as a mechanical system. Van Swieten calls the digestive functions of the human body the "nature" of the human body. Here nature is used in the same meaning as Heister used it and it is a keyword to understand Boerhaave's mechanical concept of the human body. In *Manbyô chijun* we can see clearly that Tsuboi had difficulties finding a suitable translation for the term "nature". The food we take in, writes van Swieten, is not of the same substance as the human body. In order to build the parts of the human body, which are lost due to the working of the vital functions, "nature needed to change those things (i.e. food and drink) first and to boil them and to prepare them in advance to a certain equality with the human body, that has to be nourished" (hierom heeft de natuur nodig gehad die dingen eerst te veranderen, en te kooken, en vooraf tot eene gelijkheid met het lighaam, dat gevoed moet worden, te bereiden).[43] Tsuboi translates the term "nature" as it is used in this sentence by 自然の良能 *shizen no ryônô*. *Shizen no ryônô* may tentatively be translated as "natural capacities". To clarify his statement, van Swieten repeats the above sentence but replaces the term "nature" by a more detailed definition: "the most complete tool of the created body is required to make nourishment from it (i.e. the ingested food)" (het volkomenste werktuig van het geschapen lighaam wordt vereischt, om daar uit voedsel te maken).[44] Tsuboi translates this sentence as 体中諸器ノ機能健康ニシテ.[45] Although *Edo-Haruma* and *Zufu-Haruma* respectively give 器械 and 細工道具 (both meaning "tool") as translation for "werktuig", Tsuboi interprets this word here correctly as "the various organs of the human body" and does not make any reference to the mechanical terminology.

To explain his use of the term "nature" van Swieten cites Hippocrates: if the digestive powers are insufficient, "nature itself is insufficient"(dat de natuur zelve te kort schiet: translated by Tsuboi as 良能衰弱スルノ徵ナレバナリ.[48] Note again the use of *ryônô*). Van Swieten defines this notion of nature as used by Hippocrates as "a convergence of all natural conditions, which are required for life to be continuous and durable and at the same time a very fast movement" (eene samenloop van alle natuurlijke gesteldheden, die vereischt worden, op dat het leven bestendig en duurzaam, en te gelijk eene zeer vlugge beweeging zy).[49] This is an abstract definition, from which it is difficult, especially for a *Rangakusha*, with his different intellectual background, to interpret the natural conditions as the corporeal mechanism. Tsuboi understood in any case that the term *shizen* was not appropriate. He first uses the transliteration *nachûru*. But he adds a note that the term *nachûru* can be interpreted as *honzen* and natural functions as *honzen no kinô* (本然トハ本然ノ機能ノ義ナリ).[50] *Honzen no sei* is a term used in the metaphysics of Zhu Xi to represent the pure innate properties of the human body. Its meaning stands closer to van Swieten's "nature" than Udagawa's *shizen*, but because of its Confucian philosophical connotations does not cover completely the concept of the body as an inanimate mechanism.

那去児
此二本然ト訳ス
本然之性

Van Swieten's above-mentioned definition Tsuboi translates as follows: "When using this word [i.e. *nachûru*] people refer to the functions which assure that the various organs and humours have all their innate properties so that they can maintain the vital energy and make fast and light movements" (此語ヲ用フル者ハ人身諸器諸液共ニ本然ノ稟質ヲ具有シテ能ク生気ヲ保持シ又能ク運動ヲシテ軽敏ナラシムル所ノ機能ヲ言フ).[51] Tsuboi uses here Confucian vitalistic terms such as *honzen no hinshitsu* and *seiki*, which are incompatible with the mechanical ideas of the original. Tsuboi's use of these Confucian terms probably has its origins in the vagueness of the Dutch edition. In the Dutch edition "natuur" is tautologically defined as "alle natuurlijke gesteldheden", whereas in the original Latin edition, "Natura" is defined as "aggregatum omnium conditionum physicarum" ("the aggregate of all physical conditions").

本然ノ稟質、生気

In the following sentences where van Swieten further explains his definition of "nature" by means of some examples, Tsuboi is still struggling with the interpretation of this term. Van Swieten writes: "The physicians do no injustice to the Supreme Being, when they attribute so much to Nature, because by this [i.e. Nature] they mean the structure of the created body" (De Geneesmeesters doen dan het Opperwezen geen ongelijk, wanneer zy zo veel aan de Natuur toeschryven; want door deeze verstaan zy het maaksel van het geschapen lighaam).[52] Tsuboi apparently did not understand this sentence and gave his own interpretation: "If the physicians would not use their own knowledge and would leave everything to nature, they would acquire with other words divine knowledge. When following nature it however is possible for them to acquire the fundamental knowledge of the heaven-made human body" (医者若シ能ク私知ヲ用ヒズシテ一ニ那去児ニ委任スル時ハ則チ神知ニ等シキ事ヲ得那去児ニ従フ時ハ人身天造ノ原質知了スル事ヲ得可ケレバナリ).[53] It seems that Tsuboi interpreted the word "verstaan" as "to understand" and not as "to mean", in which meaning it was used in the original sentence. Due to this misinterpretation, Tsuboi did not correctly translate the sentence that "by nature they mean the structure of the created body". As stated above, there was a debate in the seventeenth century concerning the interpretation of Hippocrates' statement that "nature cures diseases". In the above sentence, van Swieten interprets this "nature" of Hippocrates as the structure of the human body. By the words "to follow nature" as a translation for "to attribute to nature" (aan de natuur toeschrijven), Tsuboi composes a popular concept in the Edo period that the physician had to be the servant of nature. This concept was attributed by *the Rangakusha* to Hippocrates. Tsuboi was one of the admirers of Hippocrates and wrote some Chinese poems in praise of him. It seems that for lack of understanding of the above sentence, Tsuboi fell back on this concept.

那去児ニ従フ

Van Swieten continues his explanation by citing van Helmont:[54] "Van Helmont has therefore said very rightly that Nature is an order from God, by which the things are what they are and work as they are ordered to work" (Helmont heeft dan ook zeer wel gezegd, dat de Natuur is een bevel van God, waar door de dingen zyn het geenze zyn, en werken, het geen haar bevolen is te werken).[55] Tsuboi translates as follows: "Helmont has said that nature is an order from the creative process (*zôka*). The things peculiar to the human body all originate from the creative process and all their functions are what they are because the creative process makes them to be like that." (「ヘルモント」云ク那去児ハ造化ノ使令ナリ人身固有ノ諸物ハ共ニ造化ニ由テ生ズル所ニシテ其官能ハ造化ノ指揮シテ然カラシムル所ナリ).[56] I have translated *zôka* not by "Creator", but by "creative process", which is in my opinion closer to the original meaning of this term. *Zôka* was used by many *Rangakusha* to translate the terms "nature" and "Creator", but it does not completely cover the Western meaning implied in the terms. *Zôka* was used in Confucian texts to represent the creative process of Yin and Yang, the cosmic dual forces. By using *zôka* for Creator, Tsuboi again avoids the religious aspect in this statement and replaces it by Confucian terminology. Tsuboi uses the phonetic substitute *nachûru* only in the paragraph concerning the natural functions. The natural functions he first translates as *honki* (the fundamental functions), but later he switches to *seiki* (the innate functions). Throughout the text Tsuboi does not restrict the translation of "nature" to one term but uses mainly *honzen* to representing innate qualities and *shizen no ryônô* in the Hippocratic meaning of nature healing. The animal functions (dierlijke werkingen) are defined as all cognitive functions in the human body that are governed by the soul. Tsuboi first translates the animal functions as *konki*, a term used in traditional Chinese medicine to represent the positive *vitalising* energy which governs the mind, but he afterwards corrects *konki* in *shinki*. *Shin* was, as indicated above, the name for the category of the mind, which means that Tsuboi's translation is quite appropriate.

造化

本機

性機

魂機

神機

iv Conclusion

Through the analysis of the translations of Heister's and van Swieten's work by Udagawa and Tsuboi, I have tried to show that these *rangakusha* had many difficulties in overcoming the cultural and linguistic barrier and gaining an understanding of Boerhaave's mechanical concept of the human body. A literal translation, such as the one Udagawa made, was apparently not sufficient to reach a profound understanding, because the traditional Japanese terms were used in a totally different context. Udagawa's translation comes in some cases to opposite conclusions regarding the "natuur" of the human body. We have seen that Udagawa had to fall back on traditional Chinese vitalistic concepts to explain the functions of the body, in contrast to Heister's statement that the structure of the human body itself is the origin of its functions and that these functions work as an inanimate mechanism.

Tsuboi grasped the mechanical concept better as is apparent for example in his translation of the blood circulation. Contrary to Udagawa, who translates the word "natuur" in all cases by *shizen*, Tsuboi tried to use different terms to represent the different meanings the word "nature" had. His interpretations are but for a few exceptions correct, which indicates that he had a good understanding of Boerhaave's theories. However, regarding the ideological connotations of the original terms, it cannot be said that he translated them word for word with the same accuracy. Tsuboi often gives a free translation for purely mechanical terms. For the term "natuur", when it is used with mechanical connotations, he first uses a transliteration and later on *honzen*, a Confucian term. Christian terminology such as "Creator" he omits or expresses by abstract terms such as *zôka*. From these examples we may conclude that Tsuboi, although he had a good understanding of Boerhaave's mechanical concept, excluded purely mechanical and religious concepts, or transformed them into terms that fitted his own intellectual framework. It is indeed quite difficult to read the mechanical concept of the original explicitly in Tsuboi's translation.

NOTES

Sentences in classical Chinese are transposed in kakikudashibun 書下文.

1 Herman Boerhaave, *Institutiones medicae: in usus annuae exercitationis domesticos digestae* (Lugduni Batavorum: van der Linden, 1708).

2 Herman Boerhaave, *Aphorismi de cognoscendis et curandis morbis in usum doctrinae domesticae* (Lugduni Batavorum: van der Linden, 1709).

3 E. Kegel-Brinkgreve and A.M. Luyendijk-Elshout, trans., *Boerhaave's Orations*, Publications of the Sir Thomas Browne Institute, n.s., no. 4, with introduction and notes by E. Kegel-Brinkgreve and A.M. Luyendijk-Elshout (Leiden: Brill, 1983), 96.

4 Kegel-Brinkgreve and Luyendijk-Elshout, *Boerhaave's Orations*, 111.

5 Rangakusha 蘭学者: a scholar in the Edo period who studied Western sciences by means of the Dutch language.

6 Laurentius Heister, *Practicaal geneeskundig handboek; nevens een voorafgaande verhandeling van de voortreffelykheit en uitstekendheit der mechanisch-geneeskundige leerwyze* (Amsterdam: Jan Morterre, 1762).

7 Udagawa Genshin 宇田川玄真 (1769-1834): native of Ise; went to Edo to study under Udagawa Genzui 宇田川玄随, Ôtsuki Gentaku 大槻玄沢, Katsuragawa Hoshû 桂川甫周 and Sugita Genpaku 杉田玄白; succeeded the Udagawa family after Genzui's death.

8 Udagawa Genshin, trans., *Heisuteru naikasho* 「歇伊私的児内科書」, appended: *Jinshin kyûri ijutsuron* 「人身究理医術論」 (unpublished manuscript, Edo, n.d.). For my analysis I have consulted the manuscript of the Central Library of Kyoto University (Fujikawa Collection).

9 Gerard van Swieten, *Verklaaring der korte stellingen van Herman Boerhaave: over de kennis en geneezing der ziektens*, 4 vols. (Leyden: Verbeek, 1760).

10 Tsuboi Shindô 坪井信道 (1795-1848): native of Mino. After learning Chinese Medicine, Tsuboi studied *Rangaku* under Nakai Kotaku and Udagawa Genshin.

11 Tsuboi Shindô, trans., *Buruhabe manbyô-chijun* 「蒲爾花歇萬病治準」 (covering the original chapters 1-169); 「*Buruhabe naibyô-ron* 蒲爾花歇内病論」 (covering the original chapters 558-718), trans. (unpublished manuscript, Edo, 1826). I have consulted the manuscript of the Kyô-u Library of the Takeda Science Foundation in Osaka (Udagawa Collection). Both works are hereafter cited as *Manbyô-chijun*.

12 Liu An 劉安, ed., *Enanji* 「淮南子」, Chûgoku koten bungaku taikei (「中国古典文学大系」, Series of Chinese Classical Literature), vol. 6, trans. Togawa Yoshio 戸川芳郎, Kiyama Hideo 木山英雄 and Sawatani Shôji 沢谷昭次 (Tokyo: Heibonsha, 1974), 12,「形体は生の宅舎、気志は生の内実、精神は生

の主宰」. Originally published as Huainanzi.

13 Udagawa Genzui's handwritten copy of de Gorter's book and his translation notes for *Seisetsu naika sen'yô* can be found in the Kyô-u Library of the Takeda Science Foundation in Osaka (Udagawa Collection), under the title *Uda-shi Hikyû* 「宇多氏秘笈」 (Udagawa's secret wicker box), fasc. 2-8.

14 Johannes de Gorter, *Gezuiverde geneeskonst, of kort onderwys der meeste inwendige ziekten, ten nutte van chirurgyns* (Amsterdam: Isaac Tirion, 1744), 31-32.

15 Udagawa Genzui, *Seisetsu naika sen'yô* 「西説内科撰要」 (Edo, completed 1790, published 1791-1808). I have consulted the facsimile edition of Waseda University Library (Tokyo: Waseda Univ. Press, 1995).

16 Edo-Haruma Dutch-Japanese Dictionary. I have consulted *Haruma-shi Oranda jisho* 「波児馬氏和蘭辞書」, a manuscript copy ascribed to Takano Chôei and kept in the Kyô-u Library of the Takeda Science Foundation in Osaka (Udagawa Collection). This dictionary lists following equivalents for "natuur": 自然、神力ニテ造ル、造化ノ神、性質、形状、自然ノ理、欲シ好ム、自然ノ好欲、損ズル、枯ルル.

17 Egbert Buys, *Nieuw en volkomen woordenboek van konsten en weetenschappen* (Amsterdam: S.J Baalde, 1769-1778. 10 vols.), vol.VII, 627-628.

18 Hippocrates is frequently cited as an authority by Heister and his contemporaries. The high value Boerhaave attributed to Hippocrates in the clinical field, can be derived from his *Oration to Recommend the Study of Hippocrates (Oratio de Commendando Studio Hippocratico,* 1701).

19 Heister, *Verhandeling van de Voortreffelykheit en Uitstekendheit der Mechanisch-Geneeskundige Leerwyze,* 101-103.

20 Udagawa, *Jinshin kyûri ijutsu-ron*, chap. 38-39.

21 Heister, *Verhandeling van de Voortreffelykheit en Uitstekendheit der Mechanisch-Geneeskundige Leerwyze,* 106.

22 Udagawa, *Jinshin kyûri ijutsu-ron*, chap. 40.

23 Ibid.

24 Heister, *Verhandeling van de Voortreffelykheit en Uitstekendheit der Mechanisch-Geneeskundige Leerwyze,* 109-110.

25 Udagawa, *Jinshin kyûri ijutsu-ron*, chap. 41.

26 Heister, *Verhandeling van de Voortreffelykheit en Uitstekendheit der Mechanisch-Geneeskundige Leerwyze,* 106.

27 Udagawa, *Jinshin kyûri ijutsu-ron*, chap. 9.

28 Ibid., chap. 10.

29 Heister, *Verhandeling van de Voortreffelykheit en Uitstekendheit der Mechanisch-Geneeskundige Leerwyze,* 29.

30 Udagawa, *Jinshin kyûri ijutsu-ron*, chap. 11.

31 Van Swieten, *Verklaaring der korte stellingen van Herman Boerhaave*, deel 1, 1.

32 Tsuboi, *Manbyô-chijun*, chap. 1, fol.1.

33 Van Swieten, *Verklaaring der korte stellingen van Herman Boerhaave*, deel 1, 2.

34 Tsuboi, *Manbyô-chijun*, chap. 1, fol.2.

35 Gerardi van Swieten, *Commentaria in Hermanni Boerhaave aphorismos de cognoscendis et curandis morbis* (Lugduni Batavorum: Johannem et Hermannum Verbeek, 1742-1772. 5 vols.), vol.1, 2.

36 William Harvey, *Exercitatio de motu cordis et sanguinis in animalibus* (Francofurti: Guilielmi Fitzeri, 1628).

37 Richard Lower, *Tractatus de corde. Item de motu & colore sanguinis et chyli in eum transitu* (Londini: John Redmayne, 1669).

38 Van Swieten, *Verklaaring der korte stellingen van Herman Boerhaave*, deel 1, 2-3.

39 Tsuboi, *Manbyô-chijun*, chap. 1, fol.3.

40 Van Swieten, *Verklaaring der korte stellingen van Herman Boerhaave*, deel 1, 4.

41 Tsuboi, *Manbyô-chijun*, chap. 1, fol.5.

42 Van Swieten, *Verklaaring der korte stellingen van Herman Boerhaave*, deel 1, 5.

43 Ibid.

44 Ibid.

45 Tsuboi, Manbyô-chijun, chap. 1, fol.7.

46 vid. note 16.

47 Hendrik Doeff's Dutch-Japanese Dictionary popularly known as *Zufu-Haruma*. I have consulted a manuscript copy ascribed to Tsuboi in Waseda University Library.

48 Tsuboi, *Manbyô-chijun*, chap. 1, fol.8.

49 Van Swieten, *Verklaaring der korte stellingen van Herman Boerhaave*, deel 1, 6.

50 Tsuboi, *Manbyô-chijun*, chap. 1, fol.8.
51 Ibid.
52 Van Swieten, *Verklaaring der korte stellingen van Herman Boerhaave*, deel 1, 6.
53 Tsuboi, *Manbyô-chijun*, chap. 1, fol.8.
54 Van Helmont (1577-1644) was one of the pioneers of the chemiatric method in medicine. He tried to explain digestion as a series of fermentations. Although Boerhaave had more mechanistic views regarding the digestion system, van Helmont is frequently cited in van Swieten's commentary.
55 Van Swieten, *Verklaaring der korte stellingen van Herman Boerhaave*, deel 1, 6.
56 Tsuboi, *Manbyô-chijun*, chap. 1, fol.9.

ILLUSTRATIONS

ILL 1 Gerard van Swieten, *Commentaria in Hermanni Boerhaave Aphorismos, de cognoscendis et curan dis morbis.* Tomus secundus, Parisiis, Apud Guillelmum Cavelier, MDCCLXXI.

ILL 2 Joannes de Gorter, *Gezuiverde Geneeskonst, of kort onderwys der meeste inwendige ziekten, ten nutte van chirurgyns, Die ter Zee of Velde dienende, of in andere omstandigheden, zig genoodzaakt vinden dusdanige Ziekten te behandelen*, derde druk, merkelijk vermeerderd, Te Amsterdam, By Isaak Tirion, MDCCLXI.

ILL 3 *Institutiones medicae in usus annuae exercitationis domesticos Digestae ab Hermanno Boerhaave, Edicio altera*, Lugduni Batavorum, Apud Johannem vander Linden, MDCCXIII.

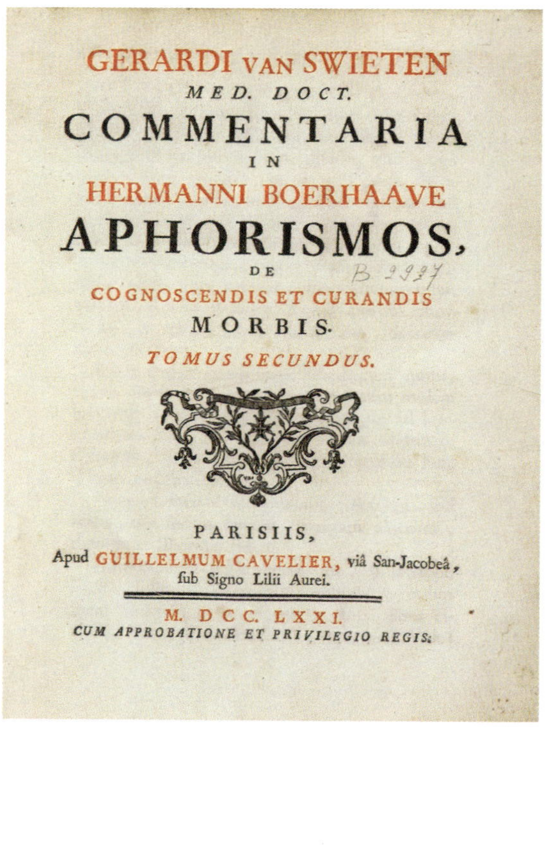

GERARDI VAN SWIETEN
MED. DOCT.
COMMENTARIA
IN
HERMANNI BOERHAAVE
APHORISMOS,
DE
COGNOSCENDIS ET CURANDIS
MORBIS.
TOMUS SECUNDUS.

PARISIIS,
Apud GUILLELMUM CAVELIER, viâ San-Jacobeâ, ſub Signo Lilii Aurei.

M. DCC. LXXI.
CUM APPROBATIONE ET PRIVILEGIO REGIS.

ILL 1

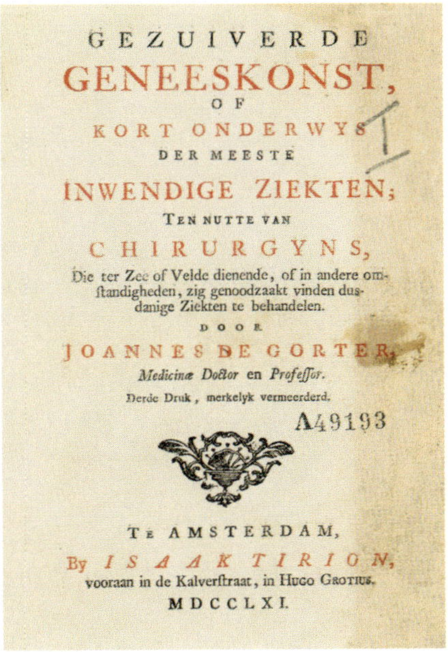
GEZUIVERDE
GENEESKONST,
OF
KORT ONDERWYS
DER MEESTE
INWENDIGE ZIEKTEN;
TEN NUTTE VAN
CHIRURGYNS,
Die ter Zee of Velde dienende, of in andere omſtandigheden, zig genoodzaakt vinden dusdanige Ziekten te behandelen.
DOOR
JOANNES DE GORTER,
Medicinæ Doctor en *Profeſſor.*
Derde Druk, merkelyk vermeerderd.

TE AMSTERDAM,
By *ISAAK TIRION,*
vooraan in de Kalverſtraat, in HUGO GROTIUS.
MDCCLXI.

ILL 2

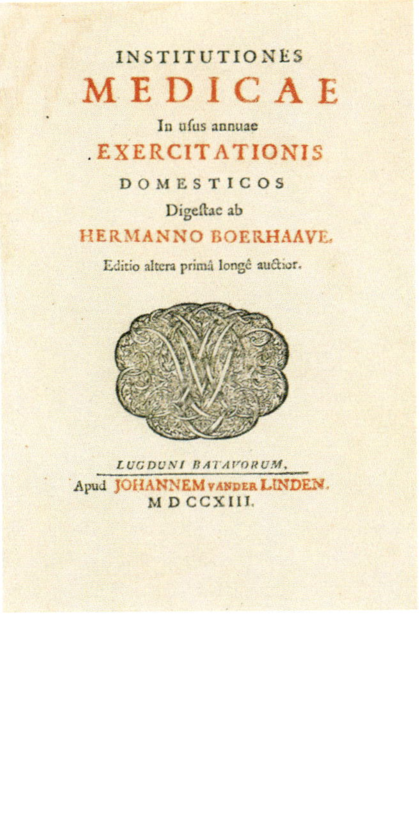
INSTITUTIONES
MEDICAE
In uſus annuae
EXERCITATIONIS
DOMESTICOS
Digeſtae ab
HERMANNO BOERHAAVE.
Editio altera primâ longê auctior.
LUGDUNI BATAVORUM,
Apud JOHANNEM VANDER LINDEN.
MDCCXIII.

ILL 3

A

B

G

H

M

P

Q

R

S

Universitaire Pers Leuven / Leuven University Press / Presses Universitaires de Louvain
Blijde Inkomststraat 5, B-3000 Leuven (Belgium)

ISBN 90 5867 179 8
D/2001/1869/87

Prepress and layout by Oskar D Creative Services, Antwerp.
Printed by Antilope Printing, Lier.